de Gruyter Textbook
Robinson · An Introduction to Abstract Algebra

Derek J. S. Robinson

An Introduction to Abstract Algebra

Walter de Gruyter
Berlin · New York 2003

Author

Derek J. S. Robinson
Department of Mathematics
University of Illinois at Urbana-Champaign
1409 W. Green Street
Urbana, Illinois 61801-2975
USA

Mathematics Subject Classification 2000: 12-01, 13-01, 16-01, 20-01

Keywords: group, ring, field, vector space, Polya theory, Steiner system, error correcting code

∞ Printed on acid-free paper which falls within the guidelines of the ANSI to ensure permanence and durability.

Library of Congress Cataloging-in-Publication Data

> Robinson, Derek John Scott.
>   An introduction to abstract algebra / Derek J. S. Robinson.
>   p.   cm. − (De Gruyter textbook)
>   Includes bibliographical references and index.
>   ISBN 3-11-017544-4 (alk. paper)
>   1. Algebra, Abstract.   I. Title.   II. Series.
>   QA162.R63   2003
>   512'.02−dc21   2002041471

ISBN 3 11 017544 4

Bibliographic information published by Die Deutsche Bibliothek

Die Deutsche Bibliothek lists this publication in the Deutsche Nationalbibliografie; detailed bibliographic data is available in the Internet at <http://dnb.ddb.de>.

© Copyright 2003 by Walter de Gruyter GmbH & Co. KG, 10785 Berlin, Germany.
All rights reserved, including those of translation into foreign languages. No part of this book may be reproduced in any form or by any means, electronic or mechanical, including photocopy, recording or any information storage and retrieval system, without permission in writing from the publisher.
Printed in Germany.
Cover design: Hansbernd Lindemann, Berlin.
Typeset using the author's TeX files: I. Zimmermann, Freiburg.
Printing and binding: Hubert & Co. GmbH & Co. KG, Göttingen.

In Memory of My Parents

Preface

The origins of algebra are usually traced back to Muhammad ben Musa al-Khwarizmi, who worked at the court of the Caliph al-Ma'mun in Baghdad in the early 9th Century. The word derives from the Arabic *al-jabr*, which refers to the process of adding the same quantity to both sides of an equation. The work of Arabic scholars was known in Italy by the 13th Century, and a lively school of algebraists arose there. Much of their work was concerned with the solution of polynomial equations. This preoccupation of mathematicians lasted until the beginning of the 19th Century, when the possibility of solving the general equation of the fifth degree in terms of radicals was finally disproved by Ruffini and Abel.

This early work led to the introduction of some of the main structures of modern abstract algebra, groups, rings and fields. These structures have been intensively studied over the past two hundred years. For an interesting historical account of the origins of algebra the reader may consult the book by van der Waerden [15].

Until quite recently algebra was very much the domain of the pure mathematician; applications were few and far between. But all this has changed as a result of the rise of information technology, where the precision and power inherent in the language and concepts of algebra have proved to be invaluable. Today specialists in computer science and engineering, as well as physics and chemistry, routinely take courses in abstract algebra.

The present work represents an attempt to meet the needs of both mathematicians and scientists who are interested in acquiring a basic knowledge of algebra and its applications. On the other hand, this is not a book on applied algebra, or discrete mathematics as it is often called nowadays.

As to what is expected of the reader, a basic knowledge of matrices is assumed and also at least the level of maturity consistent with completion of three semesters of calculus. The object is to introduce the reader to the principal structures of modern algebra and to give an account of some of its more convincing applications. In particular there are sections on solution of equations by radicals, ruler and compass constructions, Polya counting theory, Steiner systems, orthogonal latin squares and error correcting codes. The book should be suitable for students in the third or fourth year of study at a North American university and in their second or third year at a university in the United Kingdom.

There is more than enough material here for a two semester course in abstract algebra. If just one semester is available, Chapters 1 through 7 and Chapter 10 could be covered. The first two chapters contain some things that will be known to many readers and can be covered more quickly. In addition a good deal of the material in Chapter 8 will be familiar to anyone who has taken a course in linear algebra.

A word about proofs is in order. Often students from outside mathematics question the need for rigorous proofs, although this is perhaps becoming less common. One answer is that the only way to be certain that a statement is correct or that a computer program will always deliver the correct answer is to prove it. As a rule complete proofs are given and they should be read, although on a first reading some of the more complex arguments could be omitted. The first two chapters, which contain much elementary material, are a good place for the reader to develop and polish theorem proving skills. Each section of the book is followed by a selection of problems, of varying degrees of difficulty.

This book is based on courses given over many years at the University of Illinois at Urbana-Champaign, the National University of Singapore and the University of London. I am grateful to many colleagues for much good advice and lots of stimulating conversations: these have led to numerous improvements in the text. In particular I am most grateful to Otto Kegel for reading the entire text. However full credit for all errors and mis-statements belongs to me. Finally, I thank Manfred Karbe, Irene Zimmermann and the staff at Walter de Gruyter for their encouragement and unfailing courtesy and assistance.

Urbana, Illinois, November 2002 *Derek Robinson*

Contents

1 Sets, relations and functions   1
   1.1 Sets and subsets 1
   1.2 Relations, equivalence relations and partial orders 4
   1.3 Functions 9
   1.4 Cardinality 13

2 The integers   17
   2.1 Well-ordering and mathematical induction 17
   2.2 Division in the integers 19
   2.3 Congruences 24

3 Introduction to groups   31
   3.1 Permutations of a set 31
   3.2 Binary operations: semigroups, monoids and groups 39
   3.3 Groups and subgroups 44

4 Cosets, quotient groups and homomorphisms   52
   4.1 Cosets and Lagrange's Theorem 52
   4.2 Normal subgroups and quotient groups 60
   4.3 Homomorphisms of groups 67

5 Groups acting on sets   78
   5.1 Group actions and permutation representations 78
   5.2 Orbits and stabilizers 81
   5.3 Applications to the structure of groups 85
   5.4 Applications to combinatorics – counting labellings and graphs 92

6 Introduction to rings   99
   6.1 Definition and elementary properties of rings 99
   6.2 Subrings and ideals 103
   6.3 Integral domains, division rings and fields 107

7 Division in rings   115
   7.1 Euclidean domains 115
   7.2 Principal ideal domains 118
   7.3 Unique factorization in integral domains 121
   7.4 Roots of polynomials and splitting fields 127

8 Vector spaces — 134
- 8.1 Vector spaces and subspaces — 134
- 8.2 Linear independence, basis and dimension — 138
- 8.3 Linear mappings — 147
- 8.4 Orthogonality in vector spaces — 155

9 The structure of groups — 163
- 9.1 The Jordan–Hölder Theorem — 163
- 9.2 Solvable and nilpotent groups — 171
- 9.3 Theorems on finite solvable groups — 178

10 Introduction to the theory of fields — 185
- 10.1 Field extensions — 185
- 10.2 Constructions with ruler and compass — 190
- 10.3 Finite fields — 195
- 10.4 Applications to latin squares and Steiner triple systems — 199

11 Galois theory — 208
- 11.1 Normal and separable extensions — 208
- 11.2 Automorphisms of field extensions — 213
- 11.3 The Fundamental Theorem of Galois Theory — 221
- 11.4 Solvability of equations by radicals — 228

12 Further topics — 235
- 12.1 Zorn's Lemma and its applications — 235
- 12.2 More on roots of polynomials — 240
- 12.3 Generators and relations for groups — 243
- 12.4 An introduction to error correcting codes — 254

Bibliography — 267

Index of notation — 269

Index — 273

Chapter 1
Sets, relations and functions

The concepts introduced in this chapter are truly fundamental and underlie almost every branch of mathematics. Most of the material is quite elementary and will be familiar to many readers. Nevertheless readers are encouraged at least to review the material to check notation and definitions. Because of its nature the pace of this chapter is brisker than in subsequent chapters.

1.1 Sets and subsets

By a *set* we shall mean any well-defined collection of objects, which are called the *elements* of the set. Some care must be exercised in using the term "set" because of Bertrand Russell's famous paradox, which shows that not every collection can be regarded as a set. Russell considered the collection C of all sets which are not elements of themselves. If C is allowed to be a set, a contradiction arises when one inquires whether or not C is an element of itself. Now plainly there is something suspicious about the idea of a set being an element of itself, and we shall take this as evidence that the qualification "well-defined" needs to be taken seriously. A collection that is not a set is called a *proper class*.

Sets will be denoted by capital letters and their elements by lower case letters. The standard notation
$$a \in A$$
means that a is a element of the set A, (or a *belongs* to A). The negation of $a \in A$ is denoted by $a \notin A$. Sets can be defined either by writing their elements out between braces, as in $\{a, b, c, d\}$, or alternatively by giving a formal description of the elements, the general format being
$$A = \{a \mid a \text{ has property } P\},$$
i.e., A is the set of all objects with the property P. If A is a finite set, the number of its elements is written
$$|A|.$$

Subsets. Let A and B be sets. If every element of A is an element of B, we write
$$A \subseteq B$$

and say that A is a *subset* of B, or that A *is contained* in B. If $A \subseteq B$ and $B \subseteq A$, so that A and B have exactly the same elements, then A and B are said to be *equal*,
$$A = B.$$
The negation of this is $A \neq B$. The notation $A \subset B$ is used if $A \subseteq B$ and $A \neq B$; then A is a *proper* subset of B.

Special sets. A set with no elements at all is called an *empty set*. An empty set E is a subset of any set A; for if this were false, there would be an element of E that is not in A, which is certainly wrong. As a consequence there is just one empty set; for if E and E' are two empty sets, then $E \subseteq E'$ and $E' \subseteq E$, so that $E = E'$. This unique empty set is written
$$\emptyset.$$

Some further standard sets with a reserved notation are:
$$\mathbb{N}, \mathbb{Z}, \mathbb{Q}, \mathbb{R}, \mathbb{C},$$
which are respectively the sets of natural numbers $0, 1, 2, \ldots,$ integers, rational numbers, real numbers and complex numbers.

Set operations. Next we recall the familiar set operations of union, intersection and complement. Let A and B be sets. The *union* $A \cup B$ is the set of all objects which belong to A or B (possibly both); the *intersection* $A \cap B$ consists of all objects that belong to both A and B. Thus
$$A \cup B = \{x \mid x \in A \text{ or } x \in B\},$$
while
$$A \cap B = \{x \mid x \in A \text{ and } x \in B\}.$$
It should be clear how to define the union and intersection of an arbitrary collection of sets $\{A_\lambda \mid \lambda \in \Lambda\}$; these are written
$$\bigcup_{\lambda \in \Lambda} A_\lambda \quad \text{and} \quad \bigcap_{\lambda \in \Lambda} A_\lambda.$$
The *relative complement* of B in A is
$$A - B = \{x \mid x \in A \text{ and } x \notin B\}.$$
Frequently one has to deal only with subsets of some fixed set U, called the *universal set*. If $A \subseteq U$, then the *complement* of A in U is
$$\bar{A} = U - A.$$

Properties of set operations. We list for future reference the fundamental properties of union, intersection and complement.

(1.1.1) *Let A, B, C be sets. Then the following statements are valid:*

   (i) $A \cup B = B \cup A$ and $A \cap B = B \cap A$ (*commutative laws*).

   (ii) $(A \cup B) \cup C = A \cup (B \cup C)$ and $(A \cap B) \cap C = A \cap (B \cap C)$ (*associative laws*).

   (iii) $A \cap (B \cup C) = (A \cap B) \cup (A \cap C)$ and $A \cup (B \cap C) = (A \cup B) \cap (A \cup C)$ (*distributive laws*).

   (iv) $A \cup A = A = A \cap A$.

   (v) $A \cup \emptyset = A$, $A \cap \emptyset = \emptyset$.

   (vi) $A - \left(\bigcup_{\lambda \in \Lambda} B_\lambda\right) = \bigcap_{\lambda \in \Lambda}(A - B_\lambda)$ and $A - \left(\bigcap_{\lambda \in \Lambda} B_\lambda\right) = \bigcup_{\lambda \in \Lambda}(A - B_\lambda)$ (*De Morgan's Laws*).[1]

The easy proofs of these results are left to the reader as an exercise: hopefully most of these properties will be familiar.

Set products. Let $A_1, A_2, \ldots, A_n$ be sets. By an *n-tuple* of elements from $A_1, A_2, \ldots, A_n$ is to be understood a sequence of elements $a_1, a_2, \ldots, a_n$ with $a_i \in A_i$. The n-tuple is usually written $(a_1, a_2, \ldots, a_n)$ and the set of all n-tuples is denoted by

$$A_1 \times A_2 \times \cdots \times A_n.$$

This is the *set product* (or *cartesian product*) of $A_1, A_2, \ldots, A_n$. For example $\mathbb{R} \times \mathbb{R}$ is the set of coordinates of points in the plane.

The following result is a basic counting tool.

(1.1.2) *If $A_1, A_2, \ldots, A_n$ are finite sets, then $|A_1 \times A_2 \times \cdots \times A_n| = |A_1| \cdot |A_2| \ldots |A_n|$.*

Proof. In forming an n-tuple $(a_1, a_2, \ldots, a_n)$ we have $|A_1|$ choices for a_1, $|A_2|$ choices for $a_2, \ldots, |A_n|$ choices for a_n. Each choice of a_i's yields a different n-tuple. Therefore the total number of n-tuples is $|A_1| \cdot |A_2| \ldots |A_n|$. □

The power set. The *power set* of a set A is the set of all subsets of A, including the empty set and A itself; it is denoted by

$$P(A).$$

The power set of a finite set is always a larger set, as the next result shows.

---
[1] Augustus De Morgan (1806–1871)

(1.1.3) *If A is a finite set, then $|P(A)| = 2^{|A|}$.*

Proof. Let $A = \{a_1, a_2, \ldots, a_n\}$ with distinct a_i's. Also put $I = \{0, 1\}$. Each subset B of A is to correspond to an n-tuple $(i_1, i_2, \ldots, i_n)$ with $i_j \in I$. Here the rule for forming the n-tuple corresponding to B is this: $i_j = 1$ if $a_j \in B$ and $i_j = 0$ if $a_j \notin B$. Conversely every n-tuple $(i_1, i_2, \ldots, i_n)$ with $i_j \in I$ determines a subset B of A, defined by $B = \{a_j \mid 1 \leq j \leq n,\ i_j = 1\}$. It follows that the number of subsets of A equals the number of elements in $I \times I \times \cdots \times I$, (with n factors). By (1.1.2) we obtain $|P(A)| = 2^n = 2^{|A|}$. □

The power set $P(A)$, together with the operations $\cup$ and $\cap$, constitute what is known as a *Boolean[2] algebra*; such algebras have become very important in logic and computer science.

Exercises (1.1)

1. Prove as many parts of (1.1.1) as possible.

2. Let A, B, C be sets such that $A \cap B = A \cap C$ and $A \cup B = A \cup C$. Prove that $B = C$.

3. If A, B, C are sets, establish the following:
    (a) $(A - B) - C = A - (B \cup C)$.
    (b) $A - (B - C) = (A - B) \cup (A \cap B \cap C)$.

4. The *disjoint union* $A \oplus B$ of sets A and B is defined by the rule $A \oplus B = A \cup B - A \cap B$, so its elements are those that belong to exactly one of A and B. Prove the following statements:
    (a) $A \oplus A = \emptyset$, $A \oplus B = B \oplus A$.
    (b) $(A \oplus B) \oplus C = A \oplus (B \oplus C)$.
    (c) $(A \oplus B) \cap C = (A \cap C) \oplus (B \cap C)$.

1.2 Relations, equivalence relations and partial orders

In mathematics it is often not sufficient to deal with the individual elements of a set since it may be critical to understand how elements of the set are related to each other. This leads us to formulate the concept of a relation.

Let A and B be sets. Then a *relation R between A and B* is a subset of the set product $A \times B$. The definition will be clarified if we use a more suggestive notation: if $(a, b) \in R$, then a is said to be *related* to b by R and we write

$$a \, R \, b.$$

---
[2]George Boole (1815–1864)

1.2 Relations, equivalence relations and partial orders

The most important case is of a relation R between A and itself; this is called *a relation on the set* A.

Examples of relations. (i) Let A be a set and define $R = \{(a, a) \mid a \in A\}$. Thus $a_1 \, R \, a_2$ means that $a_1 = a_2$ and R is the relation of equality on A.

(ii) Let P be the set of points and L the set of lines in the plane. A relation R from P to L is defined by: $p \, R \, \ell$ if the point p lies on the line ℓ. So R is the relation of incidence.

(iii) A relation R on the set of integers $\mathbb{Z}$ is defined by: $a \, R \, b$ if $a - b$ is even.

The next result confirms what one might suspect, that a finite set has many relations.

(1.2.1) *If A is a finite set, the number of relations on A equals $2^{|A|^2}$.*

For this is the number of subsets of $A \times A$ by (1.1.2) and (1.1.3).

The concept of a relation on a set is evidently a very broad one. In practice the relations of greatest interest are those which have special properties. The most common of these are listed next. Let R be a relation on a set A.

(a) R is *reflexive* if $a \, R \, a$ for all $a \in A$.

(b) R is *symmetric* if $a \, R \, b$ always implies that $b \, R \, a$.

(c) R is *antisymmetric* if $a \, R \, b$ and $b \, R \, a$ imply that $a = b$;

(d) R is *transitive* if $a \, R \, b$ and $b \, R \, c$ imply that $a \, R \, c$.

Relations which are reflexive, symmetric and transitive are called *equivalence relations*; they are of fundamental importance. Relations which are reflexive, antisymmetric and transitive are also important; they are called *partial orders*.

Examples. (a) Equality on a set is both an equivalence relation and a partial order.

(b) A relation R on $\mathbb{Z}$ is defined by: $a \, R \, b$ if and only if $a - b$ is even. This is an equivalence relation.

(c) If A is any set, the relation of containment $\subseteq$ is a partial order on the power set $P(A)$.

(d) A relation R on $\mathbb{N}$ is defined by $a \, R \, b$ if a divides b. Here R is a partial order on $\mathbb{N}$.

Equivalence relations and partitions. The structure of an equivalence relation on a set will now be analyzed. The essential conclusion will be that an equivalence relation causes the set to split up into non-overlapping non-empty subsets.

Let E be an equivalence relation on a set A. First of all we define the E-*equivalence class* of an element a of A to be the subset

$$[a]_E = \{x \mid x \in A \text{ and } x E a\}.$$

By the reflexive law $a \in [a]_E$, so
$$A = \bigcup_{a \in A} [a]_E$$
and A is the union of all the equivalence classes.

Next suppose that the equivalence classes $[a]_E$ and $[b]_E$ both contain an integer x. Assume that $y \in [a]_E$; then $y\,E\,a$, $a\,E\,x$ and $x\,E\,b$, by the symmetric law. Hence $y\,E\,b$ by two applications of the transitive law. Therefore $y \in [b]_E$ and we have proved that $[a]_E \subseteq [b]_E$. By the same reasoning $[b]_E \subseteq [a]_E$, so that $[a]_E = [b]_E$. It follows that distinct equivalence classes are disjoint, i.e., they have no elements in common.

What has been shown so far is that the set A is the union of the E-equivalence classes and that distinct equivalence classes are disjoint. A decomposition of A into disjoint non-empty subsets is called a *partition* of A. Thus E determines a partition of A.

Conversely, suppose that a partition of A into non-empty disjoint subsets A_λ, $\lambda \in \Lambda$, is given. We would like to construct an equivalence relation on A corresponding to the partition. Now each element of A belongs to a unique subset A_λ; thus we may define $a\,E\,b$ to mean that a and b belong to the same subset A_λ. It follows immediately from the definition that the relation E is an equivalence relation; what is more, the equivalence classes are just the subsets A_λ of the original partition.

We summarize these conclusions in:

(1.2.2) (i) *If E is an equivalence relation on a set A, the E-equivalence classes form a partition of A.*

(ii) *Conversely, each partition of A determines an equivalence relation on A for which the equivalence classes are the subsets in the partition.*

Thus the concepts of equivalence relation and partition are in essence the same.

Example (1.2.1) In the equivalence relation (b) above there are two equivalence classes, the sets of even and odd integers; of course these form a partition of $\mathbb{Z}$.

Partial orders. Suppose that R is a partial order on a set A, i.e., R is a reflexive, antisymmetric, transitive relation on A. Instead of writing $a\,R\,b$ it is customary to employ a more suggestive symbol and write
$$a \preceq b.$$
The pair $(A, \preceq)$ then constitutes a *partially ordered set* (or *poset*).

The effect of a partial order is to impose a hierarchy on the set A. This can be visualized by drawing a picture of the poset called a *Hasse[3] diagram*. It consists of

---

[3] Helmut Hasse (1898-1979).

vertices and edges drawn in the plane, the vertices representing the elements of A. A sequence of upward sloping edges from a to b, as in the diagram below, indicates that $a \preceq b$, for example. Elements a, b not connected by such a sequence of edges do not satisfy $a \preceq b$ or $b \preceq a$. In order to simplify the diagram as far as possible, it is agreed that unnecessary edges are to be omitted.

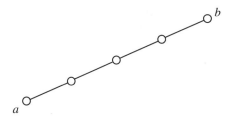

A very familiar poset is the power set of a set A with the partial order $\subseteq$, i.e. $(P(A), \subseteq)$.

Example (1.2.2) Draw the Hasse diagram of the poset $(P(A), \subseteq)$ where $A = \{1, 2, 3\}$.

This poset has $2^3 = 8$ vertices, which can be visualized as the vertices of a cube (drawn in the plane) standing on one vertex.

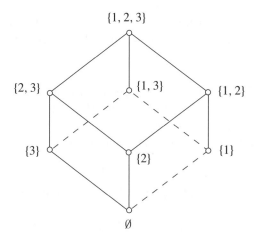

Partially ordered sets are important in algebra since they can provide a useful representation of substructures of algebraic structures such as subsets, subgroups, subrings etc..

A partial order $\preceq$ on a set A is called a *linear order* if, given $a, b \in A$, either $a \preceq b$ or $b \preceq a$ holds. Then $(A, \preceq)$ is called a *linearly ordered set* or *chain*. The Hasse diagram of a chain is a single sequence of edges sloping upwards. Obvious examples of chains are $(\mathbb{Z}, \leq)$ and $(\mathbb{R}, \leq)$ where $\leq$ is the usual "less than or equal to". Finally, a linear order on A is called a *well order* if each non-empty subset X of A contains a *least* element a, i.e., such that $a \preceq x$ for all elements $x \in X$. For example, it would seem clear that $\leq$ is a well order on the set of all positive integers, although this is actually an axiom, the Well-Ordering Law, which is discussed in Section 2.1.

Lattices. Consider a poset $(A, \preceq)$. If $a, b \in A$, then a *least upper bound* (or lub) of a and b is an element $\ell \in A$ such that $a \preceq \ell$ and $b \preceq \ell$, and if $a \preceq x$ and $b \preceq x$, with x in A, then $\ell \preceq x$. Similarly a *greatest lower bound* (or glb) of a and b is an element $g \in A$ such that $g \preceq a$ and $g \preceq b$, while $x \preceq a$ and $x \preceq b$ imply that $x \preceq g$. Part of the Hasse diagram of $(A, \preceq)$ is the lozenge shaped figure

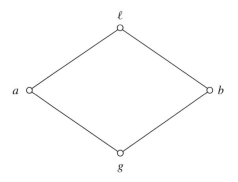

A poset in which each pair of elements has an lub and a glb is called a *lattice*. For example, $(P(S), \subseteq)$ is a lattice since the lub and glb of A and B are $A \cup B$ and $A \cap B$ respectively.

The composite of relations. Since a relation is a subset, two relations may be combined by forming their union or intersection. However there is a more useful way of combining relations called *composition*: let R and S be relations between A and B and between B and C respectively. Then the *composite relation*

$$S \circ R$$

is the relation between A and C defined by $a \: S \circ R \: c$ if there exists $b \in B$ such that $a \: R \: b$ and $b \: S \: c$.

For example, assume that $A = \mathbb{Z}$, $B = \{a, b, c\}$, $C = \{\alpha, \beta, \gamma\}$. Define relations $R = \{(1, a), (2, b), (4, c)\}$, $S = \{(a, \alpha), (b, \gamma), (c, \beta)\}$. Then $S \circ R = \{(1, \alpha), (2, \gamma), (4, \beta)\}$.

In particular one can form the composite of any two relations R and S on a set A. Notice that the condition for a relation R to be transitive can now be expressed in the form $R \circ R \subseteq R$.

A result of fundamental importance is the associative law for composition of relations.

(1.2.3) *Let R, S, T be relations between A and B, B and C, and C and D respectively. Then $T \circ (S \circ R) = (T \circ S) \circ R$.*

Proof. Let $a \in A$ and $d \in D$. Then $a \, (T \circ (S \circ R)) \, d$ means that there exists $c \in C$ such that $a \, (S \circ R) \, c$ and $c \, T \, d$, i.e., there exists $b \in B$ such that $a \, R \, b$, $b \, S \, c$ and $c \, T \, d$. Therefore $b \, (T \circ S) \, d$ and $a \, ((T \circ S) \circ R) \, d$. Thus $T \circ (S \circ R) \subseteq (T \circ S) \circ R$, and in the same way $(T \circ S) \circ R \subseteq T \circ (S \circ R)$. □

Exercises (1.2)

1. Determine whether or not each of the binary relations R defined on the sets A below is reflexive, symmetric, antisymmetric, transitive:
   (a) $A = \mathbb{R}$ and $a \, R \, b$ means $a^2 = b^2$.
   (b) $A = \mathbb{R}$ and $a \, R \, b$ means $a - b \leq 2$.
   (c) $A = \mathbb{Z} \times \mathbb{Z}$ and $(a, b) \, R \, (c, d)$ means $a + d = b + c$.
   (d) $A = \mathbb{Z}$ and $a \, R \, b$ means that $b = a + 3c$ for some integer c.

2. A relation $\sim$ on $\mathbb{R} - \{0\}$ is defined by $a \sim b$ if $ab > 0$. Show that $\sim$ is an equivalence relation and identify the equivalence classes.

3. Let $A = \{1, 2, 3, \ldots, 12\}$ and let $a \preceq b$ mean that a divides b. Show that $(A, \preceq)$ is a poset and draw its Hasse diagram.

4. Let $(A, \preceq)$ be a poset and let $a, b \in A$. Show that a and b have at most one lub and at most one glb.

5. Given linearly ordered sets $(A_i, \preceq_i)$, $i = 1, 2, \ldots, k$, suggest a way to make $A_1 \times A_2 \times \cdots \times A_k$ into a linearly ordered set.

6. How many equivalence relations are there on a set S with 1, 2, 3 or 4 elements?

7. Suppose that A is a set with n elements. Show that there are 2^{n^2-n} reflexive relations on A and $2^{n(n+1)/2}$ symmetric ones.

1.3 Functions

A more familiar object than a relation is a function. While functions are to be found throughout mathematics, they are usually first encountered in calculus, as real-valued functions of a real variable. Functions can provide convenient descriptions of complex processes in mathematics and the information sciences.

Let A and B be sets. A *function* or *mapping from A to B*, in symbols

$$\alpha : A \to B,$$

is a rule which assigns to each element a of A a unique element $\alpha(a)$ of B, called the *image of a* under α. The sets A and B are the *domain* and *codomain* of α respectively. The *image* of the function α is

$$\mathrm{Im}(\alpha) = \{\alpha(a) \mid a \in A\},$$

which is a subset of B. The set of all functions from A to B will sometimes be written Fun(A, B).

Examples of functions. (a) The functions which appear in calculus are the functions whose domain and codomain are subsets of $\mathbb{R}$. Such a function can be visualized by drawing its graph in the usual way.

(b) Given a function $\alpha : A \to B$, we may define

$$R_\alpha = \{(a, \alpha(a)) \mid a \in A\} \subseteq A \times B.$$

Then R_α is a relation between A and B. Observe that R_α is a special kind of relation since each a in A is related to a unique element of B, namely $\alpha(a)$.

Conversely, suppose that R is a relation between A and B such that each $a \in A$ is related to a unique $b \in B$. We may define a corresponding function $\alpha_R : A \to B$ by $\alpha_R(a) = b$ if and only if $a\ R\ b$. Thus functions from A to B may be regarded as special types of relation between A and B.

This observation permits us to form the composite of two functions $\alpha : A \to B$ and $\beta : B \to C$, by forming the composite of the corresponding relations: thus $\beta \circ \alpha : A \to C$ is defined by

$$\beta \circ \alpha(a) = \beta(\alpha(a)).$$

(c) *The characteristic function of a subset.* Let A be a fixed set. For each subset X of A define a function $\alpha_X : A \to \{0, 1\}$ by the rule

$$\alpha_X(a) = \begin{cases} 1 & \text{if } a \in X \\ 0 & \text{if } a \notin X. \end{cases}$$

Then α_X is called the *characteristic function* of the subset X. Conversely, every function $\alpha : A \to \{0, 1\}$ is the characteristic function of some subset of A – which subset?

(d) The *identity function* on a set A is the function id$_A : A \to A$ defined by id$_A(a) = a$ for all $a \in A$.

Injectivity and surjectivity. There are two special types of function of critical importance. A function $\alpha : A \to B$ is called *injective* (or *one-one*) if $\alpha(a) = \alpha(a')$ always implies that $a = a'$, i.e., distinct elements of A have distinct images in B under α. Next $\alpha : A \to B$ is *surjective* (or *onto*) if each element of B is the image under α of at least one element of A, i.e., Im$(\alpha) = B$. Finally, $\alpha : A \to B$ is said to be *bijective* (or a *one-one correspondence*) if it is both injective and surjective.

Example (1.3.1) (a) $\alpha : \mathbb{R} \to \mathbb{R}$ where $\alpha(x) = 2^x$ is injective but not surjective.

(b) $\alpha : \mathbb{R} \to \mathbb{R}$ where $\alpha(x) = x^3 - 4x$ is surjective but not injective. Here surjectivity is best seen by drawing the graph of $y = x^3 - 4x$. Note that any line parallel to the x-axis meets the curve at least once. But α is not injective since $\alpha(0) = 0 = \alpha(2)$.

(c) $\alpha : \mathbb{R} \to \mathbb{R}$ where $\alpha(x) = x^3$ is bijective.

(d) $\alpha : \mathbb{R} \to \mathbb{R}$ where $\alpha(x) = x^2$ is neither injective nor surjective.

Inverse functions. Functions $\alpha : A \to B$ and $\beta : B \to A$ are said to be *mutually inverse* if $\alpha \circ \beta = \mathrm{id}_B$ and $\beta \circ \alpha = \mathrm{id}_A$. If β' is another inverse for α, then $\beta = \beta'$: for by the associative law (1.2.3) we have $\beta' \circ (\alpha \circ \beta) = \beta' \circ \mathrm{id}_B = \beta' = (\beta' \circ \alpha) \circ \beta = \mathrm{id}_A \circ \beta = \beta$. Therefore α has a unique inverse if it has one at all; we write

$$\alpha^{-1} : B \to A$$

for the unique inverse of α, if it exists.

It is important to be able to recognize functions which possess inverses.

(1.3.1) *A function $\alpha : A \to B$ has an inverse if and only if it is bijective.*

Proof. Assume that $\alpha^{-1} : A \to B$ exists. If $\alpha(a_1) = \alpha(a_2)$, then, applying α^{-1} to each side, we arrive at $a_1 = a_2$, which shows that α is injective. Next, to show that α is surjective, let $b \in B$. Then $b = \mathrm{id}_B(b) = \alpha(\alpha^{-1}(b)) \in \mathrm{Im}(\alpha)$, showing that $\mathrm{Im}(\alpha) = B$ and α is surjective. Thus α is bijective.

Conversely, let α be bijective. If $b \in B$, there is precisely one element a in A such that $\alpha(a) = b$ since α is bijective. Define $\beta : B \to A$ by $\beta(b) = a$. Then $\alpha\beta(b) = \alpha(a) = b$ and $\alpha\beta = \mathrm{id}_B$. Also $\beta\alpha(a) = \beta(b) = a$; since every a in A arises in this way, $\beta\alpha = \mathrm{id}_A$ and $\beta = \alpha^{-1}$. □

The next result records some useful facts about inverses.

(1.3.2) (i) *If $\alpha : A \to B$ is a bijective function, then so is $\alpha^{-1} : B \to A$, and indeed $(\alpha^{-1})^{-1} = \alpha$.*

(ii) *If $\alpha : A \to B$ and $\beta : B \to C$ are bijective functions, then $\beta \circ \alpha : A \to C$ is bijective, and $(\beta \circ \alpha)^{-1} = \alpha^{-1} \circ \beta^{-1}$.*

Proof. (i) The equations $\alpha \circ \alpha^{-1} = \mathrm{id}_B$ and $\alpha^{-1} \circ \alpha = \mathrm{id}_A$ tell us that α is the inverse of α^{-1}.

(ii) Check directly that $\alpha^{-1} \circ \beta^{-1}$ is the inverse of $\beta \circ \alpha$, using the associative law twice: thus $(\beta \circ \alpha) \circ (\alpha^{-1} \circ \beta^{-1}) = ((\beta \circ \alpha) \circ \alpha^{-1}) \circ \beta^{-1} = (\beta \circ (\alpha \circ \alpha^{-1})) \circ \beta^{-1} = (\beta \circ \mathrm{id}_A) \circ \beta^{-1} = \beta \circ \beta^{-1} = \mathrm{id}_C$. Similarly $(\alpha^{-1} \circ \beta^{-1}) \circ (\beta \circ \alpha) = \mathrm{id}_A$. □

Application to automata. As an illustration of how the language of sets and functions may be used to describe information systems we shall briefly consider automata. An *automaton* is a theoretical device which is a basic model of a digital computer. It consists of an input tape, an output tape and a "head", which is able to read symbols on the input tape and print symbols on the output tape. At any instant the system is in one of a number of states. When the automaton reads a symbol on the input tape, it goes to another state and writes a symbol on the output tape.

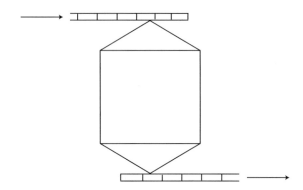

To make this idea precise we define an automaton A to be a 5-tuple

$$(I, O, S, \nu, \sigma)$$

where I and O are the respective sets of input and output symbols, S is the set of states, $\nu : I \times S \to O$ is the *output function* and $\sigma : I \times S \to S$ is the *next state function*. The automaton operates in the following manner. If it is in state $s \in S$ and an input symbol $i \in I$ is read, the automaton prints the symbol $\nu(i, s)$ on the output tape and goes to state $\sigma(i, s)$. Thus the mode of operation is determined by the three sets I, O, S and the two functions ν, σ.

Exercises (1.3)

1. Which of the following functions are injective, surjective, bijective?
   (a) $\alpha : \mathbb{R} \to \mathbb{Z}$ where $\alpha(x) = [x]$ (= the largest integer $\leq x$).
   (b) $\alpha : \mathbb{R}^{>0} \to \mathbb{R}$ where $\alpha(x) = \log(x)$. (Here $\mathbb{R}^{>0} = \{x \mid x \in \mathbb{R}, \ x > 0\}$).
   (c) $\alpha : A \times B \to B \times A$ where $\alpha((a, b)) = (b, a)$.

2. Prove that the composite of injective functions is injective, and the composite of surjective functions is surjective.

3. Let $\alpha : A \to B$ be a function between finite sets. Show that if $|A| > |B|$, then α cannot be injective, and if $|A| < |B|$, then α cannot be surjective.

4. Define $\alpha : \mathbb{R} \to \mathbb{R}$ by $\alpha(x) = \frac{x^3}{x^2+1}$. Prove that α is bijective.

5. Give an example of functions α, β on a set A such that $\alpha \circ \beta = \text{id}_A$ but $\beta \circ \alpha \neq \text{id}_A$.

6. Let $\alpha : A \to B$ be a injective function. Show that there is a surjective function $\beta : B \to A$ such that $\beta \circ \alpha = \text{id}_A$.

7. Let $\alpha : A \to B$ be a surjective function. Show that there is an injective function $\beta : B \to A$ such that $\alpha \circ \beta = \mathrm{id}_B$.

8. Describe a simplified version of an automaton with no output tape where the output is determined by the new state – this is called a *state output automaton*.

9. Let $\alpha : A \to B$ be a function. Define a relation E_α on A by the rule: $a \, E_\alpha \, a'$ means that $\alpha(a) = \alpha(a')$. Prove that E_α is an equivalence relation on A. Then show that conversely, if E is any equivalence relation on a set A, then $E = E_\alpha$ for some function α with domain A.

1.4 Cardinality

If we want to compare two sets, a natural basis for comparison is the "size" of each set. If the sets are finite, their sizes are just the numbers of elements in the set. But how can one measure the size of an infinite set? A reasonable point of view would be to hold that two sets have the same size if their elements can be paired off. Certainly two finite sets have the same number of elements precisely when their elements can be paired. The point to notice is that this idea also applies to infinite sets, making it possible to give a rigorous definition of the size of a set, its *cardinal*.

Let A and B be two sets. Then A and B are said to be *equipollent* if there is a bijection $\alpha : A \to B$: thus the elements of A and B may be paired off as $(a, \alpha(a))$, $a \in A$. It follows from (1.3.2) that equipollence is an equivalence relation on the class of all sets. Thus each set A belongs to a unique equivalence class, which will be written

$$|A|$$

and called the *cardinal* of A. Informally we think of $|A|$ as the collection of all sets with the same "size" as A. A *cardinal number* is the cardinal of some set.

If A is a finite set with exactly n elements, then A is equipollent to the set $\{0, 1, \ldots, n-1\}$ and $|A| = |\{0, 1, \ldots, n-1\}|$. It is reasonable to identify the finite cardinal $|\{0, 1, \ldots, n-1\}|$ with the non-negative integer n. For then cardinal numbers appear as infinite versions of the non-negative integers.

Let us sum up our very elementary conclusions so far.

(1.4.1) (i) *Every set A has a unique cardinal number $|A|$.*

(ii) *Two sets are equipollent if and only if they have the same cardinal.*

(iii) *The cardinal of a finite set is to be identified with the number of its elements.*

Since we plan to use cardinals to compare the sizes of sets, it makes sense to define a "less than or equal to" relation $\leq$ on cardinals as follows:

$$|A| \leq |B|$$

means that there is an injective function $\alpha : A \to B$. Of course we will write $|A| < |B|$ if $|A| \leq |B|$ and $|A| \neq |B|$.

It is important to realize that this definition of $\leq$ depends only on the cardinals $|A|$ and $|B|$, not on the sets A and B. For if $A' \in |A|$ and $B' \in |B|$, then there are bijections $\alpha' : A' \to A$ and $\beta' : B \to B'$; by composing these with $\alpha : A \to B$ we obtain the injection $\beta' \circ \alpha \circ \alpha' : A' \to B'$. Thus $|A'| \leq |B'|$.

Next we prove a famous result about inequality of cardinals.

(1.4.2) (The Cantor[4]–Bernstein[5] Theorem) *If A and B are sets such that $|A| \leq |B|$ and $|B| \leq |A|$, then $|A| = |B|$.*

The proof of (1.4.2) is our most challenging proof so far and some readers may prefer to skip it. However the basic idea behind it is not difficult.

Proof. By hypothesis there are injective functions $\alpha : A \to B$ and $\beta : B \to A$. These will be used to construct a bijective function $\gamma : A \to B$, which will show that $|A| = |B|$.

Consider an arbitrary element a in A; then either $a = \beta(b)$ for some unique $b \in B$ or else $a \notin \mathrm{Im}(\beta)$: here we use the injectivity of β. Similarly, either $b = \alpha(a')$ for a unique $a' \in A$ or else $b \notin \mathrm{Im}(\alpha)$. Continuing this process, we can trace back the "ancestry" of the element a. There are three possible outcomes: (i) we eventually reach an element of $A - \mathrm{Im}(\beta)$; (ii) we eventually reach an element of $B - \mathrm{Im}(\alpha)$; (iii) the process continues without end.

Partition the set A into three subsets corresponding to possibilities (i), (ii), (iii), and call them AA, AB, $A\infty$ respectively. In a similar fashion the set B decomposes into three disjoint subsets BA, BB, $B\infty$; for example, if $b \in BA$, we can trace b back to an element of $A - \mathrm{Im}(\beta)$.

Now we are in a position to define the function $\gamma : A \to B$. First notice that the restriction of α to AA is a bijection from AA to BA, and the restriction of α to $A\infty$ is a bijection from $A\infty$ to $B\infty$. Also, if $x \in AB$, there is a unique element $x' \in BB$ such that $\beta(x') = x$. Now define

$$\gamma(x) = \begin{cases} \alpha(x) & \text{if } x \in AA \\ \alpha(x) & \text{if } x \in A\infty \\ x' & \text{if } x \in AB. \end{cases}$$

Then γ is the desired bijection. □

Corollary (1.4.3) *The relation $\leq$ is a partial order on cardinal numbers.*

For we have proved antisymmetry in (1.4.2), while reflexivity and transitivity are clearly true by definition. In fact one can do better since $\leq$ is even a *linear* order. This is because of:

---

[4] Georg Cantor (1845–1918).
[5] Felix Bernstein (1878–1956).

(1.4.4) (The Law of Trichotomy) *If A and B are sets, then exactly one of the following must hold:*
$$|A| < |B|, \quad |A| = |B|, \quad |B| < |A|.$$

The proof of this theorem will not be given at this point since it depends on advanced material. See however (12.1.5) below for a proof.

The next result establishes the existence of arbitrarily large cardinal numbers.

(1.4.5) *If A is any set, then $|A| < |P(A)|$.*

Proof. The easy step is to show that $|A| \leq |P(A)|$. This is because the assignment $a \mapsto \{a\}$ sets up an injection from A to $P(A)$.

Next assume that $|A| = |P(A)|$, so that there is a bijection $\alpha : A \rightarrow P(A)$. Of course at this point we are looking for a contradiction. The trick is to consider the subset $B = \{a \mid a \in A, \ a \notin \alpha(a)\}$ of A. Then $B \in P(A)$, so $B = \alpha(a)$ for some $a \in A$. Now either $a \in B$ or $a \notin B$. If $a \in B$, then $a \notin \alpha(a) = B$; if $a \notin B = \alpha(a)$, then $a \in B$. This is our contradiction. □

Countable sets. The cardinal of the set of natural numbers $\mathbb{N} = \{0, 1, 2, \ldots\}$ is denoted by
$$\aleph_0$$
where $\aleph$ is the Hebrew letter aleph. A set A is said to be *countable* if $|A| \leq \aleph_0$. Essentially this means that the elements of A can be "labelled" by attaching to each element a natural number as a label. An uncountable set cannot be so labelled.

We need to explain what is meant by an infinite set for the next result to be meaningful. A set A will be called *infinite* if it has a subset that is equipollent with $\mathbb{N}$, i.e., if $\aleph_0 \leq |A|$. An *infinite cardinal* is the cardinal of an infinite set.

(1.4.6) $\aleph_0$ *is the smallest infinite cardinal.*

Proof. If A is an infinite set, then A has a subset B such that $\aleph_0 = |B|$. Hence $\aleph_0 \leq |A|$. □

It follows that *if A is a countable set, then either A is finite or $|A| = \aleph_0$*. As the final topic of the chapter we consider the cardinals of the sets $\mathbb{Q}$ and $\mathbb{R}$.

(1.4.7) (a) *The set $\mathbb{Q}$ of rational numbers is countable.*

(b) *The set $\mathbb{R}$ of real numbers is uncountable.*

Proof. (a) Each positive rational number has the form $\frac{m}{n}$ where m and n are positive integers. Arrange these rationals in rectangular array, with $\frac{m}{n}$ in the m-th row and n-th column. Of course each rational will occur infinitely often because of cancellation. Now follow the path indicated by the arrows in the diagram below: This creates a sequence in which every positive rational number appears infinitely often. Delete all

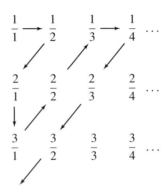

repetitions in the sequence. Insert 0 at the beginning of the sequence and insert $-r$ immediately after r for each positive rational r. Now every rational occurs exactly once in the sequence. Hence $\mathbb{Q}$ is countable.

(b) It is enough to show that the set I of all real numbers r such that $0 \leq r \leq 1$ is uncountable: this is because $|I| \leq |\mathbb{R}|$.

Assume that I is countable, so that it can be written in the form $\{r_1, r_2, r_3, \ldots\}$. Write each r_i as a decimal, say

$$r_i = 0 \cdot r_{i1} r_{i2} \ldots$$

where $0 \leq r_{ij} \leq 9$. We shall get a contradiction by producing a number in the set I which does not equal any r_i. Define

$$s_i = \begin{cases} 0 & \text{if } r_{ii} \neq 0 \\ 1 & \text{if } r_{ii} = 0 \end{cases}$$

and let s be the decimal $0 \cdot s_1 s_2 \ldots$; then certainly $s \in I$. Hence $s = r_i$ for some i, so that $s_i = r_{ii}$; but this is impossible by the definition of s_i. □

Exercises (1.4)

1. A finite set cannot be equipollent to a proper subset.

2. A set is infinite if and only if it has the same cardinal as some proper subset.

3. If there is a surjection from a set A to a set B, then $|B| \leq |A|$.
4. Show that $|\mathbb{Z}| = \aleph_0$ and $|\mathbb{Z} \times \mathbb{Z}| = \aleph_0$.

5. Let $A_1, A_2, \ldots$ be countably many countable sets. Prove that $\bigcup_{i=1,2,\ldots} A_i$ is a countable set. [Hint: write $A_i = \{a_{i0}, a_{i1}, \ldots\}$ and follow the method of the proof of (1.4.7)(a)].

6. Suggest reasonable definitions of the sum and product of two cardinal numbers. [Hint: try using the union and set product]

Chapter 2
The integers

The role of the integers is central in algebra, as it is in all parts of mathematics. One reason for this is that the set of integers $\mathbb{Z}$, together with the standard arithmetic operations of addition and multiplication, serves as a model for several of the fundamental structures of algebra, including groups and rings. In this chapter we will develop the really basic properties of the integers.

2.1 Well-ordering and mathematical induction

We begin by listing the properties of the fundamental arithmetic operations on $\mathbb{Z}$, addition and multiplication. In the following a, b, c are arbitrary integers.

(i) $a + b = b + a$, $ab = ba$ (commutative laws);

(ii) $(a + b) + c = a + (b + c)$, $(ab)c = a(bc)$ (associative laws);

(iii) $(a + b)c = ac + bc$ (distributive law);

(iv) $0 + a = a$, $1 \cdot a = a$ (existence of identities);

(v) each integer a has a *negative* $-a$ with the property $a + (-a) = 0$;

(vi) if $ab = 0$, then $a = 0$ or $b = 0$.

Next we list properties of the less than or equal to relation $\leq$ on $\mathbb{Z}$.

(vii) $\leq$ is a linear order on $\mathbb{Z}$, i.e., the relation $\leq$ is reflexive, antisymmetric and transitive; in addition, for any pair of integers a, b either $a \leq b$ or $b \leq a$;

(viii) if $a \leq b$ and $c \geq 0$, then $ac \leq bc$;

(ix) if $a \leq b$, then $-b \leq -a$.

These properties we shall take as axioms. But there is a further property of the linearly ordered set $(\mathbb{Z}, \leq)$ which is independent of the above axioms and is quite vital for the development of the elementary theory of the integers.

The Well-Ordering Law. Let k be a fixed integer and put $U = \{n \mid n \in \mathbb{Z},\ n \geq k\}$. Suppose that S is a non-empty subset of U. Then the *Well-Ordering Law* (WO) asserts that S has a smallest element. Thus $\leq$ is a well order on U in the sense of 1.2. While this may seem a harmless assumption, it cannot be deduced from axioms (i)–(ix) and must be assumed as another axiom.

The importance of WO for us is that it provides a sound basis for the method of proof by mathematical induction. This is embodied in

(2.1.1) (The Principle of Mathematical Induction) *Let k be an integer and let U be the set $\{n \mid n \in \mathbb{Z},\ n \geq k\}$. Assume that S is a subset of U such that*

(i) $k \in S$, and

(ii) *if $n \in S$, then $n + 1 \in S$.*

Then S equals U.

Once again the assertion sounds fairly obvious, but in order to prove it, we must use WO. To see how WO applies, assume that $S \neq U$, so that $S' = U - S$ is not empty. Then WO guarantees that S' has a smallest element, say s. Notice that $k < s$ since $k \in S$ by hypothesis. Thus $k \leq s - 1$ and $s - 1 \notin S'$ because s is minimal in S'. Hence $s - 1 \in S$, which by (ii) above implies that $s \in S$, a contradiction. Thus (2.1.1) is established.

The method of proof by induction. Suppose that k is a fixed integer and that for each integer $n \geq k$ there is a proposition $p(n)$, which is either true or false. Assume that the following hold:

(i) $p(k)$ is true;

(ii) if $p(n)$ is true, then $p(n + 1)$ is true.

Then we can conclude that $p(n)$ is true for all $n \geq k$.

For let S be the set of all integers $n \geq k$ for which $p(n)$ is true. Then the hypotheses of PMI (Principle of Mathematical Induction) apply to S. The conclusion is that S equals $\{n \mid n \in \mathbb{Z},\ n \geq k\}$, i.e., $p(n)$ is true for all $n \geq k$.

Here is a simple example of proof by mathematical induction.

Example (2.1.1) Use mathematical induction to show that $8^{n+1} + 9^{2n-1}$ is a multiple of 73 for all positive integers n.

Let $p(n)$ denote the statement: $8^{n+1} + 9^{2n-1}$ is a multiple of 73. Then $p(1)$ is certainly true since $8^{n+1} + 9^{2n-1} = 73$ when $n = 1$. Assume that $p(n)$ is true; we have to deduce that $p(n + 1)$ is true. Now we may rewrite $8^{(n+1)+1} + 9^{2(n+1)-1}$ in the form

$$8^{n+2} + 9^{2n+1} = 8(8^{n+1} + 9^{2n-1}) + 9^{2n+1} - 8 \cdot 9^{2n-1}$$
$$= 8(8^{n+1} + 9^{2n-1}) + 73 \cdot 9^{2n-1}.$$

Since both terms in the last expression are multiples of 73, so is the left hand side. Thus $p(n + 1)$ is true, and by PMI the statement $p(n)$ is true for all $n \geq 1$.

(2.1.2) (The Principle of Mathematical Induction – Alternate Form) *Let k be any integer and let $U = \{n \mid n \in \mathbb{Z}, n \geq k\}$. Suppose that S is a subset of U such that:*

(i) $k \in S$;

(ii) *if $m \in S$ for all integers m such that $k \leq m < n$, then $n \in S$.*

Then $S = U$.

This variant of PMI follows from WO just as the original form does. There are situations where proof by induction cannot be used but proof by the alternate form is effective. In such a case one has a proposition $p(n)$ for $n \geq k$ such that:

(i) $p(k)$ is true;

(ii) if $p(m)$ is true whenever $k \leq m < n$, then $p(m)$ is true.

The conclusion is that $p(n)$ is true for all $n \geq k$.

A good example of a proposition where this type of induction proof is successful is the Fundamental Theorem of Arithmetic – see (2.2.7).

Our approach to the integers in this section has been quite naive: we have simply stated as axioms all the properties that we needed. For an excellent axiomatic treatment of the construction of the integers, including an account of the axioms of Peano see [4].

Exercises (2.1).

1. Use induction to establish the following summation formulas for $n \geq 1$.
   (a) $1 + 2 + 3 + \cdots + n = \frac{1}{2}n(n+1)$;
   (b) $1^2 + 2^2 + 3^2 + \cdots + n^2 = \frac{1}{6}n(n+1)(2n+1)$;
   (c) $1^3 + 2^3 + 3^3 + \cdots + n^3 = \left(\frac{1}{2}n(n+1)\right)^2$.

2. Deduce the alternate form of PMI from WO.

3. Prove that $2^n > n^3$ for all integers $n \geq 10$.

4. Prove that $2^n > n^4$ for all integers $n \geq 17$.

5. Prove by mathematical induction that $6 \mid n^3 - n$ for all integers $n \geq 0$.

6. Use the alternate form of mathematical induction to show that any n cents worth of postage where $n \geq 12$ can be made up using 4 cent and 5 cent stamps. [Hint: first verify the statement for $n \leq 15$].

2.2 Division in the integers

In this section we establish the basic properties of the integers that relate to division, notably the Division Algorithm, the existence of greatest common divisors and the Fundamental Theorem of Arithmetic.

Recall that if a, b are integers, then a *divides* b, in symbols

$$a \mid b,$$

if there is an integer c such that $b = ac$. The following properties of division are simple consequences of the definition, as the reader should verify.

(2.2.1) (i) *The relation of division is a partial order on $\mathbb{Z}$.*
(ii) *If $a \mid b$ and $a \mid c$, then $a \mid bx + cy$ for all integers x, y.*
(iii) *$a \mid 0$ for all a, while $0 \mid a$ only if $a = 0$.*
(iv) *$1 \mid a$ for all a, while $a \mid 1$ only if $a = \pm 1$.*

The Division Algorithm. The first result about the integers of real significance is the *Division Algorithm*; it codefies the time-honored process of dividing one integer by another to obtain a quotient and remainder. It should be noted that the proof of the result uses WO.

(2.2.2) *Let a, b be integers with $b \neq 0$. Then there exist unique integers q (the quotient) and r (the remainder) such that $a = bq + r$ and $0 \leq r < |b|$.*

Proof. Let S be the set of all non-negative integers of the form $a - bq$, where $q \in \mathbb{Z}$. In the first place we need to observe that S is not empty. For if $b > 0$, and we choose an integer $q \leq \frac{a}{b}$, then $a - bq \geq 0$; if $b < 0$, choose an integer $q \geq \frac{a}{b}$, so that again $a - bq \geq 0$. Applying the Well-Ordering Law to the set S, we conclude that it contains a smallest element, say r. Then $r = a - bq$ for some integer q, and $a = bq + r$.

Now suppose that $r \geq |b|$. If $b > 0$, then $a - b(q + 1) = r - b < r$, while if $b < 0$, then $a - b(q - 1) = r + b < r$. In each case a contradiction is reached since we have produced an integer in S which is less than r. Hence $r < |b|$.

Finally, we must show that q and r are unique. Suppose that $a = bq' + r'$ where $q', r' \in \mathbb{Z}$ and $0 \leq r' < |b|$. Then $bq + r = bq' + r'$ and $b(q - q') = r' - r$. Thus $|b| \cdot |q - q'| = |r - r'|$. If $q \neq q'$, then $|r - r'| \geq |b|$, whereas $|r - r'| < |b|$ since $0 \leq r, r' < |b|$. Therefore $q = q'$ and it follows at once that $r = r'$. □

When $a < 0$ or $b < 0$, care must be taken to ensure that a negative remainder is not obtained. For example, if $a = -21$ and $b = -4$, then $-21 = (-4)6 + 3$, so that $q = 6$ and $r = 3$.

Greatest common divisors. Let $a_1, a_2, \ldots, a_n$ be integers. An integer c which divides every a_i is called a *common divisor* of $a_1, a_2, \ldots, a_n$. Our next goal is to establish the existence of a *greatest common divisor*.

(2.2.3) *Let $a_1, a_2, \ldots, a_n$ be integers. Then there is a unique integer $d \geq 0$ such that*

(i) *d is a common divisor of $a_1, a_2, \ldots, a_n$;*

(ii) *every common divisor of $a_1, a_2, \ldots, a_n$ divides d;*

(iii) $d = a_1 x_1 + \cdots + a_n x_n$ *for some integers x_i.*

Proof. If all of the a_i are 0, we can take $d = 0$ since this fits the description. So assume that at least one a_i is non-zero. Then the set

$$S = \{a_1 x_1 + a_2 x_2 + \cdots + a_n x_n \mid x_i \in \mathbb{Z}, \ a_1 x_1 + a_2 x_2 + \cdots + a_n x_n > 0\}$$

is non-empty. By WO there is a least element in S, say $d = a_1 x_1 + a_2 x_2 + \cdots + a_n x_n$. If an integer c divides each a_i, then $c \mid d$ by (2.2.1). It remains to show that $d \mid a_i$ for all i.

By the Division Algorithm we can write $a_i = d q_i + r_i$ where $q_i, r_i \in \mathbb{Z}$ and $0 \le r_i < d$. Then

$$r_i = a_i - d q_i = a_1(-d q_1) + \cdots + a_i(1 - d q_i) + \cdots + a_n(-d q_n).$$

If $r_i \ne 0$, then $r_i \in S$, which contradicts the minimality of d in S. Hence $r_i = 0$ and $d \mid a_i$ for all i.

Finally, we need to prove that d is unique. If d' is another integer satisfying (i) and (ii), then $d \mid d'$ and $d' \mid d$, so that $d = d'$ since $d, d' \ge 0$. □

The integer d of (2.2.3) is called the *greatest common divisor* of $a_1, a_2, \ldots, a_n$, in symbols

$$d = \gcd\{a_1, a_2, \ldots, a_n\}.$$

If $d = 1$, the integers $a_1, a_2, \ldots, a_n$ are said to be *relatively prime*; of course this means that these integers have no common divisors except ± 1.

The Euclidean[1] Algorithm. The proof of the existence of gcd's which has just been given is not constructive, i.e., it does not provide a method for calculating gcd's. Such a method is given by a well-known procedure called the *Euclidean Algorithm*.

Assume that a, b are integers with $b \ne 0$. Apply the Division Algorithm successively for division by $b, r_1, r_2, \ldots$ to get

$$\begin{aligned} a &= b q_1 + r_1, & 0 &< r_1 < |b|, \\ b &= r_1 q_2 + r_2, & 0 &< r_2 < r_1, \\ r_1 &= r_2 q_3 + r_3, & 0 &< r_3 < r_2, \\ &\ \vdots & & \\ r_{n-3} &= r_{n-2} q_{n-1} + r_{n-1}, & 0 &< r_{n-1} < r_{n-2}, \\ r_{n-2} &= r_{n-1} q_n + 0. & & \end{aligned}$$

Here r_{n-1} is the least non-zero remainder, which must exist by WO. With this notation we can state:

---
[1] Euclid of Alexandria (325–265 BC)

(2.2.4) (The Euclidean Algorithm) *The greatest common divisor of a and b equals the last non-zero remainder r_{n-1}.*

Proof. Starting with the second last equation in the system above, we can solve back for r_{n-1}, obtaining eventually an expression of the form $r_{n-1} = ax + by$, where $x, y \in \mathbb{Z}$. This shows that any common divisor of a and b must divide r_{n-1}. Also we can use the system of equations above to show successively that $r_{n-1} \mid r_{n-2}$, $r_{n-1} \mid r_{n-3}, \ldots$, and finally $r_{n-1} \mid b$, $r_{n-1} \mid a$. It now follows that $r_{n-1} = \gcd\{a, b\}$ by uniqueness of gcd's. □

Example (2.2.1) Find $\gcd(76, 60)$. We compute successively:

$$76 = 60 \cdot 1 + 16$$
$$60 = 16 \cdot 3 + 12$$
$$16 = 12 \cdot 1 + 4$$
$$12 = 4 \cdot 3 + 0$$

Hence $\gcd\{76, 60\}$ = the last non-zero remainder 4. Also by reading back from the third equation we obtain the predicted expression for the gcd, $4 = 76 \cdot 4 + 60 \cdot (-5)$.

Of course the Eucludean algorithm may also be applied to calculate gcd's of more than two integers: for example, $\gcd\{a, b, c\} = \gcd\{\gcd\{a, b\}, c\}$.

A very useful tool in working with divisibility is:

(2.2.5) (Euclid's Lemma) *Let a, b, m be integers. If m divides ab and m is relatively prime to a, then m divides b.*

Proof. By hypothesis $\gcd\{a, m\} = 1$, so by (2.2.3) there are integers x, y such that $1 = mx + ay$. Multiplying by b, we obtain $b = mbx + aby$. Since m divides ab, it divides the right side of the equation. Hence m divides b. □

Recall that a *prime number* is an integer $p > 1$ such that 1 and p are its only positive divisors. If p is a prime and a is any integer, then either $\gcd\{a, p\} = 1$ or $p \mid a$. Thus (2.2.5) has the consequence.

(2.2.6) *If a prime p divides ab where $a, b \in \mathbb{Z}$, then p divides a or b.*

The Fundamental Theorem of Arithmetic. It is a familiar and fundamental fact that every integer greater than 1 can be expressed as a product of primes. The proof of this result is an excellent example of proof by the alternate form of mathematical induction.

(2.2.7) *Every integer $n > 1$ can be expressed as a product of primes. Moreover the expression is unique up to the order of the factors.*

Proof. (i) *Existence.* We show that n is a product of primes, which is certainly true if $n = 2$. Assume that every integer m satisfying $2 \leq m < n$ is a product of primes. If n itself is a prime, there is nothing to prove. Otherwise $n = n_1 n_2$ where $1 < n_i < n$. Then n_1 and n_2 are both products of primes, whence so is $n = n_1 n_2$. The result now follows by the alternate form of mathematical induction, (2.1.2).

(ii) *Uniqueness.* In this part we have to show that n has a unique expression as a product of primes. Again this is clearly correct for $n = 2$. Assume that if $2 \leq m < n$, then m is uniquely expressible as a product of primes. Next suppose that $n = p_1 p_2 \ldots p_r = q_1 q_2 \ldots q_s$ where the p_i and q_j are primes. Then $p_1 \mid q_1 q_2 \ldots q_s$, and by (2.2.6) p_1 must divide, and hence equal, one of the q_j's; we can assume $p_1 = q_1$ by relabelling the q_j's if necessary. Now cancel p_1 to get $m = p_2 \ldots p_r = q_2 \ldots q_s$. Since $m = n/p_1 < n$, we deduce that $p_2 = q_2, \ldots, p_r = q_r$, and $r = s$, after a possible further relabelling of the q_j's. So the result is proven. □

A convenient expression for an integer $n > 1$ is

$$n = p_1^{e_1} \ldots p_k^{e_1}$$

where the p_i are *distinct* primes and $e_i > 0$. That the p_i and e_i are unique up to order is implied by (2.2.7).

Finally, we prove a famous theorem of Euclid on the infinity of primes.

(2.2.8) *There exist infinitely many prime numbers.*

Proof. Suppose this is false and let $p_1, p_2, \ldots, p_k$ be the list of all primes. The trick is to produce a prime that is not on the list. To do this put $n = p_1 p_2 \ldots p_k + 1$. Now no p_i can divide n, otherwise $p_i \mid 1$. But n is certainly divisible by at least one prime, so we reach a contradiction. □

Example (2.2.2) *If p is a prime, then $\sqrt{p}$ is not a rational number.*

Indeed assume that $\sqrt{p}$ is a rational, so that $\sqrt{p} = \frac{m}{n}$ where m, n are integers; evidently there is nothing to be lost in assuming that m and n are relatively prime since any common factor can be cancelled. Squaring both sides, we get $p = m^2/n^2$ and $m^2 = pn^2$. Hence $p \mid m^2$, and Euclid's Lemma shows that $p \mid m$. Write $m = pm_1$ for some integer m_1. Then $p^2 m_1^2 = pn^2$ and so $pm_1^2 = n^2$. Thus $p \mid n^2$, which means that $p \mid n$ and $\gcd\{m, n\} \neq 1$, a contradiction.

Exercises (2.2)

1. Prove that $\gcd\{a_1, a_2, \ldots, a_{m+1}\} = \gcd\{\gcd\{a_1, a_2, \ldots, a_m\}, a_{m+1}\}$.

2. Prove that $\gcd\{ka_1, ka_2, \ldots, ka_m\} = k \cdot \gcd\{a_1, a_2, \ldots, a_m\}$ where $k \geq 0$.

3. Use the Euclidean Algorithm to compute the following gcd's: $\gcd\{840, 410\}$, $\gcd\{24, 328, 472\}$. Then express each gcd as a linear combination of the relevant integers.

4. Suppose that a, b, c are given integers. Prove that the equation $ax + by = c$ has a solution for x, y in integers if and only if $\gcd\{a, b\}$ divides c.

5. Find all solutions in integers of the equation $6x + 11y = 1$.

6. If p and q are distinct primes, then $\sqrt{pq}$ is irrational.

7. A *least common multiple* (= lcm) of integers $a_1, a_2, \ldots, a_m$ is an integer $\ell \geq 0$ such that each a_i divides ℓ and ℓ divides any integer which is divisible by every a_i.
    (a) Let $a_i = p_1^{e_{i1}} p_2^{e_{i2}} \ldots p_n^{e_{in}}$ where the e_{ij} are integers ≥ 0 and the primes p_i are all different; show that $\text{lcm}\{a_1, \ldots, a_m\} = p_1^{f_1} p_2^{f_2} \ldots p_m^{f_m}$ with $f_j = \max\{e_{1j}, \ldots, e_{mj}\}$.
    (b) Prove that $\gcd\{a, b\} \cdot \text{lcm}\{a, b\} = ab$.

8. Let a and b be integers with $b > 0$. Prove that there are integers u, v such that $a = bu + v$ and $-\frac{b}{2} \leq v < \frac{b}{2}$. [Hint: start with the Division Algorithm].

9. Prove that $\gcd\{4n + 5, 3n + 4\} = 1$ for all integers n.

10. Prove that $\gcd\{2n + 6, n^2 + 3n + 2\} = 2$ or 4 for any integer n, and that both possibilities can occur.

11. Show that if $2^n + 1$ is prime, then n must have the form 2^l. (Such primes are called *Fermat primes*).

12. The only integer n which is expressible as $a^3(3a + 1)$ and $b^2(b + 1)^3$ with a, b relatively prime and positive is 2000.

2.3 Congruences

The notion of congruence was introduced by Gauss[2] in 1801, but it had long been implicit in ancient writings concerned with the computation of dates.

Let m be a positive integer. Then two integers a, b are said to be *congruent modulo m*, in symbols

$$a \equiv b \pmod{m},$$

if m divides $a - b$. Thus congruence modulo m is a relation on $\mathbb{Z}$ and an easy check reveals that it is an equivalence relation. Hence the set $\mathbb{Z}$ splits up into equivalence classes, which in this context are called *congruence classes modulo m*: see (1.2.2). The unique congruence class to which an integer a belongs is written

$$[a] \text{ or } [a]_m = \{a + mq \mid q \in \mathbb{Z}\}.$$

By the Division Algorithm any integer a can be written in the form $a = mq + r$ where $q, r \in \mathbb{Z}$ and $0 \leq r < m$. Thus $a \equiv r \pmod{m}$ and $[a] = [r]$. Therefore $[0], [1], \ldots, [m-1]$ are all the congruence classes modulo m. Furthermore if $[r] = [r']$ where $0 \leq r, r' < m$, then $m \mid r - r'$, which can only mean that $r = r'$. Thus we have proved:

---
[2] Carl Friedrich Gauss (1777–1855).

2.3 Congruences

(2.3.1) *Let m be any positive integer. Then there are exactly m congruence classes modulo m, namely $[0], [1], \ldots, [m-1]$.*

Congruence arithmetic. We shall write

$$\mathbb{Z}_m$$

for the set of all congruences classes modulo m. Let us show next how to introduce operations of addition and multiplication for congruence classes, thereby introducing the possibility of arithmetic in $\mathbb{Z}_m$.

The *sum* and *product* of congruence classes modulo m are defined by the rules

$$[a] + [b] = [a+b] \quad \text{and} \quad [a] \cdot [b] = [ab].$$

These definitions are surely the natural ones. However some care must be exercised in making definitions of this type. Each congruence class can be represented by any one of its elements: we need to ensure that the sum and product specified above depend only on the congruence classes themselves, not on the chosen representatives.

To this end, let $a' \in [a]$ and $b' \in [b]$. We need to show that $[a+b] = [a'+b']$ and $[ab] = [a'b']$. Now $a' = a + mu$ and $b' = b + mv$ for some $u, v \in \mathbb{Z}$. Therefore $a' + b' = (a+b) + m(u+v)$ and also $a'b' = ab + m(av + bu + muv)$; from these equations it follows that $a' + b' \equiv a + b \pmod{m}$ and $a'b' \equiv ab \pmod{m}$. Thus $[a'+b'] = [a+b]$, and $[a'b'] = [ab]$, which is what we needed to check.

Now that we know the sum and product of congruence classes to be well-defined, it is a simple task to establish the basic properties of these operations.

(2.3.2) *Let m be a positive integer and let $[a], [b], [c]$ be congruence classes modulo m. Then*

(i) $[a] + [b] = [b] + [a]$ *and* $[a] \cdot [b] = [b] \cdot [a]$;

(ii) $([a] + [b]) + [c] = [a] + ([b] + [c])$, *and* $([a][b])[c] = [a]([b][c])$;

(iii) $([a] + [b])[c] = [a][c] + [b][c]$;

(iv) $[0] + [a] = [a]$, *and* $[1][a] = [a]$;

(v) $[a] + [-a] = [0]$.

Since all of these properties are valid in $\mathbb{Z}$ as well as $\mathbb{Z}_m$ – see 2.1 – we recognize some common features of the arithmetics on $\mathbb{Z}$ and $\mathbb{Z}_m$. This can be expressed by saying that $\mathbb{Z}$ and $\mathbb{Z}_m$ are both *commutative rings with identity*, as will be explained in Chapter 6.

Fermat's[3] Theorem. Before proceeding to this well-known theorem, we shall establish a frequently used property of the binomial coefficients. If n and r are integers satisfying $0 \leq r \leq n$, the *binomial coefficient* $\binom{n}{r}$ is the number of ways of choosing r objects from a set of n distinct objects. There is the well-known formula

$$\binom{n}{r} = \frac{n!}{r!(n-r)!} = \frac{n(n-1)\ldots(n-r+1)}{r!}.$$

The property we need here is:

(2.3.3) *If p is a prime and $0 < r < p$, then $\binom{p}{r} \equiv 0 \pmod{p}$.*

Proof. Write $\binom{p}{r} = pm$ where m is the rational number

$$\frac{(p-1)(p-2)\ldots(p-r+1)}{r!}.$$

Note that each prime appearing as a factor of the numerator or denominator of m is smaller than p. Write $m = \frac{u}{v}$ where u and v are relatively prime integers. Then $v\binom{p}{r} = pmv = pu$ and by Euclid's Lemma v divides p. Now $v \neq p$, so $v = 1$ and $m = u \in \mathbb{Z}$. Hence p divides $\binom{p}{r}$. □

We are now able to prove Fermat's Theorem.

(2.3.4) *If p is a prime and x is any integer, then $x^p \equiv x \pmod{p}$.*

Proof. Since $(-x)^p \equiv -x^p \pmod{p}$, whether or not p is odd, there is no loss in assuming $x \geq 0$. We will use induction on x to show that $x^p \equiv x \pmod{p}$, which certainly holds for $x = 0$. Assume it is true for x. Then by the Binomial Theorem

$$(x+1)^p = \sum_{r=0}^{p} \binom{p}{r} x^r \equiv x^p + 1 \pmod{p}$$

since p divides $\binom{p}{r}$ if $0 < r < p$. Therefore $(x+1)^p \equiv x+1 \pmod{p}$ and the induction is complete □

Solving Congruences. Just as we solve equations for unknown real numbers, we can try to solve congruences for unknown integers. The simplest case is that of a *linear congruence* with a single unknown x; this has the form $ax \equiv b \pmod{m}$, where a, b, m are given integers.

(2.3.5) *Let a, b, m be integers with $m > 0$. Then there is a solution x of the congruence $ax \equiv b \pmod{m}$ if and only if $\gcd\{a, m\}$ divides b.*

---

[3]Pierre de Fermat (1601–1665)

2.3 Congruences

Proof. Set $d = \gcd\{a, m\}$. If x is a solution of congruence $ax \equiv b \pmod{m}$, then $ax = b + mq$ for some $q \in \mathbb{Z}$; from this it follows at once that d must divide b.

Conversely, assume that $d \mid b$. Now by (2.2.3) there are integers u, v such that $d = au + mv$. Multiplying this equation by the integer b/d, we obtain $b = a(ub/d) + m(vb/d)$. Put $x = ub/d$; then $ax \equiv b \pmod{m}$ and x is a solution of the congruence. □

The most important case occurs when $b = 1$.

Corollary (2.3.6) *Let a, m be integers with $m > 0$. Then the congruence $ax \equiv 1 \pmod{m}$ has a solution x if and only if a is relatively prime to m.*

It is worthwhile translating the last result into the language of congruence arithmetic. Given an integer $m > 0$ and a congruence class $[a]$ modulo m, there is a congruence class $[x]$ such that $[a][x] = [1]$ if and only if a is relatively prime to m. Thus we can tell which congruence classes modulo m have "inverses"; they are classes $[x]$ where $0 < x < m$ and x is relatively prime to m. The number of invertible congruence classes modulo m is denoted by

$$\phi(m).$$

Here ϕ is called *Euler's*[4] *function.*

Next let us consider systems of linear congruences.

(2.3.7) (The Chinese Remainder Theorem) *Let $a_1, a_2, \ldots, a_k$ and $m_1, m_2, \ldots, m_k$ be integers with $m_i > 0$; assume that $\gcd\{m_i, m_j\} = 1$ if $i \neq j$. Then there is a common solution x of the system of congruences*

$$\begin{cases} x \equiv a_1 \pmod{m_1} \\ x \equiv a_2 \pmod{m_2} \\ \vdots \\ x \equiv a_k \pmod{m_k}. \end{cases}$$

When $k = 2$, this striking result was discovered in the 1st Century AD by the Chinese mathematician Sun Tse.

Proof of (2.3.7). Put $m = m_1 m_2 \ldots m_k$ and $m'_i = m/m_i$. Then m_i and m'_i are relatively prime, so by (2.3.6) there exist an integer ℓ_i such that $m'_i \ell_i \equiv 1 \pmod{m_i}$. Now put $x = a_1 m'_1 \ell_1 + \cdots + a_k m'_k \ell_k$. Then

$$x \equiv a_i m'_i \ell_i \equiv a_i \pmod{m_i}$$

since $m_i \mid m'_j$ if $i \neq j$. □

As an application of (2.3.7) a well-known formula for Euler's function will be derived.

---
[4]Leonhard Euler (1707–1783).

(2.3.8) (i) *If m and n are relatively prime positive integers, then $\phi(mn) = \phi(m)\phi(n)$.*

(ii) *If $m = p_1^{l_1} p_2^{l_2} \ldots p_k^{l_k}$ with $l_i > 0$ and distinct primes p_i, then*

$$\phi(m) = \prod_{i=1}^{k} (p_i^{l_i} - p_i^{l_i - 1}).$$

Proof. (i) Let U_m denote the set of invertible congruence classes in $\mathbb{Z}_m$. Thus $|U_m| = \phi(m)$. Define a map $\alpha : U_{mn} \to U_m \times U_n$ by the rule $\alpha([a]_{mn}) = ([a]_m, [a]_n)$. First of all observe that α is well-defined. Next suppose that $\alpha([a]_{mn}) = \alpha([a']_{mn})$. Then $[a]_m = [a']_m$ and $[a]_n = [a']_n$, equations which imply that $a - a'$ is divisible by m and n, and hence by mn. Therefore α is an injective function.

In fact α is also surjective. For if $[a]_m \in U_m$ and $[b]_n \in U_n$ are given, the Chinese Remainder Theorem tells us that there is an integer x such that $x \equiv a$ (mod m) and $x \equiv b$ (mod n). Hence $[x]_m = [a]_m$ and $[x]_n = [b]_n$, so that $\alpha([x]_{mn}) = ([a]_m, [b]_n)$. Thus we have proved that α is a bijection, and consequently $|U_{mn}| = |U_m| \times |U_n|$, as required.

(ii) Suppose that p is a prime and $n > 0$. Then there are p^{n-1} multiples of p among the integers $0, 1, \ldots, p^n - 1$; therefore $\phi(p^n) = p^n - p^{n-1}$. Finally apply (i) to obtain the formula indicated. □

We end the chapter with some examples which illustrate the utility of congruences.

Example (2.3.1) Show that an integer is divisible by 3 if and only if the sum of its digits is a multiple of 3.

Let $n = a_0 a_1 \ldots a_k$ be the decimal representation of an integer n. Thus $n = a_k + a_{k-1} 10 + a_{k-2} 10^2 + \cdots + a_0 10^k$ where $0 \le a_i < 10$. The key point here is that $10 \equiv 1$ (mod 3), i.e., $[10] = [1]$. Hence $[10^i] = [10]^i = [1]^i = [1]$, i.e., $10^i \equiv 1$ (mod 3) for all $i \ge 0$. It therefore follows that $n \equiv a_0 + a_1 + \cdots + a_k$ (mod 3). The assertion is an immediate consequence of this congruence.

Example (2.3.2) (*Days of the week*) Congruences have long been used implicitly to compute dates. As an example, let us determine what day of the week September 25 in the year 2020 will be.

To keep track of the days we assign the integers $0, 1, 2, \ldots, 6$ as labels for the days of the week, say Sunday $= 0$, Monday $= 1, \ldots$, Saturday $= 6$. Suppose that we reckon from February 9, 1997, which was a Sunday. All we have to do is count the number of days from February 9, 1997 to September 25, 2020. Allowing for leap years, this number is 8629.

Now $8629 \equiv 5$ (mod 7) and 5 is the label for Friday. We conclude that September 25, 2020 will be a Friday.

Example (2.3.3) (*The Basket of Eggs Problem*) What is the smallest number of eggs a basket can contain if, when eggs are removed k at time, there is one egg left when

$k = 2, 3, 4, 5, 6$ and no eggs left when $k = 7$? (This ancient problem is mentioned in an Indian manuscript of the 7th Century).

Let x be the number of eggs in the basket. The conditions require that $x \equiv 1$ (mod k) for $k = 2, 3, 4, 5, 6$ and $x \equiv 0$ (mod k) for $k = 7$. Clearly this amounts to x satisfying the four congruences $x \equiv 1$ (mod 3), $x \equiv 1$ (mod 4), $x \equiv 1$ (mod 5) and $x \equiv 0$ (mod 7). Furthermore these are equivalent to

$$x \equiv 1 \text{ (mod 60)} \quad \text{and} \quad x \equiv 0 \text{ (mod 7)}.$$

By the Chinese Remainder Theorem there is a solution to this pair of congruences: we have to find the smallest positive solution. Applying the method of the proof of (2.3.7), we have $m = 420, m_1 = 60, m_2 = 7$, so that $m_1' = 7, m_2' = 60$; also $\ell_1 = 43$, $\ell_2 = 2$. Therefore one solution is given by $x = 1 \cdot 7 \cdot 43 + 0 \cdot 60 \cdot 2 = 301$. If y is any other solution, note that $y - x$ must be divisible by $60 \times 7 = 420$. Thus the general solution is $x = 301 + 420q, q \in \mathbb{Z}$. So the smallest positive solution is 301.

Our next example is a refinement of Euclid's Theorem on the infinity of primes.

Example (2.3.4) Prove that there are infinitely many primes of the form $3n + 2$ where n is an integer ≥ 0.

In fact the proof is a variant of Euclid's method. Suppose the result is false and let the *odd* primes of the form $3n+2$ be $p_1, p_2, \ldots, p_k$. Now consider the positive integer $m = 3p_1 p_2 \ldots p_k + 2$. Notice that m is odd and is not divisible by p_i. Therefore m is a product of odd primes different from $p_1, \ldots, p_k$. Hence m must be a product of primes of the form $3n + 1$ since every integer is of the form $3n, 3n + 1$ or $3n + 2$. But then it follows that m itself must have the form $3n + 1$ and $m \equiv 1$ (mod 3). On the other hand, $m \equiv 2$ (mod 3), so we have reached a contradiction.

Actually this exercise is a special case of a famous theorem of Dirichlet[5]: every arithmetic progression $an + b$ where $n = 0, 1, 2 \ldots$, and the integers a and b are positive and relatively prime, contains infinitely many primes.

Example (2.3.5) (*The RSA Cryptosystem*) This is a secure system for encrypting messages which has been widely used for transmitting sensitive data since its invention in 1978 by Rivest, Shamir and Adleman. It has the advantage of being a public key system in which only the deciphering function is not available to the public.

Suppose that a sensitive message is to be sent from A to B. The parameters required are two distinct large primes p and q. Put $n = pq$ and $m = \phi(n)$; therefore $m = (p-1)(q-1)$ by (2.3.8). Let a be an integer in the range 1 to m which is relatively prime to m. Then by (2.3.6) there is a unique integer b satisfying $0 < b < m$ and $ab \equiv 1$ (mod m). The sender A is assumed to know the integers a and n, while the receiver B knows b and n.

The message to be sent is first converted to an integer x satisfying $0 < x < n$. Then A encyphers x by raising it to the power a and then reducing modulo n. In this

---
[5] Johann Peter Gustav Lejeune Dirichlet (1805–1859)

form the message is transmitted to B. On receiving this, B raises the number to the power b and reduces modulo n. The result will be the original message x. Here what is being claimed is that

$$x^{ab} \equiv x \pmod{n}$$

since $0 < x < n$. To see why this holds, first write

$$ab = 1 + lm = 1 + l(p-1)(q-1)$$

with l an integer. Then

$$x^{ab} = x^{1+l(p-1)(q-1)} = x(x^{p-1})^{l(q-1)} \equiv x \pmod{p}$$

since $x^{p-1} \equiv 1 \pmod{p}$ by Fermat's Theorem. Hence p divides $x^{ab} - x$, and similarly q also divides this number. Therefore $n = pq$ divides $x^{ab} - x$ as claimed.

Even if n and a become public knowledge, it will be difficult to break the system by finding b. For this would require the computation of the inverse of $[a]$ in $\mathbb{Z}_m$. To do this using the Euclidean Algorithm, which is what lies behind (2.3.6), one would need to know the primes p and q. But for a number $n = pq$ in which p and q have several hundreds of digits, it could take hundreds of years to discover the factorization, using currently available machines. Thus the system is secure until more efficient ways of factorizing large numbers become available.

Exercises (2.3)

1. Establish the properties of congruences listed in (2.3.2).

2. In $\mathbb{Z}_{24}$ find the inverses of $[7]$ and $[13]$.

3. Show that if n is an odd integer, then $n^2 \equiv 1 \pmod{8}$.

4. Find the general solution of the congruence $6x \equiv 11 \pmod{5}$.

5. What day of the week will April 1, 2030 be?

6. Find the smallest positive solution x of the system of congruences $x \equiv 4 \pmod{3}$, $x \equiv 5 \pmod{7}$, $x \equiv 6 \pmod{11}$.

7. Prove that there are infinitely many primes of the form $4n + 3$.

8. Prove that there are infinitely many primes of the form $6n + 5$.

9. In a certain culture the festivals of the snake, the monkey and the fish occur every 6, 5 and 11 years respectively. The next festivals occur in 3, 4 and 1 years respectively. How many years must pass before all three festivals occur in the same year?

10. Prove that no integer of the form $4n + 3$ can be written as the sum of two squares of integers.

Chapter 3
Introduction to groups

Groups form one of the most important and natural structures in algebra. They also feature in other areas of mathematics such as geometry, topology and combinatorics. In addition groups arise in many areas of science, typically in situations where symmetry is important, as in atomic physics and crystallography. More general algebraic structures which have recently come to prominence due to the rise of information science include semigroups and monoids. This chapter serves as an introduction to these types of structure.

There is a continuing debate as to whether it is better to introduce groups or rings first in an introductory course in algebra: here we take the point of view that groups are logically the simpler objects since they involve only one binary operation, whereas rings have two. Accordingly rings are left until Chapter 6.

Historically the first groups to be studied consisted of *permutations*, i.e., bijective functions on a set. Indeed for most of the 19th century "group" was synonymous with "group of permutations". Since permutation groups have the great advantage that their elements are concrete and easy to compute with, we begin this chapter with a discussion of permutations.

3.1   Permutations of a set

If X is any non-empty set, a bijective function $\pi : X \to X$ is called a *permutation* of X. Thus by (1.3.1) π has a unique inverse function $\pi^{-1} : X \to X$, which is also a permutation. The set of all permutations of the set X is denoted by the symbol

$$\mathrm{Sym}(X),$$

which stands for *the symmetric group on X*.

If π and σ are permutations of X, their composite $\pi \circ \sigma$ is also a permutation; this is because it has an inverse, namely the permutation $\sigma^{-1} \circ \pi^{-1}$ by (1.3.2). In the future for the sake of simplicity we shall usually write

$$\pi \sigma$$

for $\pi \circ \sigma$. Of course id_X, the identity function on X, is a permutation.

At this point we pause to note some aspects of the set $\mathrm{Sym}(X)$: this set is "closed" with respect to forming inverses and composites, by which we mean that if $\pi, \sigma \in \mathrm{Sym}(X)$, then π^{-1} and $\pi \circ \sigma$ belong to $\mathrm{Sym}(X)$. In addition $\mathrm{Sym}(X)$ contains the identity permutation id_X, which has the property $\mathrm{id}_X \circ \pi = \pi = \pi \circ \mathrm{id}_X$. And finally, the associative law for permutations is valid, $(\pi \circ \sigma) \circ \tau = \pi \circ (\sigma \circ \tau)$. In fact what these results assert is that the pair $(\mathrm{Sym}(X), \circ)$ is a group, as defined in 3.2. Thus the permutations of a set afford a very natural example of a group.

Permutations of finite sets. We now begin the study of permutations of a finite set with n elements, $X = \{x_1, x_2, \ldots, x_n\}$. Since π is injective, $\pi(x_1), \pi(x_2), \ldots, \pi(x_n)$ are all different and therefore constitute all n elements of the set X, but in some different order from $x_1, \ldots, x_n$. So we may think of a permutation as a rearrangement of the order $x_1, x_2, \ldots, x_n$. A convenient way to denote the permutation π is

$$\pi = \begin{pmatrix} x_1 & x_2 & \ldots & x_n \\ \pi(x_1) & \pi(x_2) & \ldots & \pi(x_n) \end{pmatrix}$$

where the second row consists of the images under π of the elements of the first row.

It should be clear to the reader that nothing essential is lost if we take X to be the set $\{1, 2, \ldots, n\}$. With this choice of X, it is usual to write

$$S_n$$

for $\mathrm{Sym}(X)$; this is called the *symmetric group of degree n*.

Computations with elements of S_n are easily performed by working directly from the definitions. An example will illustrate this.

Example (3.1.1) Let $\pi = \begin{pmatrix} 1 & 2 & 3 & 4 & 5 & 6 \\ 6 & 1 & 2 & 5 & 3 & 4 \end{pmatrix}$ and $\sigma = \begin{pmatrix} 1 & 2 & 3 & 4 & 5 & 6 \\ 6 & 1 & 4 & 3 & 2 & 5 \end{pmatrix}$ be elements of S_6. Then

$$\pi\sigma = \begin{pmatrix} 1 & 2 & 3 & 4 & 5 & 6 \\ 4 & 6 & 5 & 2 & 1 & 3 \end{pmatrix}, \quad \sigma\pi = \begin{pmatrix} 1 & 2 & 3 & 4 & 5 & 6 \\ 5 & 6 & 1 & 2 & 4 & 3 \end{pmatrix}$$

while

$$\pi^{-1} = \begin{pmatrix} 1 & 2 & 3 & 4 & 5 & 6 \\ 2 & 3 & 5 & 6 & 4 & 1 \end{pmatrix}.$$

Here $\pi\sigma$ has been computed using the definition, $\pi\sigma(i) = \pi(\sigma(i))$, while π^{-1} is readily obtained by "reading up" from $1, 2, \ldots, 6$ in the second row of π to produce the second row of π^{-1}. Notice that $\pi\sigma \neq \sigma\pi$, i.e., multiplication of permutations is not commutative.

A simple count establishes the number of permutations of a finite set.

(3.1.1) *If X is a set with n elements, then $|\mathrm{Sym}(X)| = n!$.*

Proof. Consider the number of ways of constructing the second row of a permutation

$$\pi = \begin{pmatrix} x_1 & x_2 & \ldots & x_n \\ y_1 & y_2 & \ldots & y_n \end{pmatrix}$$

Cyclic permutations. Let $\pi \in S_n$, so that π is a permutation of the set $\{1, 2, \ldots, n\}$. The *support* of π is defined to be the set of all i such that $\pi(i) \neq i$, in symbols $\text{supp}(\pi)$. Now let r be an integer satisfying $1 \leq r \leq n$. Then π is called an *r-cycle* if $\text{supp}(\pi) = \{i_1, i_2, \ldots, i_r\}$, with distinct i_j, where $\pi(i_1) = i_2, \pi(i_2) = i_3, \ldots, \pi(i_{r-1}) = i_r$ and $\pi(i_r) = i_1$. So π moves the integers $i_1, i_2, \ldots, i_r$ anticlockwise around a circle, but fixes all other integers: often π is written in the form

$$\pi = (i_1 i_2 \ldots i_r)(i_{r+1}) \ldots (i_n)$$

where the presence of a 1-cycle (j) means that $\pi(j) = j$. The notation may be further abbreviated by omitting 1-cycles, although if this is done, the integer n may need to be specified.

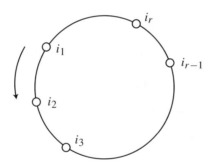

In particular a 2-cycle has the form (ij): it interchanges i and j and fixes other integers. 2-cycles are frequently called *transpositions*.

Example (3.1.2) (i) $\begin{pmatrix} 1 & 2 & 3 & 4 & 5 \\ 2 & 5 & 3 & 4 & 1 \end{pmatrix}$ is the 3-cycle $(1\,2\,5)(3)(4)$, that is, $(1\,2\,5)$.

(ii) However $\begin{pmatrix} 1 & 2 & 3 & 4 & 5 & 6 & 7 & 8 \\ 6 & 1 & 5 & 8 & 7 & 2 & 3 & 4 \end{pmatrix}$ is not a cycle. Indeed it is the composite of three cycles of length > 1, namely $(1\,6\,2) \circ (3\,5\,7) \circ (4\,8)$, as one can see by following what happens to each of the integers $1, 2, \ldots, 8$ when the permutation is applied. In fact this is an instance of an important general result, that any permutation is expressible as a composite of cycles (see (3.1.3)).

The reader should observe that there are r different ways to write an r-cycle since any element of the cycle can be the initial element: indeed $(i_1 i_2 \ldots i_r) = (i_2 i_3 \ldots i_r i_1) = \cdots = (i_r i_1 i_2 \ldots i_{r-1})$.

Disjoint permutations. Permutations π, σ in S_n are called *disjoint* if their supports are disjoint, i.e., they do not both move the same element. An important fact about disjoint permutations is that they commute, in contrast to permutations in general.

(3.1.2) *If π and σ are disjoint permutations in S_n, then $\pi\sigma = \sigma\pi$.*

Proof. Let $i \in \{1, 2, \ldots, n\}$; we will show that $\pi\sigma(i) = \sigma\pi(i)$. If $i \notin \operatorname{supp}(\pi) \cup \operatorname{supp}(\sigma)$, then plainly $\pi\sigma(i) = i = \sigma\pi(i)$. Suppose that $i \in \operatorname{supp}(\pi)$; then $i \notin \operatorname{supp}(\sigma)$ and so $\sigma(i) = i$. Thus $\pi\sigma(i) = \pi(i)$. Also $\sigma\pi(i) = \pi(i)$; for otherwise $\pi(i) \in \operatorname{supp}(\sigma)$ and so $\pi(i) \notin \operatorname{supp}(\pi)$, which leads us to $\pi(\pi(i)) = \pi(i)$. However π^{-1} can now be applied to both sides of this equation to show that $\pi(i) = i$, a contradiction since $i \in \operatorname{supp}(\pi)$. □

Powers of a permutation. Since we can form products of permutations using composition, it is natural to define powers of a permutation. Let $\pi \in S_n$ and let i be a non-negative integer. Then the i-th *power* π^i is defined recursively by the rules:

$$\pi^0 = 1, \quad \pi^{i+1} = \pi^i \pi.$$

The point to note here is that the rule allows us to compute successive powers of the permutation π as follows: $\pi^1 = \pi$, $\pi^2 = \pi\pi$, $\pi^3 = \pi^2 \pi$, etc. Powers are used in the proof of the following fundamental theorem.

(3.1.3) *Let $\pi \in S_n$. Then π is expressible as a product of disjoint cycles and the cycles appearing in the product are unique.*

Proof. We deal with the existence of the expression first. If π is the identity, then obviously $\pi = (1)(2) \ldots (n)$. Assume that $\pi \neq \operatorname{id}$ and choose an integer i_1 such that $\pi(i_1) \neq i_1$. Now the integers $i_1, \pi(i_1), \pi^2(i_1), \ldots$ belong to the finite set $\{1, 2, \ldots, n\}$ and so they cannot all be different; say $\pi^r(i_1) = \pi^s(i_1)$ where $r > s \geq 0$. Applying $(\pi^{-1})^s$ to both sides of the equation and using associativity, we find that $\pi^{r-s}(i_1) = i_1$. Hence by the Well-Ordering Law there is a least positive integer m_1 such that $\pi^{m_1}(i_1) = i_1$.

Next we will argue that the integers $i_1, \pi(i_1), \pi^2(i_1), \ldots, \pi^{m_1-1}(i_1)$ are all different. For if not and $\pi^r(i_1) = \pi^s(i_1)$ where $m_1 > r > s \geq 0$, then, just as above, we can argue that $\pi^{r-s}(i_1) = i_1$; on the other hand, $0 < r - s < m_1$, which contradicts the choice of m_1. It follows that π permutes the m_1 distinct integers $i_1, \pi(i_1), \ldots, \pi^{m_1-1}(i_1)$ in a cycle, so that we have identified the m_1-cycle $(i_1 \ \pi(i_1) \ldots \pi^{m_1-1}(i_1))$ as a component cycle of π.

If π fixes all other integers, then $\pi = (i_1 \ \pi(i_1) \ldots \pi^{m_1-1}(i_1))$ and π is an m_1-cycle. Otherwise there exists an integer $i_2 \notin \{i_1, \pi(i_1), \ldots, \pi^{m_1-1}(i_1)\}$ such that

$\pi(i_2) \neq i_2$. Just as above we identify a second cycle $(i_2 \, \pi(i_2) \ldots \pi^{m_2-1}(i_2))$ present in π. This is disjoint from the first cycle. Indeed, if the cycles had a common element, they would clearly have to coincide. It should also be clear that by a finite number of applications of this procedure we can express π as a product of disjoint cycles.

Now to establish uniqueness. Assume that there are two expressions for π as a product of disjoint cycles, say $(i_1 i_2 \ldots)(j_1 j_2 \ldots) \ldots$ and $(i'_1 i'_2 \ldots)(j'_1 j'_2 \ldots) \ldots$. By (3.1.2) disjoint cycles commute. Thus without loss of generality we can assume that i_1 occurs in the cycle $(i'_1 i'_2 \ldots)$. Since any element of a cycle can be moved up to the initial position, it can also be assumed that $i_1 = i'_1$. Then $i_2 = \pi(i_1) = \pi(i'_1) = i'_2$; similarly $i_3 = i'_3$, etc. The other cycles are dealt with in the same way. Therefore the two expressions for π are identical. □

Corollary (3.1.4) *Every element of S_n is expressible as a product of transpositions.*

Proof. Because of (3.1.3) it is sufficient to show that each cyclic permutation is a product of transpositions. That this is true follows from the easily verified identity:

$$(i_1 i_2 \ldots i_{r-1} i_r) = (i_1 i_r)(i_1 i_{r-1}) \ldots (i_1 i_3)(i_1 i_2).$$ □

Example (3.1.3) Let $\pi = \begin{pmatrix} 1 & 2 & 3 & 4 & 5 & 6 \\ 3 & 6 & 5 & 1 & 4 & 2 \end{pmatrix}$. To express π as a product of transpositions first write it as a product of disjoint cycles, following the method of the proof of (3.1.3). Thus $\pi = (1\,3\,5\,4)(2\,6)$. Also $(1\,3\,5\,4) = (1\,4)(1\,5)(1\,3)$, so that $\pi = (1\,4)(1\,5)(1\,3)(2\,6)$.

On the other hand not every permutation in S_n is expressible as a product of *disjoint* transpositions. (Why not?)

Even and odd permutations. If π is a permutation in S_n, then π replaces the natural order of integers, $1, 2, \ldots, n$ by the new order $\pi(1), \pi(2), \ldots, \pi(n)$. Thus π may cause inversions of the natural order: here we say that an *inversion* occurs if for some $i < j$ we have $\pi(i) > \pi(j)$. To count the number of inversions introduced by π it is convenient to introduce a formal device.

Consider a polynomial f in indeterminates $x_1, x_2, \ldots, x_n$. (At this point we assume an informal understanding of polynomials.) If $\pi \in S_n$, then π determines a new polynomial πf which is obtained by permuting the variables $x_1, x_2, \ldots, x_n$. Thus $\pi f(x_1, \ldots, x_n) = f(x_{\pi(1)}, \ldots, x_{\pi(n)})$. For example, if $f = x_1 - x_2 - 2x_3$ and $\pi = (1\,2)(3)$, then $\pi f = x_2 - x_1 - 2x_3$.

Now consider the special polynomial

$$f(x_1, \ldots, x_n) = \prod_{\substack{i,j=1 \\ i<j}}^{n} (x_i - x_j).$$

A typical factor in πf is $x_{\pi(i)} - x_{\pi(j)}$. Now if $\pi(i) < \pi(j)$, this is also a factor of f, while if $\pi(i) > \pi(j)$, then $-(x_{\pi(i)} - x_{\pi(j)})$ is a factor of f. Consequently $\pi f = +f$ if the number of inversions of the natural order in π is even and $\pi f = -f$ if it is odd.

Define the *sign* of the permutation π to be
$$\text{sign}(\pi) = \frac{\pi f}{f}.$$

Thus $\text{sign}(\pi) = 1$ or -1 according as the number of inversions in π is even or odd. Call π an *even permutation* if $\text{sign}(\pi) = 1$ and an *odd permutation* if $\text{sign}(\pi) = -1$.

Example (3.1.4) The even permutations in S_3 are $(1)(2)(3)$, $(1\,2\,3)$ and $(1\,3\,2)$, while the odd permutations are $(1)(2\,3)$, $(2)(1\,3)$ and $(3)(1\,2)$.

To decide if a permutation of not too large degree is even or odd a *crossover diagram* is useful. We illustrate this idea with an example.

Example (3.1.5) Is the permutation
$$\pi = \begin{pmatrix} 1 & 2 & 3 & 4 & 5 & 6 & 7 \\ 3 & 7 & 2 & 5 & 4 & 1 & 6 \end{pmatrix}$$
even or odd?

Simply join equal integers in the top and bottom rows of π and count the intersections or "crossovers", taking care to avoid multiple or unnecessary intersections. A crossover indicates the presence of an inversion of the natural order.

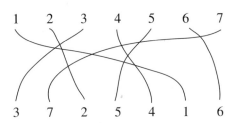

There are 11 crossovers, so $\text{sign}(\pi) = -1$ and π is an odd permutation.

The next result is very significant property of transpositions.

(3.1.5) *Transpositions are always odd.*

Proof. Consider the crossover diagram on the next page for the transposition $(i\ j)$ where $i < j$. An easy count reveals that the presence of $1 + 2(j - i - 1)$ crossovers. Since this integer is odd, $(i\ j)$ is an odd permutation. □

The basic properties of the sign function are set out next.

(3.1.6) *Let $\pi, \sigma \in S_n$. Then the following hold:*
  (i) $\text{sign}(\pi\sigma) = \text{sign}(\pi)\text{sign}(\sigma)$;
  (ii) $\text{sign}(\pi^{-1}) = \text{sign}(\pi)$.

```
1    2  ... i-1   i    i+1  ... j-1   j   j+1 ... n
|    |       |     \    |         |   /    |      |
|    |       |      \   |         |  /     |      |
|    |       |       \  |         | /      |      |
|    |       |        \ |         |/       |      |
|    |       |         X|         X        |      |
|    |       |        / |        /|        |      |
|    |       |       /  |       / |        |      |
1    2  ... i-1   j    i+1 ... j-1   i   j+1 ... n
```

Proof. Let $f = \prod_{\substack{i,j=1 \\ i<j}}^{n} (x_i - x_j)$. Then, since $\pi f = \text{sign}(\pi)f$, we have

$$\pi\sigma f(x_1, \ldots, x_n) = \pi(\sigma f(x_1, \ldots, x_n))$$
$$= \pi((\text{sign}(\sigma))f(x_1, \ldots, x_n))$$
$$= \text{sign}(\sigma)\pi f(x_1, \ldots, x_n)$$
$$= \text{sign}(\sigma)\,\text{sign}(\pi)f(x_1, \ldots, x_n).$$

But $\pi\sigma f = \text{sign}(\pi\sigma)f$. Hence $\text{sign}(\pi\sigma) = \text{sign}(\pi)\text{sign}(\sigma)$. Finally, by (i) we have $1 = \text{sign}(\text{id}) = \text{sign}(\pi\pi^{-1}) = \text{sign}(\pi)\text{sign}(\pi^{-1})$, so that $\text{sign}(\pi^{-1}) = 1/\text{sign}(\pi) = \text{sign}(\pi)$. □

Corollary (3.1.7) *A permutation π in S_n is even (odd) if and only if it is a product of an even (respectively odd) number of transpositions.*

For if $\pi = \prod_{i=1}^{k} \pi_i$ with each π_i a transposition, then

$$\text{sign}(\pi) = \prod_{i=1}^{k} \text{sign}(\pi_i) = (-1)^k$$

by (3.1.5) and (3.1.6).

The subset of all even permutations in S_n is a group denoted by

$$A_n,$$

which is called the *alternating group of degree n*. Obviously $A_1 = S_1$. For $n > 1$ exactly half of the permutations in S_n are even, as the next result shows.

(3.1.8) *If $n > 1$, there are $\frac{1}{2}n!$ even permutations and $\frac{1}{2}n!$ odd permutations in S_n.*

Proof. Define a function $\alpha : A_n \to S_n$ by the rule $\alpha(\pi) = \pi \circ (1\,2)$, observing that $\alpha(\pi)$ is odd and α is injective. Every odd permutation σ belongs to $\text{Im}(\alpha)$ since $\alpha(\pi) = \sigma$ where $\pi = \sigma \circ (1\,2) \in A_n$. Thus $\text{Im}(\alpha)$ is precisely the set of all odd permutations and $|\text{Im}(\alpha)| = |A_n|$. □

(3.1.9) (Cauchy's[1] Formula) *If π in S_n is the product of c disjoint cycles, including 1-cycles, then*

$$\text{sign}(\pi) = (-1)^{n-c}.$$

---
[1] Augustin Louis Cauchy (1789–1857).

Proof. Let $\pi = \sigma_1\sigma_2\ldots\sigma_c$ where the σ_i are disjoint cycles and σ_i has length ℓ_i. Now σ_i is expressible as a product of $\ell_i - 1$ transpositions, by the proof of (3.1.4). Hence by (3.1.6) we have $\mathrm{sign}(\sigma_2) = (-1)^{\ell_i-1}$ and

$$\mathrm{sign}(\pi) = \prod_{i=1}^{c} \mathrm{sign}(\sigma_i) = \prod_{i=1}^{c}(-1)^{\ell_i-1} = (-1)^{n-c}$$

since $\sum_{i=1}^{c} \ell_i = n$. □

Derangements. We conclude the section by discussing a special type of permutation. A permutation of a set is called a *derangement* if it fixes no elements of the set, i.e., its support is the entire set. For example, $(1234)(56)$ is a derangement in S_6. A natural question is: how many derangements does S_n contain? To answer the question we employ a famous combinatorial principle.

(3.1.10) (The Inclusion–Exclusion Principle) *If $A_1, A_2, \ldots, A_r$ are finite sets, then $|A_1 \cup A_2 \cup \cdots \cup A_r|$ equals*

$$\sum_{i=1}^{r} |A_i| - \sum_{i<j=1}^{r} |A_i \cap A_j| + \sum_{i<j<k=1}^{r} |A_i \cap A_j \cap A_k| + \cdots + (-1)^{r-1}|A_1 \cap A_2 \cap \cdots \cap A_r|.$$

Proof. We have to count the number of objects that belong to at least one A_i. Our first estimate is $\sum_{i=1}^{r}|A_i|$, but this double counts elements in more than one A_i, so we subtract $\sum_{i<j=1}^{r}|A_i \cap A_j|$. But now elements belonging to three or more A_i's have not been counted at all. So we must add $\sum_{i<j<k=1}^{r}|A_i \cap A_j \cap A_k|$. Now elements in four or more A_i's have been double counted, and so on. By a sequence of r such "inclusions" and "exclusions" we arrive at the correct formula.

It is now relatively easy to count derangements. □

(3.1.11) *The number of derangements in S_n is given by the formula*

$$d_n = n!\left(1 - \frac{1}{1!} + \frac{1}{2!} - \frac{1}{3!} + \cdots + (-1)^n \frac{1}{n!}\right).$$

Proof. Let X_i denote the set of all permutations in S_n which fix the integer i, $(1 \le i \le n)$. Then the number of derangements in S_n is

$$d_n = n! - |X_1 \cup \cdots \cup X_n|.$$

Now $|X_i| = (n-1)!$; also $|X_i \cap X_j| = (n-2)!$, $(i < j)$, and $|X_i \cap X_j \cap X_k| = (n-3)!$, $(i < j < k)$ etc. Therefore by the Inclusion–Exclusion Principle

$$d_n = n! - \left\{\binom{n}{1}(n-1)! - \binom{n}{2}(n-2)! + \binom{n}{3}(n-3)! - \cdots + (-1)^{n-1}\binom{n}{n}(n-n)!\right\}.$$

This is because there are $\binom{n}{r}$ subsets of the form $X_{i_1} \cap X_{i_2} \cap \cdots \cap X_{i_r}$ with $i_1 < i_2 < \cdots < i_r$. The required formula appears after a minor simplification of the terms in the sum. □

Notice that for large n we have $\frac{d_n}{n!} \to e^{-1} = 0.36787\ldots$. So roughly 36.8% of the permutations in S_n are derangements.

Example (3.1.6) (*The Hat Problem*) There are n people attending a party each of whom brings a different hat. All hats are checked in on arrival. Afterwards everyone picks up a hat at random. What is the probability that no one has the correct hat?

Any distribution of hats corresponds to a permutation of the original order. The permutations that are derangements give the distributions in which everyone has the wrong hat. So the probability asked for is $\frac{d_n}{n!}$ or roughly e^{-1}.

Exercises (3.1)

1. Let $\pi = \begin{pmatrix} 1 & 2 & 3 & 4 & 5 & 6 \\ 2 & 4 & 1 & 5 & 3 & 6 \end{pmatrix}$ and $\sigma = \begin{pmatrix} 1 & 2 & 3 & 4 & 5 & 6 \\ 6 & 1 & 5 & 3 & 2 & 4 \end{pmatrix}$. Compute π^{-1}, $\pi\sigma$ and $\pi\sigma\pi^{-1}$.

2. Determine which of the permutations in Exercise (3.1.1) are even and which are odd.

3. Prove that $\text{sign}(\pi\sigma\pi^{-1}) = \text{sign}(\sigma)$ for all $\pi, \sigma \in S_n$.

4. Prove that if $n > 1$, then every element of S_n is a product of *adjacent* transpositions, i.e., transpositions of the form $(i\ i+1)$. [Hint: it is enough to prove the statement for a transposition (ij) where $i < j$. Now consider the composite $(j\ j+1)(ij)(j\ j+1)$].

5. An element π in S_n satisfies $\pi^2 = \text{id}$ if and only if π is a product of disjoint transpositions.

6. How many elements π in S_n satisfy $\pi^2 = \text{id}$?

7. How many permutations in S_n contain at most one 1-cycle?

8. In the game of Rencontre there are two players. Each one deals a pack of 52 cards. If they both deal the same card, that is a "rencontre". What is the probability of a rencontre occurring?

3.2 Binary operations: semigroups, monoids and groups

Most of the structures that occur in algebra consist of a set and a number of rules for combining pairs of elements of the set. We formalize the notion of a "rule of combination" by defining a *binary operation on a set S* to be a function

$$\alpha : S \times S \to S.$$

So for each ordered pair (a, b) with a, b in S the function α produces a unique element $\alpha((a, b))$ of S. It is better notation if we write

$$a * b$$

instead of $\alpha((a, b))$ and refer to the binary operation as $*$.

Of course binary operations abound; one need think no further than addition or multiplication in sets such as $\mathbb{Z}$, $\mathbb{Q}$, $\mathbb{R}$, or composition on the set of all functions on a given set.

The first algebraic structure of interest to us is a *semigroup*, which is a pair

$$(S, *)$$

consisting of a non-empty set S and a binary operation $*$ on S which satisfies the *associative law*:

(i) $(a * b) * c = a * (b * c)$ for all $a, b, c \in S$.

If the semigroup has an *identity element*, i.e., an element e of S such that

(ii) $a * e = a = e * a$ for all $a \in S$,

then it is called a *monoid*.

Finally, a monoid is called a *group* if each element a of S has an *inverse*, i.e., an element a^{-1} of S such that

(iii) $a * a^{-1} = e = a^{-1} * a$.

Also a semigroup is said to be *commutative* if

(iv) $ab = ba$ for all elements a, b.

A commutative group is called an *abelian*[2] *group*.

Thus semigroups, monoids and groups form successively narrower classes of algebraic structures. These concepts will now be illustrated by some familiar examples.

Examples of semigroups, monoids and groups. (i) The pairs $(\mathbb{Z}, +)$, $(\mathbb{Q}, +)$, $(\mathbb{R}, +)$ are groups where $+$ is ordinary addition, 0 is an identity element and an inverse of x is its negative $-x$.

(ii) Next consider $(\mathbb{Q}^*, \cdot)$, $(\mathbb{R}^*, \cdot)$ where the dot denotes ordinary multiplication and $\mathbb{Q}^*$ and $\mathbb{R}^*$ are the sets of *non-zero* rational numbers and real numbers respectively. Here $(\mathbb{Q}^*, \cdot)$ and $(\mathbb{R}^*, \cdot)$ are groups, the identity element being 1 and the inverse of x being $\frac{1}{x}$. On the other hand, $(\mathbb{Z}^*, \cdot)$ is only a monoid since the integer 2, for example, has no inverse in $\mathbb{Z}^* = \mathbb{Z} - \{0\}$.

(iii) $(\mathbb{Z}_m, +)$ is a group where m is a positive integer. The usual addition of congruence classes is used here.

---

[2] After Niels Henrik Abel (1802–1829).

(iv) $(\mathbb{Z}_m^*, \cdot)$ is a group where m is a positive integer: here $\mathbb{Z}_m^*$ is the set of invertible congruence classes $[a]$ modulo m, i.e., such that $\gcd\{a, m\} = 1$, and multiplication of congruence classes is used. Note that $|\mathbb{Z}_m^*| = \phi(m)$ where ϕ is Euler's function.

(v) Let $M_n(\mathbb{R})$ be the set of all $n \times n$ matrices with real entries. If the usual rule of addition of matrices is used, we obtain a group.

On the other hand, $M_n(\mathbb{R})$ with matrix multiplication is only a monoid. To obtain a group we must form

$$\mathrm{GL}_n(\mathbb{R}),$$

the set of all invertible or *non-singular* matrices in $M_n(\mathbb{R})$, i.e., those with non-zero determinant. This is called the *general linear group of degree n* over $\mathbb{R}$.

(vi) For an example of a semigroup that is not a monoid we need look no further than $(E, +)$ where E is the set of all even integers. Clearly there is no identity element here.

(vii) *The monoid of functions on a set.* Let A be any non-empty set, and write $\mathrm{Fun}(A)$ for the set of all functions α on A. Then

$$(\mathrm{Fun}(A), \circ)$$

is a monoid where $\circ$ is functional composition; indeed this binary operation is associative by (1.2.3). The identity element is the identity function on A.

If we restrict attention to the bijective functions on A, i.e., to those which have inverses, we obtain the *symmetric group* on A

$$(\mathrm{Sym}(A), \circ),$$

consisting of all the permutations of A. This is one of our prime examples of groups.

(viii) *Monoids of words.* For a different type of example we consider words in an alphabet X. Here X is any non-empty set and a *word* in X is just an n-tuple of elements of X, written for convenience without parentheses as $x_1 x_2 \ldots x_n, n \geq 0$. The case $n = 0$ is the *empty word* $\emptyset$. Let $W(X)$ denote the set of all words in X.

There is a natural binary operation on X, namely juxtaposition. Thus if $w = x_1 \ldots x_n$ and $z = y_1 \ldots y_m$ are words in X, define wz to be the word $x_1 \ldots x_n z_1 \ldots z_m$. If $w = \emptyset$, then by convention $wz = z = zw$. It is clear that this binary operation is associative and $\emptyset$ is an identity element. So $W(X)$, with the operation specified, is a monoid, the so-called *free monoid generated by X*.

(ix) *Monoids and automata.* There is a somewhat unexpected connection between monoids and state output automata – see 1.3. Suppose that $A = (I, S, \nu)$ is a state output automaton with input set I, state set S and next state function $\nu : I \times S \to S$. Then A determines a monoid M_A in the following way.

Let $i \in I$ and $s \in S$; then we define $\theta_i : S \to S$ by the rule $\theta_i(s) = \nu(i, s)$. Now let M_A consist of the identity function and all composites of finite sequences of θ_i's; thus $M_A \subseteq \mathrm{Fun}(S)$. Clearly M_A is a monoid with respect to functional composition.

In fact one can go in the opposite direction as well. Suppose we start with a monoid $(M, *)$ and define an automaton $A_M = (M, M, \nu)$ where the next state function $\nu : M \times M \to M$ is defined by the rule $\nu(x_1, x_2) = x_1 * x_2$. Thus a connection between monoids and state output automata has been detected.

(x) *Symmetry groups.* As has already been remarked, groups tend to arise wherever symmetry is of importance. The size of the group can be regarded as a measure of the amount of symmetry present. Since symmetry is at heart a geometric notion, it is not surprising that geometry provides many interesting examples of groups.

A bijective function defined on 3-dimensional space or the plane is called an *isometry* if it preserves distances between points. Natural examples of isometries are translations, rotations and reflections. Now let X be a non-empty set of points in 3-space or the plane – we will refer to X as a geometric configuration. An isometry α which fixes the set X, i.e., such that

$$X = \{\alpha(x) \mid x \in X\},$$

is called a *symmetry of* X. Note that a symmetry can move the individual points of X.

It is easy to see that the symmetries of X form a group with respect to functional composition; this is the *symmetry group* $S(X)$ of X. Thus $S(X)$ is a subset of $\mathrm{Sym}(X)$, usually a proper subset.

The symmetry group of the regular n-gon. As an illustration let us analyze the symmetries of the *regular n-gon*: this is a polygon in the plane with n equal edges, ($n \geq 3$). It is convenient to label the vertices of the n-gon $1, 2, \ldots, n$, so that each symmetry may be represented by a permutation of $\{1, 2, \ldots, n\}$, i.e., by an element of S_n.

Each symmetry arises from an axis of symmetry of the figure. Of course, if we expect to obtain a group, we must include the identity symmetry, represented by $(1)(2)\ldots(n)$. There are $n-1$ anticlockwise rotations about the line perpendicular to the plane of the figure and through the centroid, through angles $i\left(\frac{2\pi}{n}\right)$, for $i = 1, 2, \ldots, n-1$. For example, the rotation through $\frac{2\pi}{n}$ is represented by the n-cycle $(1\,2\,3\ldots n)$; other rotations correspond to powers of this n-cycle.

Then there are n reflections in axes of symmetry in the plane. If n is odd, such axes join a vertex to the midpoint of the opposite edge. For example, $(1)(2\,n)(3\,n-1)\ldots$ corresponds to one such reflection. However, if n is even, there are two types of reflections, in an axis joining a pair of opposite vertices and in an axis joining midpoints of opposite edges; thus in all there are $\frac{1}{2}n + \frac{1}{2}n = n$ reflections in this case too.

Since all axes of symmetry of the n-gon have now been exhausted, we conclude that the order of its symmetry group is $1 + (n-1) + n = 2n$. This group is called the *dihedral group of order* $2n$,

$$\mathrm{Dih}(2n).$$

Notice that $\mathrm{Dih}(2n)$ is a proper subset of S_n if $2n < n!$, i.e., if $n \geq 4$. So not every permutation of the vertices arises from a symmetry when $n \geq 4$.

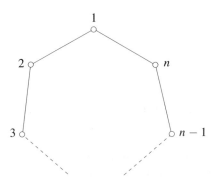

Elementary consequences of the axioms. We end this section by noting three elementary consequences of the axioms.

(3.2.1) (i) *(The Generalized Associative Law) Let $x_1, x_2, \ldots, x_n$ be elements of a semigroup $(S, *)$. If an element u is constructed by combining the n elements in the given order, using any mode of bracketing, then $u = (\cdots((x_1 * x_2) * x_3) * \cdots) * x_n$, so that u is independent of the positioning of the parentheses.*

(ii) *Every monoid has a unique identity element.*

(iii) *Every element in a group has a unique inverse.*

Proof. (i) We argue by induction on n, which can be assumed to be at least 3. If u is constructed from $x_1, x_2, \ldots, x_n$ in that order, then $u = v * w$ where v is constructed from $x_1, x_2, \ldots, x_i$ and w from $x_{i+1}, \ldots, x_n$; here $1 \leq i \leq n-1$. Then $v = (\cdots(x_1 * x_2) * \cdots * x_i)$ by induction on n. If $i = n-1$, then $w = x_n$ and the result follows at once. Otherwise $i + 1 < n$ and $w = z * x_n$ where z is constructed from $x_{i+1}, \ldots, x_{n-1}$. Then $u = v * w = v * (z * x_n) = (v * z) * x_n$ by the associative law. The result is true for $v * z$ by induction, so it is true for u.

(ii) Suppose that e and e' are two identity elements in a monoid. Then $e = e * e'$ since e' is an identity; also $e * e' = e'$ since e is an identity. Hence $e = e'$.

(iii) Let g be an element of a group and suppose g has two inverses x and x'; we claim that $x = x'$. To see this observe that $(x * g) * x' = e * x' = x'$, while also $(x * g) * x' = x * (g * x') = x * e = x$. Hence $x = x'$. □

Because of (i) above, we can without ambiguity omit all parentheses from an expression formed from elements $x_1, x_2, \ldots, x_n$ of a semigroup – an enormous gain in simplicity. Also (ii) and (iii) show that it is unambiguous to speak of *the* identity element of a monoid and *the* inverse of an element of a group.

Exercises (3.2)

1. Let S be the subset of $\mathbb{R} \times \mathbb{R}$ specified below and define $(x, y) * (x', y') = (x + x', y + y')$. Say in each case whether $(S, *)$ is a semigroup, a monoid, a group, or none of these.

(a) $S = \{(x, y) \mid x + y \geq 0\}$;
(b) $S = \{(x, y) \mid x + y > 0\}$;
(c) $S = \{(x, y) \mid |x + y| \leq 1\}$;
(d) $S = \{(x, y) \mid 2x + 3y = 0\}$.

2. Do the sets of even or odd permutations in S_n form a semigroup when functional composition is used as the binary operation?

3. Show that the set of all 2×2 real matrices with non-negative entries is a monoid but not a group when matrix addition used.

4. Let A be a non-empty set and define a binary operation $*$ on the power set $(P(A)$ by $S * T = (S \cup T) - (S \cap T)$. Prove that $(P(A), *)$ is an abelian group.

5. Define powers in a semigroup $(S, *)$ by the rules $x^1 = x$ and $x^{n+1} = x^n * x$ where $x \in S$ and n is a non-negative integer. Prove that $x^m * x^n = x^{m+n}$ and $(x^m)^n = x^{mn}$ where $m, n > 0$.

6. Let G be a monoid such that for each x in G there is a positive integer n such that $x^n = e$. Prove that G is a group.

7. Let G consist of the permutations $(1\,2)(3\,4)$, $(1\,3)(2\,4)$, $(1\,4)(2\,3)$, together with the identity permutation $(1)(2)(3)(4)$. Show that G is a group with exactly four elements in which each element is its own inverse. (This group is called the *Klein*[3] 4-group).

8. Prove that the group S_n is abelian if and only if $n \leq 2$.

9. Prove that the group $\mathrm{GL}_n(\mathbb{R})$ is abelian if and only if $n = 1$.

3.3 Groups and subgroups

From now on we shall concentrate on groups, and we start by improving the notation. In the first place it is customary not to distinguish between a group $(G, *)$ and its underlying set G, provided there is no likelihood of confusion. Then there are two standard ways of writing the group operation. In the *additive notation* we write $x + y$ for $x * y$; the identity is 0_G or 0 and the inverse of an element x is $-x$. The additive notation is often used for abelian groups, i.e., groups $(G, *)$ such that $x * y = y * x$ for all $x, y \in G$.

For non-abelian groups the *multiplicative* notation is generally used, with xy being written for $x * y$; the identity element is 1_G or 1 and the inverse of x is x^{-1}. We will employ the multiplicative notation here unless the additive notation is clearly preferable, as with a group such as $\mathbb{Z}$.

---

[3] Felix Klein (1849–1925).

3.3 Groups and subgroups

Isomorphism. It is important to decide when two groups are to be regarded as essentially the same. It is possible that two groups have very different sets elements, but their elements behave analogously with respect to the two group operations. This leads us to introduce the concept of isomorphism. Let G, H be two (multiplicative) groups. An *isomorphism* from G to H is a bijective function $\alpha : G \to H$ such that

$$\alpha(xy) = \alpha(x)\alpha(y)$$

for all $x, y \in G$. Groups G and H are said to be *isomorphic* if there exists an isomorphism from G to H, in symbols

$$G \simeq H.$$

The idea here is that, while the elements of isomorphic groups may be different, they should have the same properties in relation to their respective groups. In particular, isomorphic groups have the same order, where by the *order* of a group G we mean the cardinality of its set of elements, $|G|$.

The next result records some very useful techniques for working with group elements.

(3.3.1) *Let x, a, b be elements of a group G:*

(i) *if $xa = b$, then $x = ba^{-1}$, and if $ax = b$, then $x = a^{-1}b$.*

(ii) $(xy)^{-1} = y^{-1}x^{-1}$.

Proof. From $xa = b$ we obtain $(xa)a^{-1} = ba^{-1}$, i.e., $x(aa^{-1}) = ba^{-1}$. Since $aa^{-1} = 1$ and $x1 = x$, we obtain $x = ba^{-1}$. The second statement in (i) is dealt with similarly. By (3.2.1), in order to prove (ii) it is enough to show that $y^{-1}x^{-1}$ is an inverse of xy. This can be checked directly: $(xy)(y^{-1}x^{-1}) = x(yy^{-1})x^{-1} = x1x^{-1} = xx^{-1} = 1$; similarly $(y^{-1}x^{-1})(xy) = 1$. Consequently $(xy)^{-1} = y^{-1}x^{-1}$. □

The multiplication table of a group. Suppose that G is a group with finite order n and its elements are ordered in some fixed manner, say as $g_1, g_2, \ldots, g_n$. The group operation of G can be displayed in its *multiplication table*; this is the $n \times n$ array M whose (i, j) entry is $g_i g_j$. Thus the i-th row of M is $g_i g_1, g_i g_2, \ldots, g_i g_n$. From the multiplication table the product of any pair of group elements can be determined. Notice that all the elements in a row are different: for $g_i g_j = g_i g_k$ implies that $g_j = g_k$, i.e., $j = k$, by (3.3.1). The same is true of the columns of M. What this means is that each group element appears exactly once in each row and exactly once in each column of the array, that is, M is a *latin square*. Such configurations are studied in 10.4.

As an example, consider the group whose elements are the identity permutation 1 and the permutations $a = (12)(34)$, $b = (13)(24)$, $c = ((14)(23)$. This is the *Klein*

4-*group*, which was mentioned in Exercise (3.2.7). The multiplication table of this group is the 4×4 array

	1	a	b	c
1	1	a	b	c
a	a	1	c	b
b	b	c	1	a
c	c	b	a	1

Powers of group elements. Let x be an element of a (multiplicative) group G and let n be an integer. The *n-th power* x^n of x is defined recursively as follows:

$$x^0 = 1, \quad x^{n+1} = x^n x, \quad x^{-n} = (x^n)^{-1}$$

where $n \geq 0$. Of course if G were written additively, we would write nx instead of x^n. Fundamental for the manipulation of powers is:

(3.3.2) (The Laws of Exponents) *Let x be an element of a group G and let m, n be integers. Then*

(i) $x^m x^n = x^{m+n} = x^n x^m$;

(ii) $(x^m)^n = x^{mn}$.

Proof. (i) First we show that $x^r x^s = x^{r+s}$ where $r, s \geq 0$, using induction on s. This is clear if $s = 0$. Assuming it true for s, we have

$$x^r x^{s+1} = x^r x^s x = x^{r+s} x = x^{r+s+1},$$

thus completing the induction. Next using (3.3.1) and the definition of negative powers, we deduce from $x^r x^s = x^{r+s}$ that $x^{-r} x^{r+s} = x^s$ and $x^{-r-s} x^r = x^{-s}$. This shows that $x^{-r} x^s = x^{s-r}$ for *all* $r, s \geq 0$. In a similar way $x^r y^{-s} = x^{r-s}$ for all $r, s \geq 0$.

Finally, by inverting $x^s x^r = x^{r+s}$ where $r, s \geq 0$, we obtain $x^{-r} x^{-s} = x^{-r-s}$. Thus all cases have been covered.

(ii) When $n \geq 0$, we use induction on n: clearly it is true when $n = 0$. Assuming the statement true for n, we have $(x^m)^{n+1} = (x^m)^n x^m = x^{mn} x^m = x^{m(n+1)}$ by (i). Next $(x^m)^{-n} = ((x^m)^n)^{-1} = (x^{mn})^{-1} = x^{-mn}$, which covers the case where the second exponent is negative. □

Subgroups. Roughly speaking, a subgroup is a group contained within a larger group. To make this concept precise, we consider a group $(G, *)$ and a subset S of G. If we restrict the group operation $*$ to S, we obtain a function $*'$ from $S \times S$ to G. If $*'$ is a binary operation on S, i.e., if $x * y \in S$ whenever $x, y \in S$, and if $(S, *')$ is actually a group, then S is called a *subgroup* of G.

The first point to settle is that 1_S, the identity element of $(S, *')$, equals 1_G. Indeed $1_S = 1_S *' 1_S = 1_S * 1_S$, so $1_S * 1_S = 1_S * 1_G$. By (3.3.1) it follows that $1_S = 1_G$.

Next let $x \in S$ and denote the inverse of x in $(S, *)$ by x_S^{-1}. We want to be sure that $x_S^{-1} = x^{-1}$. Now $1_G = 1_S = x *' x_S^{-1} = x * x_S^{-1}$. Hence $x * x^{-1} = x * x_S^{-1}$ and so $x_S^{-1} = x^{-1}$. Thus inverses are the same in $(S, *')$ and in $(G, *)$.

On the basis of these conclusions we may formulate a convenient test for a subset of a group to be a subgroup.

(3.3.3) *Let S be a subset of a group G. Then S is a subgroup of G if and only if the following hold:*

(i) $1_G \in S$;

(ii) $xy \in S$ whenever $x \in S$ and $y \in S$ *(closure under products)*;

(iii) $x^{-1} \in S$ whenever $x \in S$ *(closure under inverses)*.

To indicate that S is a subgroup of a group G, we will often write

$$S \leq G,$$

and, if also $S \neq G$, we say that S is a *proper* subgroup and write

$$S < G.$$

Examples of subgroups. (i) $\mathbb{Z} \leq \mathbb{Q} \leq \mathbb{R} \leq \mathbb{C}$. These statements follow at once via (3.3.3). For the same reason $\mathbb{Q}^* \leq \mathbb{R}^* \leq \mathbb{C}^*$.

(ii) $A_n \leq S_n$. Recall that A_n is the set of even permutations in S_n. Here the point to note is that if π, σ are even permutations, then so are $\pi\sigma$ and π^{-1} by (3.1.6): of course the identity permutation is also even. However the odd permutations in S_n do not form a subgroup.

(iii) Two subgroups that are present in every group G are the *trivial* or *identity subgroup* $\{1_G\}$, which will be written 1 or 1_G, and the *improper subgroup* G itself. For some groups these are the only subgroups.

(iv) *Cyclic subgroups*. The interesting subgroups of a group are the proper nontrivial ones. An easy way to produce such subgroups is to take all the powers of a fixed element. Let G be a group and choose $x \in G$. We denote the set of all powers of the element x by

$$\langle x \rangle.$$

Using (3.3.3) and the Laws of Exponents (3.3.2), we quickly verify that $\langle x \rangle$ is a subgroup. It is called *the cyclic subgroup generated by x*. Since every subgroup of G which contains x must also contain all powers of x, it follows that $\langle x \rangle$ *is the smallest subgroup of G containing x*.

A group G is said to be *cyclic* if $G = \langle x \rangle$ for some x in G. For example, $\mathbb{Z}$ and $\mathbb{Z}_n$ are cyclic groups since, allowing for the additive notation, $\mathbb{Z} = \langle 1 \rangle$ and $\mathbb{Z}_n = \langle [1]_n \rangle$.

Next we consider intersections of subgroups.

48   3 Introduction to groups

(3.3.4) *If $\{S_\lambda \mid \lambda \in \Lambda\}$ is a set of subgroups of a group G, then $\bigcap_{\lambda \in \Lambda} S_\lambda$ is also a subgroup of G.*

This follows immediately from (3.3.3). Now suppose that X is any non-empty subset of a group G. There is at least one subgroup that contains X, namely G itself. So we may form the intersection of all the subgroups of G which contain X. This is a subgroup by (3.3.4), and it is denoted by

$$\langle X \rangle.$$

Obviously $\langle X \rangle$ is the smallest subgroup of G containing X: it is called the *subgroup generated by X*. Note that the cyclic subgroup $\langle x \rangle$ is just the subgroup generated by the singleton set $\{x\}$. Thus we have generalized the notion of a cyclic subgroup.

It is natural to enquire what form the elements of $\langle X \rangle$ take.

(3.3.5) *Let X be a non-empty subset of a group G. Then $\langle X \rangle$ consists of all elements of G of the form $x_1^{\varepsilon_1} x_2^{\varepsilon_2} \ldots x_k^{\varepsilon_k}$ where $x_i \in X$, $\varepsilon_i = \pm 1$ and $k \geq 0$, (the case $k = 0$ being interpreted as 1_G).*

Proof. Let S denote the set of all elements of the specified type. We easily check that S contains 1 and is closed under products and inversion, using (3.3.1); thus S is a subgroup. Clearly $X \subseteq S$, so that $\langle X \rangle \subseteq S$ since $\langle X \rangle$ is the smallest subgroup containing X. On the other hand, any element of the form $x_1^{\varepsilon_1} \ldots x_k^{\varepsilon_k}$ must belong to $\langle X \rangle$ since $x_i \in \langle X \rangle$. Thus $S \subseteq \langle X \rangle$ and $\langle X \rangle = S$. □

Notice that when X is a singleton $\{x\}$, we recover the fact that $\langle x \rangle$ consists of all powers of x.

The lattice of subgroups. Suppose that G is a group. Then set inclusion is a partial order on the set $S(G)$ of all subgroups of G, so that $S(G)$ is a partially ordered set. Now if H and K are subgroups of G, they have a greatest lower bound in $S(G)$, namely $H \cap K$, and also a least upper bound $\langle H \cup K \rangle$, which is usually written $\langle H, K \rangle$. This last is true because any subgroup containing H and K must also contain $\langle H, K \rangle$. This means that $S(G)$ is a lattice, in the sense of 1.2. Like any partially ordered set, $S(G)$ can be visualized by displaying its Hasse diagram; the basic component in the Hasse diagram of subgroups of a group is the "lozenge" shaped diagram.

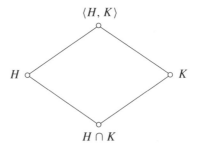

3.3 Groups and subgroups

The order of a group element. Let x be an element of a group. If the subgroup $\langle x \rangle$ has a finite number m of elements, x is said to have *finite order m*. If on the other hand $\langle x \rangle$ is infinite, x is called an element of *infinite order*. We shall write

$$|x|$$

for the order of x. The basic facts about orders of group elements are contained in:

(3.3.6) *Let x be an element of a group G.*

(i) *If x has infinite order, then all the powers of x are distinct.*

(ii) *If x has finite order m, then m is the least positive integer such that $x^m = 1$; further $\langle x \rangle$ consists of the distinct elements $1, x, \ldots, x^{m-1}$, and $\langle x \rangle$ has order m.*

(iii) *If x has finite order m, then $x^\ell = 1$ if and only if ℓ is divisible by m.*

Proof. Suppose that two powers of x are equal, say $x^i = x^j$ where $i > j$. Then $x^{i-j} = 1$ by (3.3.2). Using Well-Ordering we may choose a smallest positive integer m for which $x^m = 1$. Now let ℓ be any integer and write $\ell = mq + r$ where $q, r \in \mathbb{Z}$ and $0 \leq r < m$, using the Division Algorithm. Then by (3.3.2) again $x^\ell = (x^m)^q x^r$ and hence $x^\ell = x^r$. This shows that $\langle x \rangle = \{1, x, x^2, \ldots, x^{m-1}\}$ and x has finite order. Thus (i) is established. In addition our argument shows that $x^\ell = 1$ if and only if $x^r = 1$, i.e., $r = 0$ by the choice of m. So (iii) has also been proved.

Finally, if $x^i = x^j$ where $0 \leq j < i < m$, then $x^{i-j} = 1$. Since $0 \leq i - j < m$, it follows from the choice of m that $i = j$. Therefore $|\langle x \rangle| = m$ and x has order m, which establishes (ii). □

We will now study cyclic groups with the aim of identifying them up to isomorphism.

(3.3.7) *A cyclic group of order n is isomorphic with $\mathbb{Z}_n$. An infinite cyclic group is isomorphic with $\mathbb{Z}$.*

Proof. Let $G = \langle x \rangle$ be a cyclic group. If $|G| = n$, then $G = \{1, x, \ldots, x^{n-1}\}$. Define $\alpha : \mathbb{Z}_n \to G$ by $\alpha([i]) = x^i$, which is a well-defined function because $x^{i+nq} = x^i (x^n)^q = x^i$. Also

$$\alpha([i] + [j]) = \alpha([i + j]) = x^{i+j} = x^i x^j = \alpha([i])\alpha([j]),$$

while α is clearly bijective. Therefore, allowing for $\mathbb{Z}_n$ being written additively and G multiplicatively, we conclude that α is an isomorphism and $\mathbb{Z}_n \simeq G$. When G is infinite cyclic, the proof is similar but easier, and is left to the reader. □

There is a simple way to compute the order of an element of the symmetric group S_n.

(3.3.8) Let $\pi \in S_n$ and write $\pi = \pi_1 \pi_2 \ldots \pi_k$ where the π_i are disjoint cycles, with π_i of length ℓ_i. Then the order of π equals the least common multiple of $\ell_1, \ell_2, \ldots, \ell_k$.

Proof. Remembering that disjoint permutations commute by (3.1.2), we see that $\pi^m = \pi_1^m \pi_2^m \ldots \pi_k^m$ for any $m > 0$. Now the π_i^m's affect disjoint sets of integers, so $\pi^m = 1$, (i.e., $\pi^m = \text{id}$), if and only if $\pi_1^m = \pi_2^m = \cdots = \pi_k^m = 1$, by (3.3.6). These conditions assert that m is divisible by the orders of all the π_i. Now it is easy to see, by forming successive powers, that the order of an r-cycle is precisely r. Therefore $|\pi| = \text{lcm}\{\ell_1, \ell_2, \ldots, \ell_k\}$. (For least common multiples see Exercise 2.2.7.) □

Example (3.3.1) What is the largest possible order of an element of S_8?

Let $\pi \in S_8$ and write $\pi = \pi_1 \ldots \pi_k$ where the π_i are disjoint cycles. If π_i has length ℓ_i, then $\sum_{i=1}^{k} \ell_i = 8$ and $|\pi| = \text{lcm}\{\ell_1, \ldots, \ell_k\}$. So the question is: which positive integers $\ell_1, \ldots, \ell_k$ with sum equal to 8 have the largest least common multiple? A little experimentation will convince the reader that the answer is $\ell_1 = 3$, $\ell_2 = 5$. Hence 15 is the largest possible order of an element of S_8. For example, the permutation (1 2 3)(4 5 6 7 8) has this order.

We conclude with two more examples, including an application to number theory.

Example (3.3.2) Let G be a finite abelian group. Prove that the product of all the elements of G equals the product of all the elements of G of order 2.

The key point to notice here is that if $x \in G$, then $|x| = 2$ if and only if $x = x^{-1} \neq 1$. Since G is abelian, in the product $\prod_{g \in G} g$ we can group together elements of order greater than 2 with their inverses and then cancel each pair xx^{-1}. What is left is just the product of the elements of order 2.

Example (3.3.3) (*Wilson's[4] Theorem*) If p is a prime, then $(p-1)! \equiv -1 \pmod{p}$.

Apply Example (3.3.2) to the group $\mathbb{Z}_p^*$, the multiplicative group of non-zero congruence classes mod p. Now the only element of order 2 in $\mathbb{Z}_p^*$ is $[-1]$: for $a^2 \equiv 1 \pmod{p}$ implies that $a \equiv 1 \pmod{p}$ or $a \equiv -1 \pmod{p}$, i.e., $[a] = [1]$ or $[-1]$. It follows that $[1][2]\ldots[p-1] = [-1]$, or $(p-1)! \equiv -1 \pmod{p}$.

Exercises (3.3)

1. In each case say whether or not S is a subgroup of the group G:
    (a) $G = \text{GL}_n(\mathbb{R})$, $S = \{A \in G \mid \det(A) = 1\}$
    (b) $G = (\mathbb{R}, +)$, $S = \{x \in \mathbb{R} \mid |x| \leq 1\}$;
    (c) $G = \mathbb{R} \times \mathbb{R}$, $S = \{(x, y) \mid 3x - 2y = 1\}$: here the group operation adds the components of ordered pairs.

---

[4] John Wilson (1741–1793).

2. Let H and K be subgroups of a group G. Prove that $H \cup K$ is a subgroup if and only if $H \subseteq K$ or $K \subseteq H$.

3. Show that no group can be the union of two proper subgroups. Then exhibit a group which is the union of three proper subgroups.

4. Find the largest possible order of an element of S_{11}. How many elements of S_{11} have this order?

5. The same question for S_{12}.

6. Find the orders of the elements [3] and [7] of $\mathbb{Z}_{11}^*$.

7. Prove that a group of even order must contain an element of order 2. [Hint: assume this is false and group the non-identity elements as x, x^{-1}].

8. Assume that for each pair of elements a, b of a group G there is an integer n such that $(ab)^i = a^i b^i$ holds for $i = n, n+1, n+2$. Prove that G is abelian.

Chapter 4
Cosets, quotient groups and homomorphisms

In this chapter we probe more deeply into the nature of the subgroups of a group, and establish their critical importance in understanding the structure of groups.

4.1 Cosets and Lagrange's Theorem

Consider a group G with a fixed subgroup H. A binary relation $\sim_H$ on G is defined by the following rule: $x \sim_H y$ means that $x = yh$ for some $h \in H$. It is an easy verification that $\sim_H$ is an equivalence relation on G. Therefore by (1.2.2) the group G splits up into disjoint equivalence classes. The equivalence class to which the element x belongs is the subset

$$\{xh \mid h \in H\},$$

which is called a *left coset* of H in G and is written

$$xH.$$

Thus G is the disjoint union of the distinct left cosets of H. Notice that the only coset which is a subgroup is $1H = H$ since no other coset contains the identity element.

Next observe that the assignment $h \mapsto xh$ ($h \in H$) determines a bijection from H to xH; indeed $xh_1 = xh_2$ implies that $h_1 = h_2$. From this it follows that

$$|xH| = |H|,$$

so that each left coset of H has the cardinal of H.

Next suppose that we label the left cosets of H in some manner, say as C_λ, $\lambda \in \Lambda$, and for each λ in Λ we choose an arbitrary element t_λ from C_λ. (If Λ is infinite, we are assuming at this point a set theoretic axiom called *the axiom of choice* – for this see 12.1). Then $C_\lambda = t_\lambda H$, and since every group element belongs to some left coset of H, we have $G = \bigcup_{\lambda \in \Lambda} t_\lambda H$. Furthermore, distinct cosets being equivalence classes and therefore disjoint, each element x of G has a unique expression $x = t_\lambda h$, where $h \in H$, $\lambda \in \Lambda$. The set $\{t_\lambda \mid \lambda \in \Lambda\}$ is called a *left transversal* to H in G. Thus we have found a unique way to express elements of G in terms of the transversal and elements of the subgroup H.

In a similar fashion one can define *right cosets* of H in G; these are the equivalence classes of the equivalence relation $_H\sim$, where $x \,_H\!\sim y$ means that $x = hy$ for some h in H. The *right coset* containing x is

$$Hx = \{hx \mid h \in H\}$$

and right transversals are defined analogously.

The next result was the first significant theorem to be discovered in group theory.

(4.1.1) (Lagrange's[1] Theorem) *Let H be a subgroup of a finite group G. Then $|H|$ divides $|G|$; in fact $|G|/|H| =$ the number of left cosets of $H =$ the number of right cosets of H.*

Proof. Let ℓ be the number of left cosets of H in G. Since the number of elements in any left coset of H is $|H|$ and distinct left cosets are disjoint, a simple count of the elements of G yields $|G| = \ell \cdot |H|$; thus $\ell = |G|/|H|$. For right cosets the argument is the same. □

Corollary (4.1.2) *The order of an element of a finite group divides the group order.*

The index of a subgroup. Even in infinite groups the sets of left and right cosets of a fixed subgroup have the same cardinal. For if $H \leq G$, the assignment $xH \mapsto Hx$ clearly determines a bijection between these sets. This allows us to define the *index of H in G* to be simultaneously the cardinal of the set of left cosets and the cardinal of the set of right cosets of H; the index is written

$$|G : H|.$$

When G is finite, we have already seen that

$$|G : H| = |G|/|H|$$

by Lagrange's Theorem.

Example (4.1.1) Let G be the symmetric group S_3 and let H be the cyclic subgroup generated by $(12)(3)$.

Here $|G : H| = |G|/|H| = 6/2 = 3$, so we expect to find three left cosets and three right ones. The left cosets are:

$$H = \{\text{id}, (12)(3)\}, \quad (123)H = \{(123), (13)(2)\}, \quad (132)H = \{(132), (1)(23)\},$$

and the right cosets are

$$H = \{\text{id}, (12)(3)\}, \quad H(123) = \{(123), (1)(23)\}, \quad H(132) = \{(132), (13)(2)\}.$$

Notice that the left cosets are disjoint, as are the right ones; but left and right cosets need not be disjoint.

The next result is useful in calculations with subgroups.

---
[1] Joseph Louis Lagrange (1736–1813)

(4.1.3) *Let $H \leq K \leq G$ where G is a finite group. Then*

$$|G : H| = |G : K| \cdot |K : H|.$$

Proof. Let $\{t_\lambda \mid \lambda \in \Lambda\}$ be a left transversal to H in K, and let $\{s_\mu \mid \mu \in M\}$ be a left transversal to K in G. Thus $K = \bigcup_{\lambda \in \Lambda} t_\lambda H$ and $G = \bigcup_{\mu \in M} s_\mu K$. Hence $G = \bigcup_{\lambda \in \Lambda, \mu \in M} (s_\mu t_\lambda) H$. We claim that the elements $s_\mu t_\lambda$ belong to different left cosets of H. Indeed suppose that $s_\mu t_\lambda H = s_{\mu'} t_{\lambda'} H$; then, since $t_\lambda H \subseteq K$, we have $s_\mu K = s_{\mu'} K$, which implies that $\mu = \mu'$. Hence $t_\lambda H = t_{\lambda'} H$, which shows that $\lambda = \lambda'$. It follows that $|G : H|$, the cardinal of the set of left cosets of H in G, equals $|M \times \Lambda| = |M| \cdot |\Lambda| = |G : K| \cdot |K : H|$. □

(This result is still valid when G is an infinite group provided one defines the product of two cardinal numbers appropriately, cf. Exercise (1.4.6)).

Groups of prime order. Lagrange's Theorem is sufficiently strong to enable us to describe all groups with prime order. This is the first example of a classification theorem in group theory; it is also an indication of the importance for the structure of a group of the arithmetic properties of its order.

(4.1.4) *A group G has prime order p if and only if $G \simeq \mathbb{Z}_p$.*

Proof. Assume that $|G| = p$, and let $1 \neq x \in G$. Then $|\langle x \rangle|$ divides $|G| = p$ by (4.1.1). Hence $|\langle x \rangle| = |G|$ and $G = \langle x \rangle$, a cyclic group of order p. Thus $G \simeq \mathbb{Z}_p$ by (3.3.7). The converse is obvious. □

Example (4.1.2) Find all groups of order less than 6.

Let G be a group with $|G| < 6$. If $|G| = 1$, of course, G is a trivial group. If $|G| = 2, 3$ or 5, then (4.1.4) tells us that $G \simeq \mathbb{Z}_2, \mathbb{Z}_3$ or $\mathbb{Z}_5$ respectively. So we are left with the case where $|G| = 4$. If G contains an element x of order 4, then $G = \langle x \rangle$ and $G \simeq \mathbb{Z}_4$ by (3.3.7). Assuming that G has no elements of order 4, we may conclude from (4.1.2) that G must consist of 1 and three elements of order 2, say a, b, c.

Now ab cannot equal 1, otherwise $b = a^{-1} = a$. Also $ab \neq a$ and $ab \neq b$, as the reader should check. Hence ab must equal c; also $ba = c$ by the same argument. Similarly we can prove that $bc = a = cb$ and $ca = b = ac$.

At this point the reader should realize that G is very like the Klein 4-group $V = \{(1)(2)(3)(4), (12)(34), (13)(24), (14)(23)\}$. In fact the assignments $1_G \mapsto 1_V$, $a \mapsto (12)(34)$, $b \mapsto (13)(24)$, $c \mapsto (14)(23)$ determine an isomorphism from G to V. Our conclusion is that *up to isomorphism there are exactly six groups with order less than 6, namely* $\mathbb{Z}_1, \mathbb{Z}_2, \mathbb{Z}_3, \mathbb{Z}_4, V, \mathbb{Z}_5$.

The following application of Lagrange's Theorem gives another proof of Fermat's Theorem – see (2.3.4).

4.1 Cosets and Lagrange's Theorem   55

Example (4.1.3) Use a group theoretic argument to prove that if p is a prime and n is any integer, then $n^p \equiv n \pmod{p}$.

Apply (4.1.2) to $\mathbb{Z}_p^*$, the multiplicative group of non-zero congruence classes modulo p. If $[n] \neq [0]$, then (4.1.2) implies that the order of $[n]$ divides $|\mathbb{Z}_p^*| = p - 1$. Thus $[n]^{p-1} = [1]$, i.e., $n^{p-1} \equiv 1 \pmod{p}$. Multiply by n to get $n^p \equiv n \pmod{p}$, and observe that this also holds if $[n] = [0]$.

According to Lagrange's Theorem, the order of a subgroup of a finite group divides the group order. However the natural converse of this statement is false: there need not be a subgroup with order equal to a given divisor of the group order. This is demonstrated by the following example.

Example (4.1.4) The alternating group A_4 has order 12, but it has no subgroups of order 6.

Write $G = A_4$. First note that each non-trivial element of G is either a 3-cycle or the product of two disjoint transpositions. Also all of the latter form with the identity the Klein 4-group V. Suppose that H is a subgroup of order 6. Assume first that $H \cap V = 1$. Then there are $6 \times 4 = 24$ distinct elements of the form hv, $h \in H$, $v \in V$; for if $h_1 v_1 = h_2 v_2$ with $h_i \in H$, $v_i \in V$, then $h_2^{-1} h_1 = v_2 v_1^{-1} \in H \cap V = 1$, so that $h_1 = h_2$ and $v_1 = v_2$. Since this is impossible, we conclude that $H \cap V \neq 1$.

Let us say $H \cap V$ contains $\pi = (12)(34)$. Now H must also contain a 3-cycle since there are 8 of these in G, say $\sigma = (123) \in H$. Hence H contains $\tau = \sigma \pi \sigma^{-1} = (14)(23)$. Thus H contains $\pi \tau = (13)(24)$. It follows that $V \subseteq H$, yet $|V|$ does not divide $|H|$, a final contradiction.

Subgroups of cyclic groups. Usually a group will have many subgroups and it is can be a difficult task to find all of them. So it is of interest that the subgroups of cyclic groups are easy to describe. The basic result here is:

(4.1.5) *A subgroup of a cyclic group is cyclic.*

Proof. Let H be a subgroup of a cyclic group $G = \langle x \rangle$. If $H = 1$, then obviously $H = \langle 1 \rangle$; thus we may assume that $H \neq 1$, so that H contains some $x^m \neq 1$; since H must also contain $(x^m)^{-1} = x^{-m}$, we may as well assume that $m > 0$ here. Now choose m to be the *smallest* positive integer for which $x^m \in H$; of course we have used the Well-Ordering Law here.

Certainly it is true that $\langle x^m \rangle \subseteq H$. We will prove the reverse inclusion, which will show that $H = \langle x^m \rangle$. Let $h \in H$ and write $h = x^i$. By the Division Algorithm $i = mq + r$ where $q, r \in \mathbb{Z}$ and $0 \leq r < m$. By the Laws of Exponents (3.3.2) $x^i = x^{mq} x^r$. Then $x^r = x^{-mq} x^i$, which belongs to H since $x^m \in H$ and $x^i \in H$. By minimality of m it follows that $r = 0$ and $i = mq$. Therefore $h = x^i \in \langle x^m \rangle$. □

The next result tells us how to set about constructing the subgroups of a given cyclic group.

(4.1.6) *Let $G = \langle x \rangle$ be a cyclic group.*

(i) *If G is infinite, then each subgroup of G has the form $G_i = \langle x^i \rangle$ where $i \geq 0$. Furthermore, the G_i are all distinct and G_i has infinite order if $i > 0$.*

(ii) *If G has finite order n, then it has exactly one subgroup of order d for each positive divisor d of n, namely $\langle x^{n/d} \rangle$.*

Proof. Assume first that G is infinite and let H be a subgroup of G. By (4.1.5) H is cyclic, say $H = \langle x^i \rangle$ where $i \geq 0$. Thus $G_i = G_i$. If x^i had finite order m, then $x^{im} = 1$, which, since x has infinite order, can only mean that $i = 0$ and $H = 1$. Thus H is certainly infinite cyclic if $i > 0$. Next $G_i = G_j$ implies that $x^i \in \langle x^j \rangle$ and $x^j \in \langle x^i \rangle$, i.e., $j \mid i$ and $i \mid j$, so that $i = j$. Thus all the G_i's are different.

Next let G have finite order n and suppose d is a positive divisor of n. Now $(x^{\frac{n}{d}})^d = x^n = 1$, so $\ell = |x^{\frac{n}{d}}|$ must divide d. But also $x^{\frac{n\ell}{d}} = 1$, and hence n divides $\frac{n\ell}{d}$, i.e., d divides ℓ. It follows that $\ell = d$ and thus $K = \langle x^{n/d} \rangle$ has order exactly d.

To complete the proof, suppose that $H = \langle x^r \rangle$ is another subgroup with order d. Then $x^{rd} = 1$, so n divides rd and $\frac{n}{d}$ divides r. This shows that $H = \langle x^r \rangle \subseteq \langle x^{n/d} \rangle = K$. But $|H| = |K| = d$, from which it follows that $H = K$. Consequently there is exactly one subgroup of order d. □

Recall from 3.3 that the set of all subgroups of a group is a lattice and may be represented by a Hasse diagram.

Example (4.1.5) Display the Hasse diagram for the subgroups of a cyclic group of order 12.

Let $G = \langle x \rangle$ have order 12. By (4.1.6) the subgroups of G correspond to the positive divisors of 12, i.e., 1, 2, 3, 4, 6, 12; in fact if $i \mid 12$, the subgroup $\langle x^{12/i} \rangle$ has order i. It is now easy to draw the Hasse diagram:

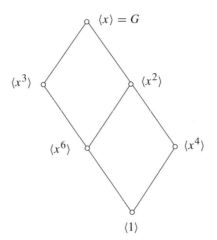

The next result is useful for calculating orders of elements in cyclic groups.

(4.1.7) *Let $G = \langle x \rangle$ be a cyclic group with finite order n. Then the order of the element x^i is*
$$\frac{n}{\gcd\{i, n\}}.$$

Proof. In the first place $(x^i)^m = 1$ if and only if $n \mid im$, i.e., $\frac{n}{d} \mid \left(\frac{i}{d}\right)m$ where $d = \gcd\{i, n\}$. Since $\frac{n}{d}$ and $\frac{i}{d}$ are relatively prime, by Euclid's Lemma this is equivalent to $\frac{n}{d}$ dividing m. Therefore $(x^i)^m = 1$ if and only if $\frac{n}{d}$ divides m, so that x^i has order $\frac{n}{d}$, as claimed. □

Corollary (4.1.8) *Let $G = \langle x \rangle$ be a cyclic group of finite order n. Then $G = \langle x^i \rangle$ if and only if $\gcd\{i, n\} = 1$.*

For G equals $\langle x \rangle$ if and only if x^i has order n, i.e., $\gcd\{i, n\} = 1$. This means that the number of generators of G equals the number of integers i satisfying $1 \leq i < n$ and $\gcd\{i, n\} = 1$. This number is $\phi(n)$ where ϕ is Euler's function, introduced in 2.3.

Every non-trivial group has at least two subgroups, itself and the trivial subgroup. Which groups have just these subgroups and no more? This question is easily answered using (4.1.7) and Lagrange's Theorem.

(4.1.9) *A group G has exactly two subgroups if and only if $G \simeq \mathbb{Z}_p$ for some prime p.*

Proof. Assume that G has exactly two subgroups, 1 and G. Let $1 \neq x \in G$; then $1 \neq \langle x \rangle \leq G$, so $G = \langle x \rangle$ and G is cyclic. Now G cannot be infinite; for then it would have infinitely many subgroups by (4.1.6). Thus G has finite order n, say. Now if n is not a prime, it has a divisor d where $1 < d < n$, and $\langle x^{n/d} \rangle$ is a subgroup of order d, which is impossible. Therefore G has prime order p and $G \simeq \mathbb{Z}_p$ by (4.1.4). Conversely, if $G \simeq \mathbb{Z}_p$, then $|G| = p$, and Lagrange's Theorem shows that G has no non-trivial proper subgroups. □

Products of subgroups. If H and K are subsets of a group G, the *product* of H and K is defined to be the subset
$$HK = \{hk \mid h \in H, k \in K\}.$$

For example, if $H = \{h\}$ and K is a subgroup, then HK is just the left coset hK. Products of more than two subsets are defined in the obvious way: $H_1 H_2 \ldots H_m = \{h_1 h_2 \ldots h_m \mid h_i \in H_i\}$.

Now even if H and K are subgroups, their product HK need not be a subgroup. For example, in S_3 take $H = \langle (12) \rangle$ and $K = \langle (13) \rangle$. Then $HK = \{\text{id}, (12), (13), (132)\}$. But HK cannot be a subgroup since 4 does not divide the order of S_3.

The following result tells us just when the product of two subgroups is a subgroup.

(4.1.10) *Let H and K be subgroups of a group G. Then HK is a subgroup if and only if $HK = KH$, and in this event $\langle H, K \rangle = HK$.*

Proof. Assume first that $HK \leq G$. Then $H \leq HK$ and $K \leq HK$, so $KH \subseteq HK$. By taking inverses on each side of this inclusion, we deduce that $HK \subseteq KH$. Hence $HK = KH$. Moreover $\langle H, K \rangle \subseteq HK$ since $H \leq HK$ and $K \leq HK$, while $HK \subseteq \langle H, K \rangle$ is always true. Therefore $\langle H, K \rangle = HK$.

Conversely, assume that $HK = KH$; we will verify that HK is a subgroup by using (3.3.3). Let $h_1, h_2 \in H$ and $k_1, k_2 \in K$. Then $(h_1 k_1)^{-1} = k_1^{-1} h_1^{-1} \in KH = HK$. Also $(h_1 k_1)(h_2 k_2) = h_1(k_1 h_2)k_2$; now $k_1 h_2 \in KH = HK$, so that $k_1 h_2 = h_3 k_3$ where $h_3 \in H, k_3 \in K$. Thus $(h_1 k_1)(h_2 k_2) = (h_1 h_3)(k_3 k_2) \in HK$. Obviously $1 \in HK$. Since we have proved that the subset HK is closed under products and inversion, it is a subgroup. □

It is customary to say that the subgroups H and K *permute* if $HK = KH$. The next result is frequently used in calculations with subgroups.

(4.1.11) (Dedekind's[2] Modular Law) *Let H, K, L be subgroups of a group and assume that $K \subseteq L$. Then*

$$(HK) \cap L = (H \cap L)K.$$

Proof. In the first place $(H \cap L)K \subseteq L$ since $K \subseteq L$; therefore $(H \cap L)K \subseteq (HK) \cap L$. To prove the converse, let $x \in (HK) \cap L$ and write $x = hk$ where $h \in H$, $k \in K$. Hence $h = xk^{-1} \in LK = L$, from which it follows that $h \in H \cap L$ and $x = hk \in (H \cap L)K$. Thus $(HK) \cap L \subseteq (H \cap L)K$. □

The reader may recognize that (4.1.11) is a special case of the distributive law $\langle H, K \rangle \cap L = \langle H \cap L, K \cap L \rangle$. However this law is false in general, (see Exercise (4.1.1) below).

Frequently one wants to count the elements in a product of finite subgroups, which makes the next result useful.

(4.1.12) *If H and K are finite subgroups of a group, then*

$$|HK| = \frac{|H| \cdot |K|}{|H \cap K|}.$$

Proof. Define a function $\alpha : H \times K \to HK$ by $\alpha((h, k)) = hk$ where $h \in H$, $k \in K$; evidently α is surjective. Now $\alpha((h_1, k_1)) = \alpha((h_2, k_2))$ holds if and only if $h_1 k_1 = h_2 k_2$, i.e., $h_2^{-1} h_1 = k_2 k_1^{-1} = d \in H \cap K$. Thus $h_2 = h_1 d^{-1}$ and $k_2 = dk_1$. It follows that the elements of $H \times K$ which have the same image under α as (h_1, k_1) are those of the form $(h_1 d^{-1}, dk_1)$ where $d \in H \cap K$. Now compute the number of

---
[2] Richard Dedekind (1831–1916)

the elements of $H \times K$ by counting their images under α and allowing for the number of elements with the same image. This gives $|H \times K| = |HK| \cdot |H \cap K|$. But of course $|H \times K| = |H| \cdot |K|$, so the result follows. □

The final result of this section provides important information about the index of the intersection of two subgroups.

(4.1.13). (Poincaré[3]) *Let H and K be subgroups of finite index in a group G. Then*

$$|G : H \cap K| \leq |G : H| \cdot |G : K|,$$

with equality if $|G : H|$ and $|G : K|$ are relatively prime.

Proof. To each left coset $x(H \cap K)$ assign the pair of left cosets (xH, xK). This is a well-defined function; for, if we were to replace x by xd with $d \in H \cap K$, then $xH = xdH$ and $xK = xdK$. The function is also injective; indeed $(xH, xK) = (yH, yK)$ implies that $xH = yH$ and $xK = yK$, i.e., $y^{-1}x \in H \cap K$, so that $x(H \cap K) = y(H \cap K)$. It follows that the number of left cosets of $H \cap K$ in G is at most $|G : H| \cdot |G : K|$.

Now assume that $|G : H|$ and $|G : K|$ are relatively prime. Since by (4.1.3) $|G : H \cap K| = |G : H| \cdot |H : H \cap K|$, we see that $|G : H|$ divides $|G : H \cap K|$, as does $|G : K|$ for a similar reason. But $|G : H|$ and $|G : K|$ are relatively prime, which means that $|G : H \cap K|$ is divisible by $|G : H| \cdot |G : K|$. This shows that $|G : H \cap K|$ must equal $|G : H| \cdot |G : K|$. □

Exercises (4.1)

1. Show that the distributive law for subgroups,

$$\langle H, K \rangle \cap L = \langle H \cap L, K \cap L \rangle$$

is false in general.

2. If H is a subgroup of a finite group, show that there are $|H|^{|G:H|}$ left transversals to H in G and the same number of right transversals.

3. Let H be a subgroup of a group G such that $G - H$ is finite. Prove that either $H = G$ or G is finite.

4. Display the Hasse diagram for the subgroup lattices of the following groups: $\mathbb{Z}_{18}$; $\mathbb{Z}_{24}$; V (the Klein 4-group); S_3.

5. Let G be a group with exactly three subgroups. Show that $G \simeq \mathbb{Z}_{p^2}$ where p is a prime. [Hint: first prove that G is cyclic].

---

[3] Henri Poincaré (1854–1912)

6. Let H and K be subgroups of a group G with relatively prime indexes in G. Prove that $G = HK$. [Hint: use (4.1.12) and (4.1.13)].

7. If the product of subsets are used as the binary operation, show that the set of all non-empty subsets of a group is a monoid.

8. Let H and K be subgroups of a finite group with relatively prime orders. Show that $H \cap K = 1$ and $|HK|$ divides the order of $\langle H, K \rangle$.

9. Let $G = \langle x \rangle$ be an infinite cyclic group and put $H = \langle x^i \rangle$, $K = \langle x^j \rangle$. Prove that $H \cap K = \langle x^\ell \rangle$ and $\langle H, K \rangle = \langle x^d \rangle$ where $d = \gcd\{i, j\}$ and $\ell = \text{lcm}\{i, j\}$.

10. Let G be a finite group with order n and let d be the minimum number of generators of G. Prove that $n \geq 2^d$, so that $d \leq [\log_2 n]$.

11. By applying Lagrange's Theorem to the group $\mathbb{Z}_n^*$, prove that $x^{\phi(n)} \equiv 1 \pmod{n}$ where n is any positive integer and x is an integer relatively prime to n. Here ϕ is Euler's function. (This is a generalization of Fermat's theorem (2.3.4)).

4.2 Normal subgroups and quotient groups

We focus next on a special type of subgroup called a *normal subgroup*. Such subgroups are important since they can be used to construct new groups, the so-called *quotient groups*. Normal subgroups are characterized in the following result.

(4.2.1) *Let H be a subgroup of a group G. Then the following statements about H are equivalent:*

(i) $xH = Hx$ for all x in G;

(ii) $xhx^{-1} \in H$ whenever $h \in H$ and $x \in G$.

The element xhx^{-1} is called the *conjugate* of h by x.

Proof of (4.2.1). First assume that (i) holds and let $x \in G$ and $h \in H$. Then $xh \in xH = Hx$, so $xh = h_1 x$ for some $h_1 \in H$; hence $xhx^{-1} = h_1 \in H$, which establishes (ii). Now assume that (ii) holds. Again let $x \in G$ and $h \in H$. Then $xhx^{-1} = h_1 \in H$, so $xh = h_1 x \in Hx$, and therefore $xH \subseteq Hx$. Next $x^{-1}hx = x^{-1}h(x^{-1})^{-1} = h_2 \in H$, which shows that $hx = xh_2 \in xH$ and $Hx \subseteq xH$. Thus $xH = Hx$. □

A subgroup H of a group G with the equivalent properties of (4.2.1) is called a *normal subgroup* of G. The notation

$$H \triangleleft G$$

is used to denote the fact that H is a normal subgroup of a group G.

Examples of normal subgroups. (i) Obvious examples of normal subgroups are: $1 \triangleleft G$ and $G \triangleleft G$, and it is possible that 1 and G are the only normal subgroups present. If $G \neq 1$ and 1 and G are the only normal subgroups, the group G is said to be a *simple group*. This is one of the great mis-nomers of mathematics since simple groups can have extremely complicated structure!

(ii) $A_n \triangleleft S_n$.

For, if $\pi \in A_n$ and $\sigma \in S_n$, then according to (3.1.6) we have

$$\text{sign}(\sigma \pi \sigma^{-1}) = \text{sign}(\sigma) \, \text{sign}(\pi)(\text{sign}(\sigma))^{-1} = (\text{sign}(\sigma))^2 = 1.$$

so that $\sigma \pi \sigma^{-1} \in A_n$.

(iii) In an abelian group G every subgroup is normal.

This is because $xhx^{-1} = hxx^{-1} = h$ for all $h \in H$, $x \in G$.

(iv) Recall that $\text{GL}_n(\mathbb{R})$ is the group of all non-singular $n \times n$ real matrices. The subset of matrices in $\text{GL}_n(\mathbb{R})$ with determinant equal to 1 is denoted by

$$\text{SL}_n(\mathbb{R}).$$

First observe that this is a subgroup, the so-called *special linear group* of degree n over $\mathbb{R}$; for if $A, B \in \text{SL}_n(\mathbb{R})$, then $\det(AB) = \det(A)\det(B) = 1$ and $\det(A^{-1}) = (\det(A))^{-1} = 1$. In addition

$$\text{SL}_n(\mathbb{R}) \triangleleft \text{GL}_n(\mathbb{R}) :$$

for if $A \in \text{SL}_n(\mathbb{R})$ and $B \in \text{GL}_n(\mathbb{R})$,

$$\det(BAB^{-1}) = \det(B)\det(A)(\det(B))^{-1} = \det(A) = 1.$$

In these computations two standard results about determinants have been used: $\det(XY) = \det(X)\det(Y)$ and $\det(X^{-1}) = (\det(X))^{-1}$.

(v) On the other hand, $\langle (12)(3) \rangle$ is a subgroup of S_3 that is *not* normal.

(vi) *The normal closure*. Let X be a non-empty subset of a group G. The *normal closure*

$$\langle X^G \rangle$$

of X in G is the subgroup generated by all the conjugates gxg^{-1} with $g \in G$ and $x \in X$. Clearly this is the smallest normal subgroup of G which contains X.

(vii) Finally, we introduce two important normal subgroups that can be formed in any group G. The *center* of G,

$$Z(G),$$

consists of all x in G such that $xg = gx$ for every g in G. The reader should check that $Z(G) \triangleleft G$. Plainly a group G is abelian if and only if $G = Z(G)$.

Next, if x, y are elements of a group G, their *commutator* is the element $xyx^{-1}y^{-1}$, which is written $[x, y]$. The significance of commutators arises from the fact that

$[x, y] = 1$ if and only if $xy = yx$, i.e., x and y commute. The *derived subgroup* or *commutator subgroup* of G is the subgroup generated by all the commutators,

$$G' = \langle [x, y] \mid x, y \in G \rangle.$$

An easy calculation reveals that $z[x, y]z^{-1} = [zxz^{-1}, zyz^{-1}]$; therefore $G' \triangleleft G$. Clearly a group G is abelian if and only if $G' = 1$.

Quotient groups. Next we must explain how to form a new group from a normal subgroup N of a group G. This is called the *quotient group* of N in G,

$$G/N.$$

The elements of G/N are the cosets $xN = Nx$, while the group operation is given by the natural rule

$$(xN)(yN) = (xy)N,$$

where $x, y \in G$.

Our first concern is to check that this binary operation on G/N is properly defined; it should depend on the two cosets xN and yN, not on the choice of coset representatives x and y. To prove this, let $x_1 \in xN$ and $y_1 \in yN$, so that $x_1 = xa$ and $y_1 = yb$ where $a, b \in N$. Then

$$x_1 y_1 = xayb = xy(y^{-1}ayb) \in (xy)N$$

since $y^{-1}ay = y^{-1}a(y^{-1})^{-1} \in N$ by normality of N. Thus $(xy)N = (x_1 y_1)N$.

It is now straightforward to verify that the binary operation just defined is associative. Also the role of the identity in G/N is played by $1N = N$, while $x^{-1}N$ is the inverse of xN, as is readily checked. It follows that G/N is a group. Note that the elements of G/N are *subsets*, not elements, of G, so that G/N is not a subgroup of G. If G is finite, then so is G/N and its order is given by

$$|G/N| = |G : N| = |G|/|N|.$$

Examples of quotient groups. (i) $G/1$ is the set of all $x1 = \{x\}$, i.e., one-element subsets of G. Also $\{x\}\{y\} = \{xy\}$. In fact this is not really a new group since $G \simeq G/1$ via the isomorphism in which $x \mapsto \{x\}$. Another trivial example of a quotient group is G/G, which is a group of order 1, with the single element G.

(ii) Let n be a positive integer. Then $\mathbb{Z}/n\mathbb{Z} = \mathbb{Z}_n$. For, allowing for the additive notation, the coset of $n\mathbb{Z}$ containing x is $x + n\mathbb{Z} = \{x + nq \mid q \in \mathbb{Z}\}$, which is just the congruence class of x modulo n.

(iii) If G is any group, the quotient group G/G' is an abelian group: indeed $(xG')(yG') = xyG' = (yx)(x^{-1}y^{-1}xy)G' = yxG' = (yG')(xG')$. If G/N is any other abelian quotient group, then

$$(xy)N = (xN)(yN) = (yN)(xN) = (yx)N :$$

thus $[x^{-1}, y^{-1}] = x^{-1}y^{-1}xy \in N$ for all $x, y \in N$. Since G' is generated by all commutators $[x^{-1}, y^{-1}]$, it follows that $G' \leq N$. So we have shown that G/G' is the "largest" abelian quotient group of G.

(iv) *The circle group.* Let r denote any real number and suppose that the plane is rotated through angle $2r\pi$ in an anti-clockwise direction about an axis through the origin and perpendicular to the plane. This results in a symmetry of the unit circle, which we will call r'.

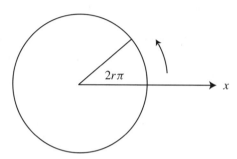

Now define $G = \{r' \mid r \in \mathbb{R}\}$, a subset of the symmetry group of the unit circle. Note that $r'_1 \circ r'_2 = (r_1 + r_2)'$ and $(r')^{-1} = (-r)'$. This shows that G is actually a subgroup of the symmetry group of the circle; in fact it is the subgroup of all rotations. Our aim is to identify G as a quotient group.

It is claimed that the assignment $r + \mathbb{Z} \mapsto r'$ determines a function $\alpha : \mathbb{R}/\mathbb{Z} \to G$: here we need to make sure that the function is well-defined. To this end let n be an integer and observe that $(r+n)' = r' \circ n' = r'$ since n' is the identity rotation. Clearly α is surjective; it is also injective because $r'_1 = r'_2$ implies that $2r_1\pi = 2r_2\pi + 2n\pi$ for some integer n and hence $r_1 + \mathbb{Z} = r_2 + \mathbb{Z}$. Thus α is a bijection. Finally $\alpha((r_1 + \mathbb{Z}) + (r_2 + \mathbb{Z})) = \alpha((r_1 + r_2) + \mathbb{Z})$, which equals

$$(r_1 + r_2)' = r'_1 \circ r'_2 = \alpha(r_1 + \mathbb{Z}) \circ \alpha(r_2 + \mathbb{Z}).$$

Therefore, allowing for the difference in notation for the groups, we conclude that α is an isomorphism from the quotient group $\mathbb{R}/\mathbb{Z}$ to the circle group G. Hence

$$G \simeq \mathbb{R}/\mathbb{Z}.$$

Subgroups of quotient groups. Suppose that N is a normal subgroup of a group G; it is natural to ask what subgroups of the quotient group G/N look like. They must surely be related in some simple fashion to the subgroups of G.

Assume that H is a subgroup of G/N and define a corresponding subset of G,

$$\overline{H} = \{x \in G \mid xN \in H\}.$$

Then it is easy to verify that $\overline{H}$ is a subgroup of G. Also the definition of $\overline{H}$ shows that $N \subseteq \overline{H}$.

Now suppose we start with a subgroup K of G which contains N. Since $N \triangleleft G$ implies that $N \triangleleft K$, we may form the quotient group K/N, which is evidently a subgroup of G/N. Notice that if $N \leq K_1 \leq G$, then $K/N = K_1/N$ implies that $K = K_1$. Thus the assignment $K \mapsto K/N$ determines an injective function from the set of subgroups of G that contain N to the set of subgroups of G/N. But the function is also surjective since $\bar{H} \mapsto H$ in the notation of the previous paragraph; therefore it is a bijection.

These arguments establish the following fundamental theorem.

(4.2.2) (The Correspondence Theorem) *Let N be a normal subgroup of a group G. Then the assignment $K \mapsto K/N$ determines a bijection from the set of subgroups of G which contain N to the set of subgroups of G/N. Furthermore, $K/N \triangleleft G/N$ if and only if $K \triangleleft G$.*

All of this is already proven except for the last statement, which follows at once from the observation $(xN)(kN)(xN)^{-1} = (xkx^{-1})N$ where $k \in K$ and $x \in G$.

Example (4.2.1) Let n be a positive integer. Then subgroups of $\mathbb{Z}_n = \mathbb{Z}/n\mathbb{Z}$ are of the form $K/n\mathbb{Z}$ where $n\mathbb{Z} \leq K \leq \mathbb{Z}$. Now by (4.1.5) there is an integer $m > 0$ such that $K = \langle m \rangle = m\mathbb{Z}$, and further m divides n since $n\mathbb{Z} \leq K$. Thus the Correspondence Theorem tells us that the subgroups of $\mathbb{Z}_n$ correspond to the positive divisors of n, a fact we already know from (4.1.6).

Example (4.2.2) Let N be a normal subgroup of a group G. Call N a *maximal normal subgroup of G* if $N \neq G$ and if $N < L \triangleleft G$ implies that $L = G$. In short "maximal normal" means "maximal *proper* normal". We deduce from the Correspondence Theorem that G/N is a simple group if and only if there are no normal subgroups of G lying properly between N and G, i.e., N is maximal normal in G. Thus maximal normal subgroups lead in a natural way to simple groups.

Direct products. Consider two normal subgroups H and K of a group G such that $H \cap K = 1$. Let $h \in H$ and $k \in K$. Then $[h, k] = (hkh^{-1})k^{-1} \in K$ since $K \triangleleft G$; also $[h, k] = h(kh^{-1}k^{-1}) \in H$ since $H \triangleleft G$. But $H \cap K = 1$, so $[h, k] = 1$, i.e., $hk = kh$. Thus elements of H commute with elements of K.

Now assume that we have $G = HK$ in addition to $H \cap K = 1$. Then G is said to be the *(internal) direct product* of H and K, in symbols

$$G = H \times K.$$

Each element of G is *uniquely* expressible in the form hk, ($h \in H$, $k \in K$). For if $hk = h'k'$ with $h' \in H$, $k' \in K$, then $(h')^{-1}h = k'k^{-1} \in H \cap K = 1$, so that $h = h'$, $k = k'$. Notice also the form taken by the group operation,

$$(h_1 k_1)(h_2 k_2) = (h_1 h_2)(k_1 k_2) \quad (h_i \in H, k_i \in K),$$

since $k_1 h_2 = h_2 k_1$.

For example, consider the Klein 4-group $V = \{\text{id}, (12)(34), (13)(24), (14)(23)\}$: here
$$V = A \times B = B \times C = A \times C$$
where $A = \langle (12)(34) \rangle$, $B = \langle (13)(24) \rangle$, $C = \langle (14)(23) \rangle$.

The direct product concept may be extended to any finite set of normal subgroups $H_1, H_2, \ldots, H_m$ of a group G where
$$H_i \cap \langle H_j \mid j = 1, 2, \ldots, m, \ j \neq i \rangle = 1$$
and
$$G = H_1 H_2 \ldots H_m.$$

In this case elements from different H_i's commute and every element of G has a unique expression of the form $h_1 h_2 \ldots h_m$ where $h_i \in H_i$. (The reader should supply a proof.) We denote the direct product by
$$G = H_1 \times H_2 \times \cdots \times H_m.$$

External direct products. So far a direct product can only be formed within a given group. We show next how to form the direct product of groups that are not necessarily subgroups of the same group. Let $H_1, H_2, \ldots, H_m$ be any finite collection of groups. First we form the set product
$$D = H_1 \times H_2 \times \cdots \times H_m,$$
consisting of all m-tuples $(h_1, h_2, \ldots, h_m)$ with $h_i \in H_i$. Then define a binary operation on D by the rule:
$$(h_1, h_2, \ldots, h_m)(h_1', h_2', \ldots, h_m') = (h_1 h_1', h_2 h_2', \ldots, h_m h_m')$$
where $h_i, h_i' \in H_i$. With this operation D becomes a group, with identity element $(1_{H_1}, 1_{H_2}, \ldots, 1_{H_m})$ and inverses given by
$$(h_1, h_2, \ldots, h_m)^{-1} = (h_1^{-1}, h_2^{-1}, \ldots, h_m^{-1}).$$

We call D the *external direct product* of $H_1, H_2, \ldots, H_m$.

Now the group H_i is not a subgroup of D, but there is a subgroup of D which is isomorphic with H_i. Let $\overline{H}_i$ consist of all elements $h(i) = (1_{H_1}, 1_{H_2}, \ldots, h, \ldots, 1_{H_m})$ where $h \in H_i$ appears as the i-th entry of $h(i)$. Then clearly $H_i \simeq \overline{H}_i \leq D$. Also, if $x = (h_1, h_2, \ldots, h_m)$ is any element of D, then $x = h_1(1) h_2(2) \ldots h_m(m)$, by the product rule in D. Hence $D = \overline{H}_1 \overline{H}_2 \ldots \overline{H}_m$. It is easy to verify that $\overline{H}_i \triangleleft D$ and
$$\overline{H}_i \cap \langle \overline{H}_j \mid j = 1, \ldots, m, \ j \neq i \rangle = 1.$$

Hence D is the internal direct product
$$D = \overline{H}_1 \times \overline{H}_2 \times \cdots \times \overline{H}_m$$
of subgroups isomorphic with $H_1, H_2, \ldots, H_m$. Thus every external direct product can be regarded as an internal direct product.

Example (4.2.3) Let $C_1, C_2, \ldots, C_k$ be finite cyclic groups of orders $n_1, n_2, \ldots, n_k$ where the n_i are pairwise relatively prime. Form the external direct product
$$D = C_1 \times C_2 \times \cdots \times C_k.$$
Therefore we have $|D| = n_1 n_2 \ldots n_k = n$, say. Now let C_i be generated by x_i and put $x = (x_1, x_2, \ldots, x_m) \in D$. We claim that x generates D, so that D is a cyclic group of order n. To prove this statement we show that an arbitrary element $x_1^{u_1} \ldots x_k^{u_k}$ of D is of the form x^r for some r. This amounts to proving that there is a solution r of the system of linear congruences $r \equiv u_i \pmod{n_i}$, $i = 1, 2, \ldots, k$. This is true by the Chinese Remainder Theorem (2.3.7) since $n_1, n_2, \ldots, n_k$ are relatively prime.

For example, let n be a positive integer and write $n = p_1^{e_1} \ldots p_k^{e_k}$ where the p_i are distinct primes and $e_i > 0$. Then the preceding discussion shows that
$$\mathbb{Z}_{p_1^{e_1}} \times \mathbb{Z}_{p_2^{e_2}} \times \cdots \times \mathbb{Z}_{p_k^{e_k}}$$
is a cyclic group of order n, and hence is isomorphic with $\mathbb{Z}_n$.

Exercises (4.2)

1. Identify all the normal subgroups of the groups S_3, Dih(8) and A_4.

2. Let H be a subgroup of a group G with index 2. Prove that $H \triangleleft G$. Is this true if 2 is replaced by 3?

3. Let $H \triangleleft K \leq G$ and $L \leq G$. Show that $H \cap L \triangleleft K \cap L$. Also, if $L \triangleleft G$, prove that $HL/L \triangleleft KL/L$.

4. Let $H \leq G$ and $N \triangleleft G$. Prove that HN is a subgroup of G.

5. Assume that $H \leq K \leq G$ and $N \triangleleft G$. If $H \cap N = K \cap N$ and $HN = KN$, prove that $H = K$.

6. Show that normality is not a transitive relation, i.e., $H \triangleleft K \triangleleft G$ need not imply that $H \triangleleft G$. [Hint: Dih(8)].

7. If H, K, L are arbitrary groups, prove that
$$H \times (K \times L) \simeq H \times K \times L \simeq (H \times K) \times L.$$

8. Let $G = H \times K$ where $H, K \leq G$. Prove that $G/H \simeq K$ and $G/K \simeq H$.

9. Let $G = \langle x \rangle$ be a cyclic group of order n. If $d \geq 0$, prove that $G/\langle x^d \rangle$ is cyclic with order $\gcd\{n, d\}$.

10. Prove that $Z(S_n) = 1$ if $n \neq 2$.

11. Prove that $S_n' \neq S_n$ if $n \neq 1$.

12. Prove that the center of the group $\mathrm{GL}_n(\mathbb{R})$ of all $n \times n$ non-singular real matrices is the subgroup of all scalar matrices.

4.3 Homomorphisms of groups

A homomorphism between two groups is a special kind of function which links the operations of the groups. More precisely, a function α from a group G to a group H is called a *homomorphism* if

$$\alpha(xy) = \alpha(x)\alpha(y)$$

for all $x, y \in G$. The reader will recognize that a bijective homomorphism is just what we have been calling an *isomorphism*.

Examples of homomorphisms. (i) $\alpha : \mathbb{Z} \to \mathbb{Z}_n$ where $\alpha(x) = [x]_n$. Here n is a positive integer. Allowing for the additive notation, what is being claimed here that $\alpha(x + y) = \alpha(x) + \alpha(y)$, i.e. $[x + y]_n = [x]_n + [y]_n$; this is just the definition of addition of congruence classes.

(ii) The determinant function $\det : \mathrm{GL}_n(\mathbb{R}) \to \mathbb{R}^*$, in which $A \mapsto \det(A)$, is a homomorphism; this is because of the identity $\det(AB) = \det(A)\det(B)$.

(iii) The sign function $\mathrm{sign} : S_n \to \{\pm 1\}$, in which $\pi \mapsto \mathrm{sign}(\pi)$, is a homomorphism since $\mathrm{sign}(\pi\sigma) = \mathrm{sign}(\pi)\mathrm{sign}(\sigma)$, by (3.1.6).

(iv) *The canonical homomorphism.* This example provides the first evidence of a link between homomorphisms and normal subgroups. Let N be a normal subgroup of a group G and define a function

$$\alpha : G \to G/N$$

by the rule $\alpha(x) = xN$. Then $\alpha(xy) = \alpha(x)\alpha(y)$, i.e., $(xy)N = (xN)(yN)$ since this is just the definition of the group operation in G/N. Thus α is a homomorphism.

(v) For any pair of groups G, H there is always at least one homomorphism from G to H, namely the *trivial homomorphism* in which $x \mapsto 1_H$ for all x in G. Another obvious example is the *identity homomorphism* from G to G, which is just the identity function on G.

Next come two very basic properties that all homomorphism possess.

(4.3.1) *Let $\alpha : G \to H$ be a homomorphism of groups. Then:*

(i) $\alpha(1_G) = 1_H$;

(ii) $\alpha(x^n) = (\alpha(x))^n$ *for all $n \in \mathbb{Z}$.*

Proof. Applying α to the equation $1_G 1_G = 1_G$, we obtain $\alpha(1_G)\alpha(1_G) = \alpha(1_G)$, which on cancellation yields $\alpha(1_G) = 1_H$.

If $n > 0$, an easy induction on n shows that $\alpha(x^n) = (\alpha(x))^n$. Next $xx^{-1} = 1_G$, so that $\alpha(x)\alpha(x^{-1}) = \alpha(1_G) = 1_H$; from this it follows that $\alpha(x^{-1}) = (\alpha(x))^{-1}$. Finally, if $n \geq 0$, we have $\alpha(x^{-n}) = \alpha((x^n)^{-1}) = (\alpha(x^n))^{-1} = ((\alpha(x))^n)^{-1} = (\alpha(x))^{-n}$, which completes the proof of (ii). □

Image and kernel. Let $\alpha : G \to H$ be a group homomorphism. Then of course the *image* of α is the subset $\mathrm{Im}(\alpha) = \{\alpha(x) \mid x \in G\}$. Another very significant subset associated with α is the *kernel*, which is defined by

$$\mathrm{Ker}(\alpha) = \{x \in G \mid \alpha(x) = 1_H\}.$$

The fundamental properties of image and kernel are given in the next result.

(4.3.2) *If $\alpha : G \to H$ is a homomorphism, the image $\mathrm{Im}(\alpha)$ is a subgroup of H and the kernel $\mathrm{Ker}(\alpha)$ is a normal subgroup of G.*

Thus another connection between homomorphisms are normal subgroups appears.

Proof of (4.3.2). By (4.3.1) $1_H \in \mathrm{Im}(\alpha)$. Let $x, y \in G$; then $\alpha(x)\alpha(y) = \alpha(xy)$ and $(\alpha(x))^{-1} = \alpha(x^{-1})$. These equations tell us that $\mathrm{Im}(\alpha)$ is a subgroup of H.

Next, if $x, y \in \mathrm{Ker}(\alpha)$, then $\alpha(xy) = \alpha(x)\alpha(y) = 1_H 1_H = 1_H$, and $\alpha(x^{-1}) = (\alpha(x))^{-1} = 1_H^{-1} = 1_H$; thus $\mathrm{Ker}(\alpha)$ is a subgroup of G. Finally, we prove that $\mathrm{Ker}(\alpha)$ is normal in G. Let $x \in \mathrm{Ker}(\alpha)$ and $g \in G$; then

$$\alpha(gxg^{-1}) = \alpha(g)\alpha(x)\alpha(g)^{-1} = \alpha(g)1_H \alpha(g)^{-1} = 1_H,$$

so that $gxg^{-1} \in \mathrm{Ker}(\alpha)$, as required. □

Examples (4.3.1) Let us compute the image and kernel of some of the homomorphisms mentioned above.

(i) $\det : \mathrm{GL}_n(\mathbb{R}) \to \mathbb{R}^*$. Here the kernel is $\mathrm{SL}_n(\mathbb{R})$, the special linear group, and the image is $\mathbb{R}^*$ (since each non-zero real number is the determinant of some $n \times n$ diagonal matrix in $\mathrm{GL}_n(\mathbb{R})$).

(ii) sign: $S_n \to \{\pm 1\}$. Here the kernel is the alternating group A_n, and the image is the entire group $\{\pm 1\}$, unless $n = 1$.

(iii) The kernel of the canonical homomorphism from G to G/N is, as one would expect, the normal subgroup N. The image is G/N.

Clearly we can tell from the image of a homomorphism whether it is surjective. It is a fact that the kernel of a homomorphism will tell us if it is injective.

(4.3.3) *Let $\alpha : G \to H$ be a group homomorphism. Then:*

(i) *α is surjective if and only if $\text{Im}(\alpha) = H$;*

(ii) *α is injective if and only if $\text{Ker}(\alpha) = 1_G$;*

(iii) *α is an isomorphism if and only if $\text{Im}(\alpha) = H$ and $\text{Ker}(\alpha) = 1_G$;*

(iv) *If α is an isomorphism, then so is $\alpha^{-1} : H \to G$.*

Proof. Of course (i) is true by definition. As for (ii), if α is injective and $x \in \text{Ker}(\alpha)$, then $\alpha(x) = 1_H = \alpha(1_G)$, so that $x = 1_G$ by injectivity of α. Conversely, assume that $\text{Ker}(\alpha) = 1_G$. If $\alpha(x) = \alpha(y)$, then $\alpha(xy^{-1}) = \alpha(x)(\alpha(y))^{-1} = 1_H$, which means that $xy^{-1} \in \text{Ker}(\alpha) = 1_G$ and $x = y$. Thus (ii) is proven, while (iii) follows at once from (i) and (ii).

To establish (iv) all we need to prove is that $\alpha^{-1}(xy) = \alpha^{-1}(x)\alpha^{-1}(y)$. Now $\alpha(\alpha^{-1}(xy)) = xy$, whereas

$$\alpha(\alpha^{-1}(x)\alpha^{-1}(y)) = \alpha(\alpha^{-1}(x))\alpha(\alpha^{-1}(y)) = xy.$$

By injectivity of α we may conclude that $\alpha^{-1}(xy) = \alpha^{-1}(x)\alpha^{-1}(y)$. □

The Isomorphism Theorems. We come now to three fundamental results about homomorphisms and quotient groups which are traditionally known as the *Isomorphism Theorems*.

(4.3.4) (First Isomorphism Theorem) *If $\alpha : G \to H$ is a homomorphism of groups, then $G/\text{Ker}(\alpha) \simeq \text{Im}(\alpha)$ via the mapping $x \text{Ker}(\alpha) \mapsto \alpha(x)$).*

Thus the image of a homomorphism may be regarded as a quotient group. Conversely, every quotient group is the image of the associated canonical homomorphism. What this means is that up to isomorphism quotient groups and homomorphic images are the same objects.

Proof of (4.3.4). Let $K = \text{Ker}(\alpha)$. We would like to define a function $\theta : G/K \to \text{Im}(\alpha)$ by the rule $\theta(xK) = \alpha(x)$, but first we need to check that this makes sense. Now if $k \in K$, then $\alpha(xk) = \alpha(x)\alpha(k) = \alpha(x)$; therefore $\theta(xK)$ depends only on the coset xK and not on the choice of x from xK. So at least θ is a well-defined function.

Next $\theta((xy)K) = \alpha(xy) = \alpha(x)\alpha(y) = \theta(xK)\theta(yK)$; thus θ is a homomorphism. It is obvious that $\text{Im}(\theta) = \text{Im}(\alpha)$. Finally, $\theta(xK) = 1_H$ if and only if $\alpha(x) = 1_H$, i.e., $x \in K$ or equivalently $xK = K = 1_{G/K}$. Therefore $\text{Ker}(\theta)$ is the identity subgroup of G/K and θ is an isomorphism from G/K to $\text{Im}(\alpha)$. □

(4.3.5) (Second Isomorphism Theorem) *Let G be a group with a subgroup H and a normal subgroup N. Then $HN \leq G$, $H \cap N \triangleleft H$ and $HN/N \simeq H/H \cap N$.*

Proof. We begin by defining a function $\theta : H \to G/N$ by the rule $\theta(h) = hN, h \in H$. It is easy to check that θ is a homomorphism. Also $\operatorname{Im}(\theta) = \{hN \mid h \in H\} = HN/N$, which is a subgroup of G/N; therefore $HN \leq G$ by (4.3.2). Next $h \in \operatorname{Ker}(\theta)$ if and only if $hN = N$, i.e., $h \in H \cap N$. Therefore $\operatorname{Ker}(\theta) = H \cap N$ and $H \cap N \triangleleft H$ by (4.3.2). Now apply the First Isomorphism Theorem to the homomorphism θ to obtain $H/H \cap N \simeq HN/N$. □

(4.3.6) (Third Isomorphism Theorem) *Let M and N be normal subgroups of a group G such that $N \subseteq M$. Then $M/N \triangleleft G/N$ and $(G/N)/(M/N) \simeq G/M$.*

Proof. Define $\theta : G/N \to G/M$ by the rule $\theta(xN) = xM$; the reader should now verify that θ is a well-defined homomorphism. Also $\operatorname{Im}(\theta) = G/M$ and $\operatorname{Ker}(\theta) = M/N$; the result follows via (4.3.4). □

Thus a quotient group of a quotient group of G is essentially just a quotient group of G, which represents a considerable simplification. Next these theorems are illustrated by some examples.

Example (4.3.2) Let m, n be positive integers. Then (4.3.5) tells us that

$$m\mathbb{Z} + n\mathbb{Z}/n\mathbb{Z} \simeq m\mathbb{Z}/m\mathbb{Z} \cap n\mathbb{Z}.$$

What does this tell us about the integers m, n? Fairly obviously $m\mathbb{Z} \cap n\mathbb{Z} = \ell\mathbb{Z}$ where ℓ is the least common multiple of m and n. Now $m\mathbb{Z} + n\mathbb{Z}$ consists of all $ma + nb$, where $a, b \in \mathbb{Z}$. From (2.2.3) we see that this is just $d\mathbb{Z}$ where $d = \gcd\{m, n\}$. So the assertion is that $d\mathbb{Z}/n\mathbb{Z} \simeq m\mathbb{Z}/\ell\mathbb{Z}$. Now $d\mathbb{Z}/n\mathbb{Z} \simeq \mathbb{Z}/(\frac{n}{d})\mathbb{Z}$ via the mapping $dx + n\mathbb{Z} \mapsto x + \frac{n}{d}\mathbb{Z}$. Similarly $m\mathbb{Z}/\ell\mathbb{Z} \simeq \mathbb{Z}/(\frac{\ell}{m})\mathbb{Z}$. Thus $\mathbb{Z}/(\frac{n}{d})\mathbb{Z} \simeq \mathbb{Z}/(\frac{\ell}{m})\mathbb{Z}$. Since isomorphic groups have the same order, it follows that $\frac{n}{d} = \frac{\ell}{m}$ or $mn = d\ell$. So what (4.3.5) implies is that

$$\gcd\{m, n\} \cdot \operatorname{lcm}\{m, n\} = mn,$$

(see also Exercise (2.2.7)).

Example (4.3.3) Consider the determinantal homomorphism $\det : \operatorname{GL}_n(\mathbb{R}) \to \mathbb{R}^*$, which has kernel is $\operatorname{SL}_n(\mathbb{R})$ and image $\mathbb{R}^*$. Then (4.3.4) tells us that

$$\operatorname{GL}_n(\mathbb{R})/\operatorname{SL}_n(\mathbb{R}) \simeq \mathbb{R}^*.$$

Automorphisms. An *automorphism* of a group G is an isomorphism from G to itself. Thus an automorphism of G is a permutation of the set of group elements which is also a homomorphism. The set of all automorphisms of G,

$$\operatorname{Aut}(G),$$

is a subset of the symmetric group $\operatorname{Sym}(G)$. Our first observation is:

4.3 Homomorphisms of groups

(4.3.7) *If G is a group, then $\mathrm{Aut}(G)$ is a subgroup of $\mathrm{Sym}(G)$.*

Proof. The identity permutation is certainly an automorphism. Also, if $\alpha \in \mathrm{Aut}(G)$, then $\alpha^{-1} \in \mathrm{Aut}(G)$ by (4.3.3)(iv). Finally, if $\alpha, \beta \in \mathrm{Aut}(G)$, then $\alpha\beta$ is certainly a permutation of G, while $\alpha\beta(xy) = \alpha(\beta(x)\beta(y)) = \alpha\beta(x)\alpha\beta(y)$, which shows that $\alpha\beta \in \mathrm{Aut}(G)$ and $\mathrm{Aut}(G)$ is a subgroup. □

In fact $\mathrm{Aut}(G)$ is usually quite a small subgroup of $\mathrm{Sym}(G)$, as will be seen in some of the ensuing examples.

Example (4.3.4) Let A be any additively written abelian group and define $\alpha : A \to A$ by $\alpha(x) = -x$. Then α is an automorphism since

$$\alpha(x+y) = -(x+y) = -x - y = \alpha(x) + \alpha(y),$$

while $\alpha^{-1} = \alpha$.

Now suppose we take A to be $\mathbb{Z}$, and let β be any automorphism of A. Now $\beta(1) = n$ for some integer n. Notice that β is completely determined by n since $\beta(m) = \beta(m1) = m\beta(1) = mn$ by (4.3.1)(ii). Also $\beta(x) = 1$ for some integer x since β is surjective. Furthermore $1 = \beta(x) = \beta(x1) = x\beta(1) = xn$. It follows that $n = \pm 1$, and hence there are just two possibilities for β, namely the identity and the automorphism α of the last paragraph. Therefore $|\mathrm{Aut}(\mathbb{Z})| = 2$. On the other hand, as the reader should verify, the group $\mathrm{Sym}(\mathbb{Z})$ is uncountable.

Inner automorphisms. An easy way to construct an automorphism is to use a fixed element of the group to form conjugates. If g is an element of a group G, define a function $\tau(g)$ on G by the rule

$$\tau(g)(x) = gxg^{-1} \quad (x \in G).$$

Recall that gxg^{-1} is the *conjugate* of x by g. Since

$$\tau(g)(xy) = g(xy)g^{-1} = (gxg^{-1})(gyg^{-1}) = (\tau(g)(x))(\tau(g)(y)),$$

we see that $\tau(g)$ is a homomorphism. Now $\tau(g^{-1})$ is clearly the inverse of $\tau(g)$; therefore $\tau(g)$ is an automorphism of G, known as the *inner automorphism induced by g*. Thus we have discovered a function

$$\tau : G \to \mathrm{Aut}(G).$$

Our next observation is that τ is a homomorphism, called the *conjugation homomorphism*; indeed $\tau(gh)$ sends x to $(gh)x(gh)^{-1} = g(hxh^{-1})g^{-1}$, which is precisely the image of x under the composite $\tau(g)\tau(h)$. Thus $\tau(gh) = \tau(g)\tau(h)$ for all $g, h \in G$.

The image of τ is the set of all inner automorphisms of G, which is denoted by

$$\text{Inn}(G).$$

This is a subgroup of $\text{Aut}(G)$ by (4.3.2). What can be said about $\text{Ker}(\tau)$? An element g belongs to $\text{Ker}(\tau)$ if and only if $\tau(g)(x) = x$ for all x in G, i.e., $gxg^{-1} = x$, or $gx = xg$. Thus the kernel of τ is the center of G,

$$\text{Ker}(\tau) = Z(G),$$

which consists of the elements which commute with every element of G.

These conclusions are summed up in the following important result.

(4.3.8) *Let G be a group and let $\tau : G \to \text{Aut}(G)$ be the conjugation homomorphism. Then $\text{Ker}(\tau) = Z(G)$ and $\text{Im}(\tau) = \text{Inn}(G)$. Hence $\text{Inn}(G) \simeq G/Z(G)$.*

The final statement follows on applying the First Isomorphism Theorem to the homomorphism τ.

Usually a group possesses non-inner automorphisms. For example, if A is an (additively written) abelian group, every inner automorphism is trivial since $\tau(g)(x) = g + x - g = g - g + x = x$. On the other hand, the assignment $x \mapsto -x$ determines an automorphism of A which is not trivial unless $2x = 0$ for all x in A.

(4.3.9) *In any group G the subgroup $\text{Inn}(G)$ is normal in $\text{Aut}(G)$.*

Proof. Let $\alpha \in \text{Aut}(G)$ and $g \in G$; we claim that $\alpha\tau(g)\alpha^{-1} = \tau(\alpha(g))$, which will establish normality. For if $x \in G$, we have

$$\tau(\alpha(g))(x) = \alpha(g)x(\alpha(g))^{-1} = \alpha(g)x\alpha(g^{-1}),$$

which equals

$$\alpha(g\alpha^{-1}(x)g^{-1}) = \alpha(\tau(g)(\alpha^{-1}(x))) = (\alpha\tau(g)\alpha^{-1})(x),$$

as required. □

By (4.3.9) we may now proceed to form the quotient group

$$\text{Out}(G) = \text{Aut}(G)/\text{Inn}(G),$$

which is termed the *outer automorphism group* of G, (even although its elements are not actually automorphisms). Thus all automorphisms of G are inner precisely when $\text{Out}(G) = 1$.

Finally, we point out that the various groups and homomorphisms introduced above fit neatly together in a sequence of groups and homomorphisms:

$$1 \to Z(G) \xrightarrow{\iota} G \xrightarrow{\tau} \text{Aut}(G) \xrightarrow{\nu} \text{Out}(G) \to 1.$$

Here ι is the inclusion map, τ is the conjugation homomorphism and ν is the canonical homomorphism associated with the normal subgroup Inn(G). Of course $1 \to Z(G)$ and Out(G) $\to 1$ are trivial homomorphisms. The sequence above is an *exact sequence*, in the sense that at each group in the interior of the sequence the image of the homomorphism on the left equals the kernel of the homomorphism on the right. For example at Aut(G) we have Im(τ) = Inn(G) = Ker(ν).

In general it can be a tricky business to determine the automorphism group of a given group. A useful aid in the process of deciding which permutations of the group are actually automorphisms is the following simple fact.

(4.3.10) *Let G be a group, $g \in G$ and $\alpha \in$ Aut(G). Then g and $\alpha(g)$ have the same order.*

Proof. By (4.3.1) $\alpha(g^m) = \alpha(g)^m$. Since α is injective, it follows that $\alpha(g)^m = 1$ if and only if $g^m = 1$. Hence $|g| = |\alpha(g)|$. □

The automorphism group of a cyclic group. As a first example we consider the automorphism group of a cyclic group $G = \langle x \rangle$. If G is infinite, then $G \simeq \mathbb{Z}$ and we have already seen in Example (4.3.4) that Aut(G) $\simeq \mathbb{Z}_2$. So assume from now on that G has finite order m.

First of all notice that α is completely determined by $\alpha(x)$ since $\alpha(x^i) = \alpha(x)^i$. Also $|\alpha(x)| = |x| = m$ by (4.3.10). Thus (4.1.7) shows that $\alpha(x) = x^i$ where $1 \le i < m$ and i is relatively prime to m. Consequently $|$Aut(G)$| \le \phi(m)$, where ϕ is Euler's function, since $\phi(m)$ is the number of such integers i.

Conversely suppose that i is an integer satisfying $1 \le i < m$ and gcd$\{i, m\} = 1$. Then the assignment $g \mapsto g^i$, ($g \in G$), determines a homomorphism $\alpha_i : G \to G$ because $(g_1 g_2)^i = g_1^i g_2^i$, the group G being abelian. Since $|x^i| = m$, the element x^i generates G and this homomorphism is surjective. But G is finite, so we may conclude that α_i must also be injective and hence $\alpha_i \in$ Aut(G). It follows that $|$Aut(G)$| = \phi(m)$, the number of such i's.

The structure of the group Aut(G) is not hard to determine. Recall that $\mathbb{Z}_m^*$ is the multiplicative group of congruence classes $[a]_m$ where a is relatively prime to m. Now there is a natural function $\theta : \mathbb{Z}_m^* \to$ Aut(G) given by $\theta([i]_m) = \alpha_i$ where α_i is as defined above; this is well-defined since $\alpha_{i+\ell m} = \alpha_i$ for all ℓ. Also θ is a homomorphism because $\alpha_{ij} = \alpha_i \alpha_j$, and the preceding discussion shows that θ is surjective and hence bijective. We have therefore established:

(4.3.11) *Let $G = \langle x \rangle$ be a cyclic group of order m. Then $\mathbb{Z}_m^* \simeq$ Aut(G) via the assignment $[i]_m \mapsto (g \mapsto g^i)$.*

In particular this establishes:

Corollary (4.3.12) *The automorphism group of a cyclic group is abelian.*

The next example is more challenging.

Example (4.3.5) Show that the order of the automorphism group of the dihedral group Dih($2p$) where p is an odd prime is $p(p-1)$.

Recall that Dih($2p$) is the symmetry group of a regular p-gon – see 3.2. First we need a good description of the elements of $G = $ Dih($2p$). If the vertices of the p-gon are labelled $1, 2, \ldots, p$, then G contains the p-cycle $\sigma = (1\ 2\ \ldots\ p)$, which corresponds to an anticlockwise rotation through angle $\frac{2\pi}{p}$. It also contains the permutation $\tau = (1)(2\ p)(3\ p-1)\ldots(\frac{p+1}{2}\ \frac{p+3}{2})$, which represents a reflection in the line through the vertex 1 and the midpoint of the opposite edge.

The elements $\sigma^r, \sigma^r\tau$, where $r = 0, 1, \ldots, p-1$, are clearly all different and there are $2p$ of them. Since $|G| = 2p$, we conclude that

$$G = \{\sigma^r, \sigma^r\tau \mid r = 0, 1, \ldots, p-1\}.$$

Note that $\sigma^r\tau$ is a reflection, so it has order 2, while σ^r is a rotation of order 1 or p.

Now let $\alpha \in $ Aut(G). Then by (4.3.10) $\alpha(\sigma)$ has order p, and hence $\alpha(\sigma) = \sigma^r$ where $1 \leq r < p$; also $\alpha(\tau)$ has order 2 and so must equal $\sigma^s\tau$ where $0 \leq s < p$. Observe that α is determined by its effect on σ and τ since $\alpha(\sigma^i) = \alpha(\sigma)^i$ and $\alpha(\sigma^i\tau) = \alpha(\sigma)^i\alpha(\tau)$. It follows that there are at most $p(p-1)$ possibilities for α and hence that $|$Aut(G)$| \leq p(p-1)$.

To show that $p(p-1)$ is the order of the automorphism group we have to construct some automorphisms. Now it is easy to see that $Z(G) = 1$; thus by (4.3.8) Inn(G) $\simeq G/Z(G) \simeq G$. Therefore $|$Inn(G)$| = 2p$, and since Inn(G) $\leq$ Aut(G), it follows from Lagrange's Theorem that p divides $|$Aut(G)$|$.

Next for $0 < r < p$ we define an automorphism α_r of G by the rules

$$\alpha_r(\sigma) = \sigma^r \quad \text{and} \quad \alpha_r(\tau) = \tau.$$

To verify that α_r is a homomorphism one needs to check that $\alpha_r(xy) = \alpha_r(x)\alpha_r(y)$; this is not hard, but it does involve some case distinctions, depending on the form of x and y. Now α_r is clearly surjective because σ^r generates $\langle\sigma\rangle$; thus it is an automorphism. Notice also that $\alpha_r\alpha_s = \alpha_{rs}$, so that $[r]_p \mapsto \alpha_r$ determines a homomorphism from $\mathbb{Z}_p^*$ to $H = \{\alpha_r \mid 1 \leq r < p\}$. This mapping is clearly surjective, while if $\alpha_r = 1$, then $r \equiv 1 \pmod{p}$, i.e., $[r]_p = [1]_p$. Hence $[r]_p \mapsto \alpha_r$ is an isomorphism from $\mathbb{Z}_p^*$ to H. Therefore H has order $p-1$ and $p-1$ divides $|$Aut(G)$|$. Consequently $p(p-1)$ divides the order of Aut(G) and hence $|$Aut(G)$| = p(p-1)$.

Since $|$Inn(G)$| = |G| = 2p$, we also see that $|$Out(G)$| = p(p-1)/2p = \frac{p-1}{2}$. Thus $|$Out(G)$| = 1$ if and only if $p = 3$; then of course $G = $ Dih(6) $\simeq S_3$.

A group G is said to be *complete* if the conjugation homomorphism $\tau : G \to $ Aut(G) is an isomorphism: this is equivalent to $Z(G) = 1$ and Out(G) $= 1$. Thus Dih($2p$) is complete if and only if $p = 3$. We shall prove in Chapter 5 that the symmetric group S_n is always complete unless $n = 2$ or 6.

4.3 Homomorphisms of groups

Semidirect products. Suppose that G is a group with a normal subgroup N and a subgroup H such that
$$G = NH \quad \text{and} \quad N \cap H = 1.$$
Then G is said to be the (*internal*) *semidirect product* of N and H. As a simple example, consider the alternating group $G = A_4$; this has a normal subgroup of order 4, namely the Klein 4-group V, and also the subgroup $H = \langle (123)(4) \rangle$ of order 3. Thus $V \cap H = 1$ and $|VH| = |V| \cdot |H| = 12$, so that $G = VH$ and G is the semidirect product of V and H.

Now let us analyze the structure of a semidirect product $G = NH$. In the first place each element $g \in G$ has a unique expression $g = nh$ with $n \in N$ and $h \in H$. For if $g = n'h'$ is another such expression, then $(n')^{-1}n = h'h^{-1} \in N \cap H = 1$, which shows that $n = n'$ and $h = h'$. Secondly, conjugation in N by an element h of H produces an automorphism of N, say $\theta(h)$. Furthermore it is easily verified that $\theta(h_1 h_2) = \theta(h_1)\theta(h_2)$, $h_i \in H$; thus $\theta : H \to \text{Aut}(N)$ is a homomorphism.

Now let us see whether, on the basis of the preceding discussion, we can reconstruct the semidirect product from given groups N and H and a given homomorphism $\theta : H \to \text{Aut}(N)$. This will be the so-called *external semidirect product*. The underlying set of this group is to be the set product of N and H, i.e.,
$$G = \{(n, h) \mid n \in N, h \in H\}.$$
A binary operation on G is defined by the rule
$$(n_1, h_1)(n_2, h_2) = (n_1 \theta(h_1)(n_2), h_1 h_2).$$
The motivation for this rule is the way we form products in an internal semidirect product NH, which is $(n_1 h_1)(n_2 h_2) = n_1(h_1 n_2 h_1^{-1}) h_1 h_2$. The identity element of G is to be $(1_N, 1_H)$ and the inverse of (n, h) is to be $(\theta(h^{-1})(n^{-1}), h^{-1})$: the latter is motivated by the fact that in an internal semidirect product NH inverses are formed according to the rule $(nh)^{-1} = h^{-1} n^{-1} = (h^{-1} n^{-1} h) h^{-1}$. We omit the entirely routine verification of the group axioms for G.

Next we look for subgroups of G which resemble the original groups N and H. Here there are natural candidates:
$$\bar{N} = \{(n, 1_H) \mid n \in N\} \quad \text{and} \quad \bar{H} = \{(1_N, h) \mid h \in H\}.$$
It is straightforward to show that these are subgroups isomorphic with N and H respectively. The group operation of G shows that
$$(n, 1_H)(1_N, h) = (n\theta(1_H)(1_N), h) = (n, h) \in \bar{N}\bar{H}$$
since $\theta(1_H)$ is the identity automorphism of N. It follows that $G = \bar{N}\bar{H}$, while it is obvious that $\bar{N} \cap \bar{H} = 1$. To show that G is the semidirect product of $\bar{N}$ and $\bar{H}$, it is only necessary to check normality of $\bar{N}$ in G. Let $n, n_1 \in N$ and $h \in H$. Then by definition $(n, h)(n_1, 1_H)(n, h)^{-1}$ equals
$$(n, h)(n_1, 1_H)(\theta(h^{-1})(n^{-1}), h^{-1}) = (n\theta(h)(n_1), h)(\theta(h^{-1})(n^{-1}), h^{-1}),$$

which equals

$$(n\theta(h)(n_1)\theta(h)\theta(h^{-1})(n^{-1}), 1_H) = (n\theta(h)(n_1)n^{-1}, 1_H) \in \bar{N}.$$

In particular conjugation in $\bar{N}$ by $(1_N, h)$ maps $(n_1, 1_H)$ to $(\theta(h)(n_1), 1_H)$. Thus $(1_N, h)$ "induces" the automorphism θ in N by conjugation.

In the special case where θ is chosen to be the trivial homomorphism, elements of $\bar{N}$ and $\bar{H}$ commute, so that G becomes the direct product $N \times H$. Thus the semidirect product is a generalization of the direct product of two groups. Semidirect products provide an important means of constructing new groups.

Example (4.3.6) Let $N = \langle n \rangle$ and $H = \langle h \rangle$ be cyclic groups with respective orders 3 and 4. Suppose we wish to form a semidirect product G of N and H. For this purpose we select a homomorphism θ from H to $\text{Aut}(N)$; there is little choice here since N has only one non-identity automorphism, namely $n \mapsto n^{-1}$. Accordingly define $\theta(h)$ to be this automorphism. The resulting group G is known as the *dicyclic group of order* 12. Observe that this group is not isomorphic with A_4 or $\text{Dih}(12)$ since, unlike these groups, G has an element of order 4.

Exercises (4.3)

1. Let $H \triangleleft K \leq G$ and assume that $\alpha : G \to G_1$ is a homomorphism. Show that $\alpha(H) \triangleleft \alpha(K) \leq G_1$ where $\alpha(H) = \{\alpha(h) \mid h \in H\}$.

2. If G and H are groups with relatively prime orders, then the only homomorphism from G to H is the trivial one.

3. Let G be a simple group. Show that if $\alpha : G \to H$ is a homomorphism, then either α is trivial or H has a subgroup isomorphic with G.

4. Prove that $\text{Aut}(V) \simeq S_3$ where V is the Klein 4-group.

5. Prove that $\text{Aut}(\mathbb{Q}) \simeq \mathbb{Q}^*$ where $\mathbb{Q}^*$ is the multiplicative group of non-zero rationals. [Hint: an automorphism is determined by its effect on 1].

6. Let G and A be groups with A additively written and abelian. Let $\text{Hom}(G, A)$ denote the set of all homomorphisms from G to A. Define a binary operation $+$ on $\text{Hom}(G, A)$ by $\alpha + \beta(x) = \alpha(x) + \beta(x)$ $(x \in G)$. Prove that with this operation $\text{Hom}(G, A)$ becomes an abelian group. Then show that $\text{Hom}(\mathbb{Z}, A) \simeq A$.

7. Let $G = \langle x \rangle$ have order 8. Write down all the automorphisms of G and verify that $\text{Aut}(G) \simeq V$: conclude that the automorphism group of a cyclic group need not be cyclic.

8. If G and H are finite groups of relatively prime orders, show that $\text{Aut}(G \times H) \simeq \text{Aut}(G) \times \text{Aut}(H)$.

9. Use the previous exercise to prove that $\phi(mn) = \phi(m)\phi(n)$ where ϕ is Euler's function and m, n are relatively prime integers. (A different proof was given in (2.3.8)).

10. An $n \times n$ matrix is called a *permutation matrix* if each row and each column contains a single 1 and if all other entries are 0.

(a) If $\pi \in S_n$, form a permutation matrix $M(\pi)$ by defining $M(\pi)_{ij}$ to be 1 if $\pi(j) = i$ and 0 otherwise. Prove that the assignment $\pi \mapsto M(\pi)$ determines an injective homomorphism from S_n to $\mathrm{GL}_n(\mathbb{Q})$.

(b) Deduce that the $n \times n$ permutation matrices form a group which is isomorphic with S_n.

(c) How can one tell from $M(\pi)$ whether π is even or odd?

11. (a) Show that each of the groups $\mathrm{Dih}(2n)$ and S_4 is a semidirect product of groups of smaller orders.

(b) Use the groups $\mathbb{Z}_3 \times \mathbb{Z}_3$ and $\mathbb{Z}_2$ to form three non-isomorphic groups of order 18 each with a normal subgroup of order 9.

Chapter 5
Groups acting on sets

Until the end of the 19th century, groups were almost always regarded as permutation groups, so that their elements acted in a natural way on a set. While group theory has now become more abstract, it remains true that groups are at their most useful when their elements act on a set. In this chapter we develop the basic theory of group actions and illustrate its utility by giving applications to both group theory and combinatorics.

5.1 Group actions and permutation representations

Let G be a group and X a non-empty set. A *left action* of G on X is a function

$$\alpha : G \times X \to X,$$

written for convenience $\alpha((g, x)) = g \cdot x$, such that

(i) $g_1 \cdot (g_2 \cdot x) = (g_1 g_2) \cdot x$,

and

(ii) $1_G \cdot x = x$

for all $g_1, g_2 \in G$ and $x \in X$. Here we should think of a group element g as operating or "acting" on a set element x to produce the set element $g \cdot x$.

There is a corresponding definition of a *right action* of G on X as a function $\beta : X \times G \to X$, with $\beta((x, g))$ written $x \cdot g$, such that $x \cdot 1_G = x$ and $(x \cdot g_1) \cdot g_2 = x \cdot (g_1 g_2)$ for all $x \in X$ and $g_1, g_2 \in G$.

For example, suppose that G is a subgroup of the symmetric group $\mathrm{Sym}(X)$ – in which event G is called *a permutation group on X*. We can define $\pi \cdot x$ to be $\pi(x)$ where $\pi \in G$ and $x \in X$; then this is a left action of G on X. There may of course be many ways for a group to act on a set, so we are dealing here with a wide generalization of a permutation group.

Permutation representations. Let G be a group and X a non-empty set. Then a homomorphism

$$\sigma : G \to \mathrm{Sym}(X)$$

is called a *permutation representation* of G on X. Thus the homomorphism σ represents elements of the abstract group G by concrete objects, namely permutations of X. A permutation representation provides a useful way of visualizing the elements in an abstract group.

What is the connection between group actions and permutation representations? In fact the two concepts are essentially identical. To see why, suppose that a permutation representation $\sigma : G \to \mathrm{Sym}(X)$ is given; then there is a corresponding left action of G on X defined by

$$g \cdot x = \sigma(g)(x),$$

where $g \in G$, $x \in X$; it is easy to check that this is indeed an action.

Conversely, if we start with a left action of G on X, say $(g, x) \mapsto g \cdot x$, there is a corresponding permutation representation $\sigma : G \to \mathrm{Sym}(X)$ defined by

$$\sigma(g)(x) = g \cdot x,$$

where $g \in G$, $x \in X$. Again it is an easy verification that the mapping σ is a homomorphism, and hence is a permutation representation of G on X. This discussion makes the following result clear.

(5.1.1) *Let G be a group and X a non-empty set. Then there is a bijection from the set of left actions of G on X to the set of permutation representations of G on X.*

If σ is a permutation representation of a group G on a set X, then

$$G/\mathrm{Ker}(\sigma) \simeq \mathrm{Im}(\sigma)$$

by the First Isomorphism Theorem (4.3.4). Thus $G/\mathrm{Ker}(\sigma)$ is isomorphic with a permutation group on X. If $\mathrm{Ker}(\sigma) = 1$, then G is itself isomorphic with a permutation group on X; in this case the representation σ is called *faithful*. Of course the term faithful can also be applied to a group action by means of its associated permutation representation.

Examples of group actions. There are several natural ways in which a group can act on a set.

(a) *Action on group elements by multiplication.* A group G can act on its underlying set G by left multiplication, that is to say

$$g \cdot x = gx,$$

where $g, x \in G$; this is an action since $1_G \cdot x = 1_G x = x$ and

$$g_1 \cdot (g_2 \cdot x) = g_1(g_2 x) = (g_1 g_2)x = (g_1 g_2) \cdot x.$$

This is the *left regular action* of G: the corresponding permutation representation

$$\lambda : G \to \mathrm{Sym}(G)$$

is given by $\lambda(g)(x) = gx$ and is called the *left regular representation* of G. Observe that $\lambda(g) = 1$ if and only if $gx = x$ for all $x \in G$, i.e., $g = 1$. Thus $\mathrm{Ker}(\lambda) = 1$ and λ is a faithful permutation representation.

It follows at once that G is isomorphic with $\mathrm{Im}(\lambda)$, which is a subgroup of $\mathrm{Sym}(G)$. We have therefore proved the following result, which demonstrates in a striking fashion the significance of permutation groups.

(5.1.2) (Cayley's[1] Theorem). *An arbitrary group G is isomorphic with a subgroup of $\mathrm{Sym}(G)$ via the assignment $g \mapsto (x \mapsto gx)$ where $x, g \in G$.*

(b) *Action on cosets*. For our next example of an action we take a fixed subgroup H of a group G and let $\mathcal{L}$ be the set of all left cosets of H in G. A left action of G on $\mathcal{L}$ is defined by the rule
$$g \cdot (xH) = (gx)H,$$
where $g, x \in G$. Again it is simple to verify that this is a left action.

Now consider the corresponding permutation representation $\lambda : G \to \mathrm{Sym}(\mathcal{L})$. Then $g \in \mathrm{Ker}(\lambda)$ if and only if $gxH = xH$ for all x in G, i.e., $x^{-1}gx \in H$ or $g \in xHx^{-1}$. It follows that
$$\mathrm{Ker}(\lambda) = \bigcap_{x \in G} xHx^{-1}.$$

Thus we have:

(5.1.3) *The kernel of the permutation representation of G on the set of left cosets of H by left multiplication is $\bigcap_{x \in G} xHx^{-1}$: this is the largest normal subgroup of G contained in H.*

For the final statement in (5.1.3), note that the intersection is normal in G. Also, if $N \triangleleft G$ and $N \leq H$, then $N \leq xHx^{-1}$ for all $x \in G$. The normal subgroup $\bigcap_{x \in G} xHx^{-1}$ is called the *normal core* of H in G. Here is an application of the action on left cosets.

(5.1.4) *Suppose that H is a subgroup of a finite group G such that $|G : H|$ equals the smallest prime dividing $|G|$. Then $H \triangleleft G$. In particular, a subgroup of index 2 is always normal in G.*

Proof. Let $|G : H| = p$, and let K be the kernel of the permutation representation of G arising from the left action of G on the set of left cosets of H. Then $K \leq H < G$ and $p = |G : H|$ divides $|G : K|$ by (4.1.3). Now G/K is isomorphic with a subgroup of the symmetric group S_p, so $|G : K|$ divides $|S_p| = p!$. But $|G : K|$ divides $|G|$ and so it cannot be divisible by a smaller prime than p. Therefore $|G : K| = p = |G : H|$, and $H = K \triangleleft G$. □

(c) *Action by conjugation.* Another natural way in which a group G can act on its underlying set is by conjugation. Define
$$g \cdot x = gxg^{-1},$$

---

[1] Arthur Cayley (1821–1895).

where $g, x \in G$; by another simple check this is an action. Again we ask what is the kernel of the action. An element g belongs to the kernel if and only if $gxg^{-1} = x$ for all $x \in G$, i.e., $gx = xg$ or g belongs to $Z(G)$, the center of G. It follows that $Z(G)$ is the kernel of the conjugation representation.

A group G can also act on its set of subgroups by conjugation; thus if $H \leq G$, define
$$g \cdot H = gHg^{-1} = \{ghg^{-1} \mid h \in H\}.$$
In this case the kernel consists of all group elements g such that $gHg^{-1} = H$ for all $H \leq G$. This normal subgroup is called the *norm* of G; clearly it contains the center $Z(G)$.

Exercises (5.1)

1. Complete the proof of (5.1.1).

2. Let $(x, g) \mapsto x \cdot g$ be a *right* action of a group G on a set X. Define $\rho : G \to \text{Sym}(X)$ by $\rho(g)(x) = x \cdot g^{-1}$. Prove that ρ is a permutation representation of G on X. Why is the inverse necessary here?

3. Establish a bijection between the set of right actions of a group G on a set X and the set of permutation representations of G on X.

4. A right action of a group G on its underlying set is defined by $x \cdot g = xg$. Verify that this is an action and describe the corresponding permutation representation of G, (this is called the *right regular representation* of G).

5. Prove that a permutation representation of a simple group is either faithful or trivial.

6. The left regular representation of a finite group is surjective if and only if the group has order 1 or 2.

7. Define a "natural" right action of a group G on the set of right cosets of a subgroup H, and then identify the kernel of the associated representation.

8. Show that the number of (isomorphism types of) groups of order n is at most $(n!)^{[\log_2 n]}$ [Hint: a group of order n can be generated by $[\log_2 n]$ elements by Exercise (4.1.10). Now apply Cayley's Theorem.]

5.2 Orbits and stabilizers

We now proceed to develop the theory of group actions, introducing the fundamental concepts of *orbit* and *stabilizer*.

Let G be a group and X a non-empty set, and suppose that a left action of G on X is given. Then a binary relation $\underset{G}{\sim}$ on X is defined by the rule:
$$a \underset{G}{\sim} b \text{ if and only if } g \cdot a = b$$

for some $g \in G$. A simple verification shows that $\underset{G}{\sim}$ is an equivalence relation on the set X. The $\underset{G}{\sim}$-equivalence class containing a is evidently

$$G \cdot a = \{g \cdot a \mid g \in G\},$$

which is called the *G-orbit* of a. Thus X is the union of all the distinct G-orbits and distinct G-orbits are disjoint, statements which follow from general facts about equivalence relations – see (1.2.2).

If X is the only G-orbit, the action of G on X – and the corresponding permutation representation of G – is called *transitive*. Thus the action of G is transitive if for each pair of elements a, b of X, there exists a g in G such that $g \cdot a = b$. For example, the left regular representation is transitive and so is the left action of a group on the left cosets of a given subgroup.

Another important concept is that of a stabilizer. The *stabilizer* in G of an element $a \in X$ is defined to be

$$St_G(x) = \{g \in G \mid g \cdot x = x\}.$$

It is easy to verify that $St_G(a)$ is a subgroup of G. If $St_G(a) = 1$ for all $a \in X$, the action is called *semiregular*. An action which is both transitive and semiregular is termed *regular*.

We illustrate these concepts by examining the group actions introduced in 5.1.

Example (5.2.1) Let G be any group.

(i) The left regular action of G is regular. Indeed $(yx^{-1})x = y$ for any $x, y \in G$, so it is transitive, while $gx = x$ implies that $g = 1$, and regularity follows.

(ii) In the conjugation action of G on G the stabilizer of x in G consists of all g in G such that $gxg^{-1} = x$, i.e., $gx = xg$. This subgroup is called the *centralizer* of x in G, the notation being

$$C_G(x) = \{g \in G \mid gx = xg\}.$$

(iii) In the conjugation action of G on its underlying set the G-orbit of x is $\{gxg^{-1} \mid g \in G\}$, i.e., the set of all conjugates of x in G. This is called the *conjugacy class* of x. The number of conjugacy classes of G is known as the *class number*.

(iv) In the action of G by conjugation on its set of subgroups, the G-orbit of $H \leq G$ is just the set of all conjugates of H in G, i.e., $\{gHg^{-1} \mid g \in G\}$. The stabilizer of H in G is an important subgroup termed the *normalizer* of H in G,

$$N_G(H) = \{g \in G \mid gHg^{-1} = H\}.$$

Both centralizers and normalizers feature extensively throughout group theory.

Next we will prove two basic theorems on group actions. The first one counts the number of elements in an orbit.

(5.2.1) *Let G be a group acting on a set X and let $x \in X$. Then the assignment $g\,St_G(x) \mapsto g \cdot x$ determines a bijection from the set of left cosets of $St_G(x)$ in G to the orbit $G \cdot x$. Hence $|G \cdot x| = |G : St_G(x)|$.*

Proof. In the first place $g\,St_G(x) \mapsto g \cdot x$ determines a well-defined function. For if $s \in St_G(x)$, then $gs \cdot x = g \cdot (s \cdot x) = g \cdot x$. Next $g_1 \cdot x = g_2 \cdot x$ implies that $g_2^{-1}g_1 \cdot x = x$, so $g_2^{-1}g_1 \in St_G(x)$, i.e., $g_1\,St_G(x) = g_2\,St_G(x)$. So the function is injective, while it is obviously surjective. □

Corollary (5.2.2) *Let G be a finite group acting on a finite set X. If the action is transitive, then $|X|$ divides $|G|$, while if the action is regular, $|X| = |G|$.*

Proof. If the action is transitive, X is the only G-orbit and so $|X| = |G : St_G(x)|$ for any $x \in X$, by (5.2.1); hence this divides $|G|$. If the action is regular, then in addition $St_G(x) = 1$ and thus $|X| = |G|$. □

The corollary tells us that if G is a transitive permutation group of degree n, then n divides $|G|$, and $|G| = n$ if G is regular.

The second main theorem on actions counts the number of orbits and has many applications. If a group G acts on a set X and $g \in G$, the *fixed point set* of g is defined to be

$$\text{Fix}(g) = \{x \in X \mid g \cdot x = x\}.$$

(5.2.3) (Burnside's[2] Lemma). *Let G be a finite group acting on a finite set X. Then the number of G-orbits equals*

$$\frac{1}{|G|} \sum_{g \in G} |\text{Fix}(g)|,$$

i.e., the average number of fixed points of elements of G.

Proof. Consider how often an element x of X is counted in the sum $\sum_{g \in G} |\text{Fix}(g)|$. This happens once for each g in $St_G(x)$. Thus the element x contributes $|St_G(x)| = |G|/|G \cdot x|$ to the sum, by (5.2.1). The same is true of each element of the orbit $|G \cdot x|$, so that the total contribution of the orbit to the sum is

$$(|G|/|G \cdot x|) \cdot |G \cdot x| = |G|.$$

It follows that $\sum_{g \in G} |\text{Fix}(g)|$ must equal $|G|$ times the number of orbits, and the result is proven. □

We illustrate Burnside's Lemma by a simple example.

---
[2] William Burnside (1852–1927).

Example (5.2.2) The group $G = \{\text{id}, (1\,2)(3)(4), (1)(2)(3\,4), (1\,2)(3\,4)\}$ acts on the set $X = \{1, 2, 3, 4\}$ as a permutation group. There are two G-orbits here, namely $\{1, 2\}$ and $\{3, 4\}$. Now it is easy to count the fixed points of the elements of G by looking for 1-cycles. Thus the four elements of the group have respective numbers of fixed points 4, 2, 2, 0. Therefore the number of G-orbits should be

$$\frac{1}{|G|}\left(\sum_{g \in G} |\text{Fix}(g)|\right) = \frac{1}{4}(4 + 2 + 2 + 0) = 2,$$

which is the correct answer.

Example (5.2.3) Show that the average number of fixed points of an element of S_n is 1.

Here the symmetric group S_n acts on the set $\{1, 2, \ldots, n\}$ in the natural way and the action is clearly transitive. By Burnside's Lemma the average number of fixed points is simply the number of S_n-orbits, which is 1 by transitivity of the action.

Exercises (5.2)

1. If g is an element of a finite group G, the number of conjugates of g divides $|G : \langle g \rangle|$.

2. If H is a subgroup of a finite group G, the number of conjugates of H divides $|G : H|$.

3. Let $G = \langle (1\,2\ldots p), (1)(2\,p)(3\,p-1)\ldots \rangle$ be a dihedral group of order $2p$ where p is an odd prime. Verify Burnside's Lemma for the group G acting on the set $\{1, 2, \ldots, p\}$ as a permutation group.

4. Let G be a finite group acting as a finite set X. If the action is semiregular, prove that $|G|$ divides $|X|$.

5. Prove that the class number of a finite group G is given by the formula

$$\frac{1}{|G|}\left(\sum_{x \in G} |C_G(x)|\right).$$

6. The class number of the direct product $H \times K$ equals the product of the class numbers of H and K.

7. Let G be a finite group acting transitively on a finite set X where $|X| > 1$. Prove that G contains at least $|X| - 1$ *fixed-point-free elements*, i.e., elements g such that $\text{Fix}(g)$ is empty.

8. Let H be a proper subgroup of a finite group G. Prove that $G \neq \bigcup_{x \in G} xHx^{-1}$. [Hint: Consider the action of G on the set of left cosets of H by multiplication. The action is transitive, so Exercise (5.2.7) may be applied.]

9. Let X be a subset of a group G. Define the *centralizer* $C_G(X)$ of X in G to be the set of elements of G that commute with every element of X. Prove that $C_G(X)$ is a subgroup and then show that $C_G(C_G(C_G(X))) = C_G(X)$.

10. Let G be a finite group with class number h. A group element is chosen at random from G and replaced. Then another group element is chosen. Prove that the probability that the two elements commute is $\frac{h}{|G|}$. What will the answer be if the first group element is *not* replaced? [Hint: use Exercise (5.2.5).]

5.3 Applications to the structure of groups

Our aim in this section is to show that group actions can be a highly effective tool for investigating the structure of groups. The first result provides important arithmetic information about the conjugacy classes of a finite group.

(5.3.1) (The class equation) *Let G be a finite group and let $C_1, \ldots, C_h$ be the distinct conjugacy classes of G. Then*

(i) $|C_i| = |G : C_G(x_i)|$ *for any x_i in C_i; thus $|C_i|$ divides $|G|$.*

(ii) $|G| = |C_1| + |C_2| + \cdots + |C_h|$.

This follows on applying (5.2.1) to the conjugation action of G on its underlying set. For in this action the G-orbit of x is its conjugacy class, while the stabilizer of x is the centralizer $C_G(x)$; thus $|G \cdot x| = |G : St_G(x)| = |G : C_G(x)|$.

There are other ways to express the class equation. Choose any $x_i \in C_i$ and put $n_i = |C_G(x_i)|$. Then $|C_i| = |G|/n_i$. On division by $|G|$, the class equation becomes

$$\frac{1}{n_1} + \frac{1}{n_2} + \cdots + \frac{1}{n_h} = 1,$$

an interesting diophantine equation for the orders of the centralizers.

Next observe that the one-element set $\{x\}$ is a conjugacy class of G if and only if x is its only conjugate in G, i.e., x belongs to the center of the group G. Now suppose we order the conjugacy classes in such a way that $|C_i| = 1$ for $i = 1, 2, \ldots, r$ and $|C_i| > 1$ if $r < i \leq h$. Then the class equation becomes:

(5.3.2) $|G| = |Z(G)| + |C_{r+1}| + \cdots + |C_h|$.

Here is a natural question: what are the conjugacy classes of the symmetric group S_n? First note that any two r-cycles in S_n are conjugate. For

$$\pi(i_1 \, i_2 \, \ldots \, i_r)\pi^{-1} = (j_1 \, j_2 \, \ldots \, j_r)$$

where π is any permutation in S_n such that $\pi(i_1) = j_1, \pi(i_2) = j_2, \ldots, \pi(i_r) = j_r$. From this and (3.1.3) it follows that *any two permutations which have the same cycle type are conjugate in S_n*. Here "cycle type" refers to the numbers of 1-cycles, 2-cycles, etc. present in the disjoint cycle decomposition. It is also easy to see that conjugate permutations must have the same cycle type. So we have the answer to our question.

(5.3.3) *The conjugacy classes of the symmetric group S_n are the sets of permutations with the same cycle type.*

It follows that the class number of S_n equals the number of different cycle types; this is just $\lambda(n)$, *the number of partitions of n*, i.e., the number of ways of writing the positive integer n as a sum of positive integers when order is not significant. This is a well-known number theoretic function which has been studied intensively.

Example (5.3.1) S_6 has 7 conjugacy classes.
For $\lambda(6) = 7$, as is seen by writing out the partitions, $6 = 5 + 1 = 4 + 1 + 1 = 4 + 2 = 3 + 1 + 1 + 1 = 3 + 2 + 1 = 3 + 3$.

As a deeper application of our knowledge of the conjugacy classes of S_n we shall prove next:

(5.3.4) *The symmetric group S_n has no non-inner automorphisms if $n \neq 6$.*

Proof. Since S_2 has only the trivial automorphism, we can assume that $n > 2$, as well as $n \neq 6$. First a general remark is in order: in any group G the automorphism group $\text{Aut}(G)$ permutes the conjugacy classes of G. Indeed, if $\alpha \in \text{Aut}(G)$, then $\alpha(xgx^{-1}) = \alpha(x)\alpha(g)(\alpha(x))^{-1}$, so α maps the conjugacy class of g to that of $\alpha(g)$.

Now let C_1 denote the conjugacy class consisting of all the 2-cycles in S_n. If π is a 2-cycle, then $\alpha(\pi)$ has order 2 and so it is a product of, say, k disjoint 2-cycles. Then $\alpha(C_1) = C_k$ say, where C_k is the conjugacy class of all (disjoint) products of k 2-cycles. The first step in the proof is to show by a counting argument that $k = 1$, i.e., α maps 2-cycles to 2-cycles. Assume to the contrary that $k \geq 2$.

Clearly $|C_1| = \binom{n}{2}$, and more generally

$$|C_k| = \binom{n}{2k} \frac{(2k)!}{(2!)^k k!}.$$

For, in order to form a product of k disjoint 2-cycles, we first choose the $2k$ integers from $1, 2, \ldots, n$ in $\binom{n}{2k}$ ways. Then we divide these $2k$ elements into k pairs, with order of pairs unimportant; this can be done in $\frac{(2k)!}{(2!)^k k!}$ ways. Forming the product, we obtain the formula for $|C_k|$.

Since $\alpha(C_1) = C_k$, we must have $|C_1| = |C_k|$, and so

$$\binom{n}{2} = \binom{n}{2k} \frac{(2k)!}{(2!)^k k!}.$$

After cancellation this becomes $(n-2)(n-3)\ldots(n-2k+1) = 2^{k-1}(k!)$. Now this is impossible if $k = 2$, while if $k = 3$ it can only hold if $n = 6$, which is forbidden. Therefore $k > 3$. Now clearly $n \geq 2k$, so that $(n-2)(n-3)\ldots(n-2k+1) \geq (2k-2)!$. This leads to $(2k-2)! \leq 2^{k-1}(k!)$, which implies that $k = 3$, a contradiction.

The argument so far has established that $k = 1$ and $\alpha(C_1) = C_1$. Write

$$\alpha((a\ b)) = (b'\ b'') \quad \text{and} \quad \alpha((a\ c)) = (c'\ c'').$$

If b', b'', c', c'' are all different, then $(a\,c)(a\,b) = (a\,b\,c)$, which has order 3 and maps to $(c'\,c'')(b'\,b'')$, which has order 2: but this is impossible. Therefore two of b', b'', c', c'' are equal and

$$\alpha((a\ b)) = (a'\,b') \quad \text{and} \quad \alpha((a\ c)) = (a'\,c').$$

Next suppose there is a d such that $\alpha((a\ d)) = (b'\,c')$ with b', $c' \neq a'$. Then $(a\,c)(a\,d)(a\,b) = (a\,b\,d\,c)$, whereas $(a'\,c')(b'\,c')(a'\,b') = (a')(b'\,c')$, another contradiction.

This argument shows that for each a there is a unique a' such that $\alpha((a\,b)) = (a'\,b')$ for all b and some b'. Therefore α determines a permutation $\pi \in S_n$ such that $\pi(a) = a'$ for all a. Then $\alpha((a\ b)) = (a'\,b') = (\pi(a)\ \pi(b))$, which equals the conjugate $\pi\,(a\,b)\,\pi^{-1}$ since the latter interchanges a' and b' and fixes all other integers. Since S_n is generated by 2-cycles by (3.1.4), it follows that α is simply conjugation by π, so that it is an inner automorphism. □

Recall that a group is called *complete* if the conjugation homomorphism $G \to \mathrm{Aut}(G)$ is an isomorphism, i.e., $Z(G) = 1$ and $\mathrm{Aut}(G) = \mathrm{Inn}(G)$ by (4.3.8). Now $Z(S_n) = 1$ if $n \neq 2$ – see Exercise (4.2.10). Hence we obtain:

Corollary (5.3.5) *The symmetric group S_n is complete if $n \neq 2$ or 6.*

Of course S_2 is not complete since it is abelian. Also it is known that the group S_6 has a non-inner automorphism, so it too is not complete.

Finite p-groups. If p is a prime number, a finite group is called a *p-group* if its order is a power of p. Finite p-groups form an important, and highly complex, class of finite groups. A first indication that these groups have special features is provided by the following result.

(5.3.6) *If G is a non-trivial finite p-group, then $Z(G) \neq 1$.*

This contrasts with arbitrary finite groups, which can easily have trivial center; for example, $Z(S_3) = 1$.

Proof of (5.3.6). Consider the class equation of G in the form

$$|G| = |Z(G)| + |C_{r+1}| + \cdots + |C_h|,$$

(see (5.3.2)). Here $|C_i|$ divides $|G|$ and hence is a power of p; also $|C_i| > 1$. If $Z(G) = 1$, then it will follow that $|G| \equiv 1 \pmod{p}$. But this is impossible because $|G|$ is a power of p. Therefore $Z(G) \neq 1$. □

Corollary (5.3.7) *If p is a prime, all groups of order p^2 are abelian.*

Proof. Let G be a group of order p^2. Then $|Z(G)| = p$ or p^2 by (5.3.6) and Lagrange's Theorem. If $|Z(G)| = p^2$, then $G = Z(G)$ is abelian. Thus we can assume that $|Z(G)| = p$, so that $|G/Z(G)| = p$. By (4.1.4) both $G/Z(G)$ and $Z(G)$ are cyclic, say $G/Z(G) = \langle a Z(G) \rangle$ and $Z(G) = \langle b \rangle$. Then every element of G has the form $a^i b^j$ where i, j are integers. But

$$(a^i b^j)(a^{i'} b^{j'}) = a^{i+i'} b^{j+j'} = (a^{i'} b^{j'})(a^i b^j)$$

since $b \in Z(G)$. This shows that G is abelian and $Z(G) = G$, a contradiction.

On the other hand, there are non-abelian groups of order $2^3 = 8$, for example Dih(8). Thus (5.3.7) does not generalize to groups of order p^3. □

Sylow's [3] Theorem. Group actions will now be used to give a proof of *Sylow's Theorem*, which is probably the most celebrated and frequently used result in elementary group theory.

Let G be a finite group and p a prime, and write $|G| = p^a m$ where p does not divide the integer m. Thus p^a is the highest power of p dividing $|G|$. Lagrange's Theorem guarantees that the order of a p-subgroup of G is at most p^a. That p-subgroups of this order actually occur is the first part of Sylow's Theorem. A subgroup of G with order p^a is called a *Sylow p-subgroup*.

(5.3.8) (Sylow's Theorem) *Let G be a finite group and let p^a denote largest power of the prime p dividing $|G|$. Then:*

(i) *every p-subgroup of G is contained in some subgroup of order p^a: in particular, Sylow p-subgroups exist;*

(ii) *if n_p is the number of Sylow p-subgroups, then $n_p \equiv 1 \pmod{p}$;*

(iii) *any two Sylow p-subgroups are conjugate in G.*

Proof. Write $|G| = p^a m$ where p does not divide m. Three group actions will be used during the course of the proof.

(a) Let **S** be the set of all *subsets* of G with exactly p^a elements. Thus **S** has s elements where

$$s = \binom{p^a m}{p^a} = \frac{m(p^a m - 1) \ldots (p^a m - p^a + 1)}{1 \cdot 2 \ldots (p^a - 1)}.$$

First we prove that p does not divide s. For this purpose consider the rational number $\frac{p^a m - i}{i}$ where $1 \leq i < p^a$. If $p^j \mid i$, then $j < a$ and hence $p^j \mid p^a m - i$. On the other hand, if $p^j \mid p^a m - i$, then $j < a$ since otherwise $p^a \mid i$. Therefore $p^j \mid i$. It follows that the integers $p^a m - i$ and i involve the same highest power of p, which can of

---
[3] Peter Ludwig Mejdell Sylow (1832–1918).

course be cancelled in $\frac{p^a m - i}{i}$; thus no p's occur in this rational number. It follows that p does not divide s, as claimed.

Now we introduce the first group action. The group G acts on the set $\mathbf{S}$ via left multiplication, i.e., $g \cdot X = gX$ where $X \subseteq G$ and $|X| = p^a$. Thus $\mathbf{S}$ splits up into disjoint G-orbits. Since $|\mathbf{S}| = s$ is not divisible by p, there must be at least one G-orbit $\mathbf{S}_1$ such that $|\mathbf{S}_1|$ is not divisible by p. Choose $X \in \mathbf{S}_1$ and put $P = St_G(X)$, which is a subgroup of course. Then $|G : P| = |\mathbf{S}_1|$, from which it follows that p does not divide $|G : P|$. However p^a divides $|G|$, and hence p^a must divide $|P|$.

Now fix x in X; then the number of elements gx with $g \in P$ equals $|P|$. Also $gx \in X$; hence $|P| \leq |X| = p^a$ and consequently $|P| = p^a$. Thus P is a Sylow p-subgroup of G and we have shown that Sylow p-subgroups exist.

(b) Let $\mathbf{T}$ denote the set of all conjugates of the Sylow p-subgroup P constructed in (a). We will argue next that $|\mathbf{T}| \equiv 1 \pmod{p}$.

The group P acts on the set $\mathbf{T}$ *by conjugation*, i.e., $g \cdot Q = gQg^{-1}$ where $g \in P$ and $Q \in \mathbf{T}$; clearly $|gQg^{-1}| = |P| = p^a$. In this action $\{P\}$ is a P-orbit since $gPg^{-1} = P$ if $g \in P$. Suppose that $\{P_1\}$ is another one-element P-orbit. Then $P_1 \lhd \langle P, P_1 \rangle$; for $xP_1x^{-1} = P_1$ if $x \in P \cup P_1$, so $N_{\langle P, P_1 \rangle}(P_1) = \langle P, P_1 \rangle$. By (4.3.5) PP_1 is a subgroup and its order is $|PP_1| = |P| \cdot |P_1|/|P \cap P_1|$, which is certainly a power of p. But $P \subseteq PP_1$ and P already has the maximum order allowed for a p-subgroup. Therefore $P = PP_1$, so $P_1 \subseteq P$ and thus $P_1 = P$ because $|P_1| = |P|$.

Consequently there is only one P-orbit of $\mathbf{T}$ with a single element. Every other P-orbit has order a power of p greater than 1. Therefore $|\mathbf{T}| \equiv 1 \pmod{p}$.

(c) Finally, let P_2 be an arbitrary p-subgroup of G. We aim to show that P_2 is contained in some conjugate of the Sylow p-subgroup P; this will complete the proof of Sylow's Theorem.

Let P_2 act on $\mathbf{T}$ by conjugation, where as before $\mathbf{T}$ is the set of all conjugates of P. Assume that P_2 is not contained in any member of $\mathbf{T}$. If $\{P_3\}$ is a one-element P_2-orbit of $\mathbf{T}$, then, arguing as in (b), we see that P_2P_3 is a p-subgroup containing P_3, so $P_3 = P_2P_3$ because $|P_3| = p^a$. Thus $P_2 \subseteq P_3 \in \mathbf{T}$, contrary to assumption. It follows that there are no one-element P_2-orbits in $\mathbf{T}$; this means that $|\mathbf{T}| \equiv 0 \pmod{p}$, which contradicts (b). □

An important special case of Sylow's Theorem is:

(5.3.9) (Cauchy's Theorem) *If the order of a finite group G is divisible by a prime p, then G contains an element of order p.*

Proof. Let P be a Sylow p-subgroup of G. Then $P \neq 1$ since p divides $|G|$. Choose $1 \neq g \in P$; then $|g| = |\langle g \rangle|$ divides $|P|$, and hence $|g| = p^m$ where $m > 0$. Then $g^{p^{m-1}}$ has order p, as required. □

While Sylow's Theorem does not tell us the exact number of Sylow p-subgroups, it does provide valuable information, which may be sufficient to determine it. Let us review what is known. Suppose P is a Sylow p-subgroup of a finite group G. Then, since every Sylow p-subgroup is a conjugate of P, the number of Sylow p-subgroups of G equals the number of conjugates of P, which by (5.2.1) is

$$n_p = |G : N_G(P)|,$$

where $N_G(P)$ is the normalizer of P in G – see 5.2. Hence n_p divides $|G : P|$ since $P \leq N_G(P)$. Also of course

$$n_p \equiv 1 \pmod{p}. \qquad \square$$

Example (5.3.2) Find the numbers of Sylow p-subgroups of the alternating group A_5.

Let $G = A_5$. We can assume that p divides $|G|$, so that $p = 2$, 3 or 5. Note that a non-trivial element of G has one of the cycle types of even permutations,

$$(**)(**)(*), \quad (***)(*)(*), \quad (*****)$$

If $p = 2$, then $n_2 \mid \frac{60}{4} = 15$ and $n_1 \equiv 1 \pmod{2}$, so $n_2 = 1, 3, 5$ or 15. There are $5 \times 3 = 15$ elements of order 2 in G, with three of them in each Sylow 2-subgroup. Hence $n_2 \geq 5$. If $n_2 = 15$, then $P = N_G(P)$ where P is a Sylow 2-subgroup, since $P \leq N_G(P) \leq G$ and $|G : N_G(P)| = 15 = |G : P|$. But this is wrong since P is normalized by a 3-cycle – note that the Klein 4-group is normal in A_4. Consequently $n_2 = 5$.

Next $n_3 \mid \frac{60}{3} = 20$ and $n_3 \equiv 1 \pmod{3}$. Thus $n_3 = 1, 4$ or 10. Now G has $\binom{5}{3} \times 2 = 20$ elements of order 3, which shows that $n_3 > 4$. Hence $n_3 = 10$. Finally, $n_5 \mid 12$ and $n_5 \equiv 1 \pmod{5}$, so $n_5 = 6$ since $n_5 = 1$ would give only 4 elements of order 5.

The next result provides some very important information about the group A_5.

(5.3.10) *The alternating group A_5 is simple.*

Proof. Let $G = A_5$ and suppose N is a proper non-trivial normal subgroup of G. Now the order of an element of G is 1, 2, 3, or 5. If N contains an element of order 3, then it contains a Sylow 3-subgroup of G, and by normality it contains all such. Hence N contains all 3-cycles. Now the easily verified equations $(ab)(ac) = (acb)$ and $(ac)(bd) = (abc)(abd)$, together with the fact that every permutation in G is a product of an *even* number of transpositions, shows that G is generated by 3-cycles. Therefore $N = G$, which is a contradiction.

Next suppose G has an element of order 5; then N contains a Sylow 5-subgroup and hence all Sylow subgroups; thus N contains all 5-cycles. But $(12345)(12543) = (132)$, which delivers the contradiction that N contains a 3-cycle.

The argument so far tells us that each element of N has order a power of 2, which implies that $|N|$ is a power of 2 by Cauchy's Theorem. Since $|N|$ divides $|G| = 60$, this order must be 2 or 4. We leave it to the reader to disprove these possibilities. This final contradiction shows that G is a simple group. $\qquad \square$

5.3 Applications to the structure of groups

More generally, A_n is simple for all $n \geq 5$ – see (9.1.7). We will show in Chapter 11 that the simplicity of A_5 is intimately connected with the insolvability of polynomial equations of degree 5 by radicals.

Example (5.3.3) Find all groups of order 21.

Let G be a group of order 21. Then G has elements a, b with orders 7, 3 respectively by (5.3.9). Now the order of $\langle a \rangle \cap \langle b \rangle$ divides both 7 and 3. Thus $\langle a \rangle \cap \langle b \rangle = 1$, and $|\langle a \rangle \langle b \rangle| = |a| \cdot |b| = 21$, which means that $G = \langle a \rangle \langle b \rangle$. Next $\langle a \rangle$ is a Sylow 7-subgroup of G, and $n_7 \equiv 1 \pmod{7}$ and $n_7 \mid 3$. Hence $n_7 = 1$, so that $\langle a \rangle \triangleleft G$ and $bab^{-1} = a^i$ where $1 \leq i < 7$. If $i = 1$, then G is abelian and $|ab| = 21$. In this case $G = \langle ab \rangle$ and $G \simeq Z_{21}$.

Next assume $i \neq 1$. Now $b^3 = 1$ and $bab^{-1} = a^i$, with $2 \leq i < 7$, imply that $a = b^3 a b^{-3} = a^{i^3}$. Hence $7 \mid i^3 - 1$, which shows that $i = 2$ or 4. Now $[2]_7 = [4]_7^{-1}$ since $8 \equiv 1 \pmod{7}$. Since we can replace b by b^{-1} if necessary, there is nothing to be lost in assuming $i = 2$.

Thus far we have discovered that $G = \{a^u b^v \mid 0 \leq u < 7, 0 \leq v < 3\}$ and that $a^7 = 1 = b^3$, $bab^{-1} = a^2$. But is there really such a group? An example is easily found using permutations. Put $\pi = (1\,2\,3\,4\,5\,6\,7)$ and $\sigma = (2\,3\,5)(4\,7\,6)$. Then one quickly verifies that $\pi^7 = 1 = \sigma^3$ and $\sigma \pi \sigma^{-1} = \pi^2$. A brief computation reveals that the assignments $a \mapsto \pi, b \mapsto \sigma$ determine an isomorphism from G to the group $\langle \pi, \sigma \rangle$. Therefore we have shown that *up to isomorphism there are exactly two groups of order* 21.

Example (5.3.4) Show that there are no simple groups of order 300.

Suppose that G is a simple group of order 300. Since $300 = 2^2 \cdot 3 \cdot 5^2$, a Sylow 5-subgroup P has order 25. Now $n_5 \equiv 1 \pmod{5}$ and n_5 divides $300/25 = 12$. Thus $n_5 = 1$ or 6. Now $n_5 = 1$ implies that $P \triangleleft G$, which is impossible. Hence $n_5 = 6$ and $|G : N_G(P)| = 6$. But the left action of G on the set of left cosets of $N_G(P)$ (see 5.1) leads to a homomorphism θ from G to S_6. Also $\text{Ker}(\theta) = 1$ since G is simple. Thus θ is injective and $|G|$ divides S_6, which is false.

Exercises (5.3)

1. A finite p-group cannot be simple unless it has order p.

2. Let G be a group of order pq where p and q are primes such that $p \not\equiv 1 \pmod{q}$ and $q \not\equiv 1 \pmod{p}$. Prove that G is cyclic.

3. Show that if p is a prime, a group of order p^2 is isomorphic with Z_{p^2} or $Z_p \times Z_p$.

4. Let P be a Sylow p-subgroup of a finite group G and let $N \triangleleft G$. Prove that $P \cap N$ and PN/N are Sylow p-subgroups of N and G/N respectively.

5. Show that there are no simple groups of orders 312, 616 or 1960.

6. Show that every group of order 561 is cyclic. [Hint: show that there is a cyclic normal subgroup $\langle x \rangle$ of order 11×17; then use the fact that 3 does not divide $|\text{Aut}(\langle x \rangle)|$.]

7. Let G be a group of order $2m$ where m is odd. Prove that G has a normal subgroup of order m. [Hint: let λ be the left regular representation of G. By Cauchy's Theorem there is an element g of order 2 in G. Now argue that $\lambda(g)$ must be an odd permutation.]

8. Find all finite groups with class number at most 2.

9. A group of order 10 is isomorphic with $\mathbb{Z}_{10}$ or Dih(10). [Follow the method of Example (5.3.3).]

10. Show that up to isomorphism there are three groups of order 55.

11. If H is a proper subgroup of a finite p-group G, prove that $H < N_G(H)$. [Hint: use induction on $|G| > 1$, noting that $H \triangleleft HZ(G)$.]

12. Let P be a Sylow p-subgroup of a finite group G and let H be a subgroup of G containing $N_G(P)$. Prove that $H = N_G(H)$. [Hint: if $g \in N_G(H)$, then P and gPg^{-1} are conjugate in H.]

13. Let G be a finite group and suppose it is possible to choose one element from each conjugacy class in such a way that all the selected elements commute. Prove that G is abelian. [Hint: use (5.3.2)].

5.4 Applications to combinatorics – counting labellings and graphs

Group actions can be used very effectively to solve certain types of counting problem. To illustrate this suppose that we wish to color the six faces of a cube and five colors are available. How many different coloring schemes can one come up with? At first thought one might answer 5^6 since each of the six faces can be colored in five different ways. However this answer is not correct since by merely rotating the cube one can pass from one coloring scheme to another. Clearly these two coloring schemes are not really different. So not all of the 5^6 colorings schemes are distinct.

Let us pursue further the idea of rotating the cube. The group of rotations of the cube acts on the set of all possible coloring schemes. If two colorings belong to the same orbit, they should be regarded as identical since one arises from the other by a suitable rotation of the cube. So what we really want to do is count the number of orbits, and for this purpose Burnside's Lemma (5.2.3) is ideally suited.

Labelling problems. Our problem is really about the labelling of sets. Let X and L be two non-empty sets, with L referred to as the set of *labels*. Suppose we have to assign a label to each element of the set X; then we need to specify a function

$$\alpha : X \to L.$$

5.4 Applications to combinatorics – counting labellings and graphs   93

Call such a function α a *labelling* of X by L. Thus the set of all such labellings of X by L is

$$\text{Fun}(X, L).$$

Now suppose that G is a group acting on the set X. Then G can be made to act on the set of labellings $\text{Fun}(X, L)$ by the rule

$$g \cdot \alpha(x) = \alpha(g^{-1} \cdot x),$$

where $g \in G$, $x \in X$ and $\alpha \in \text{Fun}(X, L)$. What this equation asserts is that the labelling $g \cdot \alpha$ assigns to the set element $g \cdot x$ the same label as α assigns to x. The example of the cube should convince the reader that this is the correct action.

First we must verify that this really is an action of G on $\text{Fun}(X, L)$. To do this let $g_1, g_2 \in G$, $x \in X$ and $\alpha \in \text{Fun}(X, L)$; then

$$(g_1 \cdot (g_2 \cdot \alpha))(x) = g_2 \cdot \alpha(g_1^{-1} \cdot x)$$

$$= \alpha(g_2^{-1} \cdot (g_1^{-1} \cdot x)) = \alpha((g_1 g_2)^{-1} \cdot x) = ((g_1 g_2) \cdot \alpha)(x).$$

Hence $g_1 \cdot (g_2 \cdot \alpha) = (g_1 g_2) \cdot \alpha$. Also $1_G \cdot \alpha(x) = \alpha(1_G \cdot x) = \alpha(x)$, so that $1_G \cdot \alpha = \alpha$. Therefore we have an action of G on $\text{Fun}(X, L)$.

Our goal is to count the number of G-orbits in $\text{Fun}(X, L)$. This is done in the following fundamental result.

(5.4.1) (Polya's [4] Theorem) *Let G be a finite group acting on a finite set X, and let L be a finite set of labels. Then the number of G-orbits of labellings of X by L is*

$$\frac{1}{|G|} \left(\sum_{g \in G} \ell^{m(g)} \right)$$

where $\ell = |L|$ and $m(g)$ is the number of disjoint cycles in the permutation of X corresponding to g.

Proof. Burnside's Lemma shows that the number of G-orbits of labellings is

$$\frac{1}{|G|} \left(\sum_{g \in G} |\text{Fix}(g)| \right)$$

where $\text{Fix}(g)$ is the set of labellings fixed by g. Thus we have to count such labellings. Now $\alpha \in \text{Fix}(g)$ if and only if $g \cdot \alpha(x) = \alpha(x)$, i.e., $\alpha(g^{-1} \cdot x) = \alpha(x)$ for all $x \in X$. This equation asserts that α is constant on the $\langle g \rangle$-orbit $\langle g \rangle \cdot x$. Now the $\langle g \rangle$-orbits arise from the disjoint cycles involved in the permutation of X corresponding to g. Therefore, to construct a labelling in $\text{Fix}(g)$ all we need to do is assign a label to

---
[4] George Polya (1887–1985).

5 Groups acting on sets

each cycle of g. This can be done in $\ell^{m(g)}$ ways where $m(g)$ is the number of cycles; consequently $|\operatorname{Fix}(g)| = \ell^{m(g)}$ and we have our formula. □

Now let us use Polya's Theorem to solve some counting problems.

Example (5.4.1) How many ways are there to design a necklace with 11 beads if c different colors of beads are available?

Here it is assumed that the beads are identical apart from color. We can visualize the necklace as a regular 11-gon with the beads as the vertices. The labels are the c colors and one color has to be assigned to each vertex. Now a symmetry of the 11-gon can be applied without changing the design of the necklace. Recall that the group of symmetries G is a dihedral group $\operatorname{Dih}(22)$ – see 3.2. It consists of the identity, rotations through $\left(\frac{2\pi}{11}\right)i$, for $i = 1, 2, \ldots, 10$, and reflections in a line joining a vertex to the midpoint of the opposite edge.

For each element $g \in G$ count the number of $\langle g \rangle$-orbits in the set of vertices $X = \{1, 2, \ldots, 11\}$, so that Polya's formula can be applied. The results of the count are best displayed in tabular form.

Type of element	Cycle type of permutation	Number of elements of type	m
identity	eleven 1-cycles	1	11
rotation through $\frac{2\pi i}{11}$	one 11-cycle	10	1
reflection	one 1-cycle and five 2 cycles	11	6

From the table and Polya's formula we deduce that the number of different designs is

$$\frac{1}{22}(c^{11} + 11c^6 + 10c) = \frac{1}{22}c(c^5 + 1)(c^5 + 10).$$

Next let us tackle the cube-coloring problem with which the section began.

Example (5.4.2) How many ways are there to color the faces of a cube using c colors?

In this problem the relevant group is the rotation group G of the cube since this acts on the set of colorings. In fact $G \simeq S_4$: the easiest way to see this is to observe that the rotations permute the four diagonals of the cube, but this isomorphism is not really needed to solve the problem.

It is a question of computing the number of G-orbits of $\operatorname{Fun}(X, L)$ where the set X consists of the six faces of the cube and $|L| = c$. To identify the rotations in G, we examine the various axes of symmetry of the cube. For each element of G record the number of cycles in the corresponding permutation of X. Again the results are conveniently displayed in a table.

5.4 Applications to combinatorics – counting labellings and graphs

Type of element	Cycle type of permutation	Number of elements of type	m
identity	six 1-cycles	1	6
rotation about line through centroids of opposite faces through			
$\pi/2$	two 1-cycles and one 4-cycle	3	3
π	two 1-cycles, two 2-cycles	3	4
$\frac{3\pi}{2}$	two 1-cycles and one 4-cycle	3	3
rotation about a diagonal through			
$\frac{2\pi}{3}$	two 3-cycles	4	2
$\frac{4\pi}{3}$	two 3-cycles	4	2
rotation about a line joining midpoints of opposite edges through π	three 2-cycles	6	3

This confirms that $|G| = 24$, and Polya's formula gives the answer

$$\frac{1}{24}(c^6 + 3c^3 + 3c^4 + 3c^3 + 4c^2 + 4c^2 + 6c^3) = \frac{1}{24}(c^6 + 3c^4 + 12c^3 + 8c^2).$$

When $c = 5$, the formula yields 800. So there are 800 different ways to color the faces of a cube using 5 colors.

From these examples the reader can see that Polya's method allows us to solve some complex combinatorial problems that would otherwise be intractable.

Counting graphs. We conclude the chapter by describing how Polya's methods can be used to count graphs. First some brief remarks about graphs.

A *graph* Γ consists of a non-empty set V of *vertices* and a relation E on V which is symmetric and *irreflexive*, i.e., $v \not\mathrel{E} v$ for all $v \in V$. If $u \mathrel{E} v$, call the 2-element set $\{u, v\}$ an *edge* of Γ. Since E is symmetric, we can identify E with the set of all edges of Γ.

A graph $\Gamma = (V, E)$ may be visualized by representing the vertices by points in the plane and the edges by lines joining the vertices. Simple examples of graphs are:

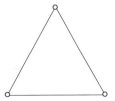 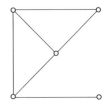 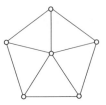

Graph theory has many applications outside mathematics, for example to transportation systems, telephone networks and electrical circuits.

Two graphs $\Gamma_i = (V_i, E_i), i = 1, 2$, are said to be *isomorphic* if there is a bijection $\theta : V_1 \to V_2$ such that $\{u, v\} \in E_1$ if and only if $\{\theta(u), \theta(v)\} \in E_2$. For example, the graphs

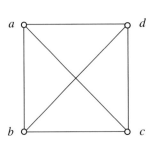 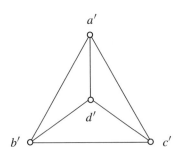

are isomorphic because of the bijection $a \mapsto a', b \mapsto b', c \mapsto c', d \mapsto d'$.

Here we are interested in finding the number of non-isomorphic graphs on a given set of n vertices. For this purpose we may confine ourselves to graphs with the vertex set $V = \{1, 2, \ldots, n\}$. The first step is to observe that a graph $\Gamma = (V, E)$ is determined by its *edge function*

$$\alpha_\Gamma : V^{[2]} \to \{0, 1\}$$

where $V^{[2]}$ is the set of all 2-element sets $\{u, v\}$, with $u \neq v$ in V, and

$$\alpha_\Gamma(\{u, v\}) = \begin{cases} 0 & \text{if } (u, v) \notin E \\ 1 & \text{if } (u, v) \in E. \end{cases}$$

Now the symmetric group S_n acts on the vertex set $V = \{1, 2, \ldots, n\}$ in the natural way and this leads to an action of S_n on $V^{[2]}$ in which

$$\pi \cdot \{u, v\} = \{\pi(u), \pi(v)\}$$

where $\pi \in S_n$. It follows that S_n acts on the set of possible edge functions for V,

$$F = \text{Fun}(V^{[2]}, \{0, 1\}).$$

It is a consequence of the definition of an isomorphism of graphs that graphs $\Gamma_1 = (V, E_1)$ and $\Gamma_2 = (V, E_2)$ are isomorphic if and only if there exists a $\pi \in S_n$

5.4 Applications to combinatorics – counting labellings and graphs

such that $\pi \cdot \alpha_{\Gamma_1} = \alpha_{\Gamma_2}$, i.e., α_{Γ_1} and α_{Γ_2} belong to the same S_n-orbit of F. So we have to count the S_n-orbits of F. Of course (5.4.1) applies to this situation and we can therefore deduce:

(5.4.2) *The number of non-isomorphic graphs with n vertices is given by*

$$g(n) = \frac{1}{n!}\left(\sum_{\pi \in S_n} 2^{m(\pi)}\right)$$

where $m(\pi)$ is the number of disjoint cycles present in the permutation of $V^{[2]}$ induced by π.

To use this result one must be able to compute $m(\pi)$. While formulas for $m(\pi)$ are available, we shall be content to calculate these numbers directly for small values of n.

Example (5.4.3) Show that there are exactly 11 non-isomorphic graphs with 4 vertices.

All we have to do is compute $m(\pi)$ for π of each cycle type in S_4. Note that $|V^{[2]}| = \binom{4}{2} = 6$. Of course $m(1) = 6$. If π is a 4-cycle, say $(1\,2\,3\,4)$, there are two cycles in the permutation of $V^{[2]}$ produced by π, namely $(\{1, 2\}, \{2, 3\}, \{3, 4\}, \{4, 1\})$ and $(\{1, 3\}, \{2, 4\})$; thus $m(\pi) = 2$. Also there are six 4-cycles in S_4.

If π is a 3-cycle, say $(1\,2\,3)(4)$, there are two cycles, $(\{1, 2\}, \{2, 3\}, \{1, 3\})$ and $(\{1, 4\}, \{2, 4\}, \{3, 4\})$, so $m(\pi) = 2$; there are eight such 3-cycles.

If π involves two 2-cycles, say $\pi = (1\,2)(3\,4)$, there are four cycles $(\{1, 2\})$, $(\{3, 4\}), (\{1, 3\}, \{2, 4\}), (\{1, 4\}, \{2, 3\})$; so $m(\pi) = 4$. There are three such π's.

Finally, there are six transpositions π and it is easy to see that $m(\pi) = 4$. The formula in (5.4.2) therefore yields

$$g(4) = \frac{1}{4!}\left(2^6 + 6 \cdot 2^2 + 8 \cdot 2^2 + 3 \cdot 2^4 + 6 \cdot 2^4\right) = 11.$$

This result can be verified by actually enumerating the graphs.

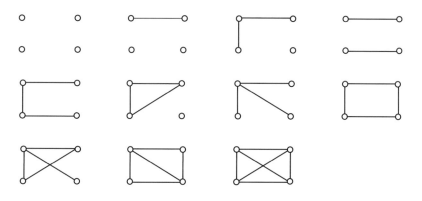

Exercises (5.4)

1. Show that there are $\frac{1}{10}c(c^2+1)(c^2+4)$ ways to label the vertices of a regular pentagon using c labels.

2. The same problem for the *edges* of the pentagon.

3. A baton has n bands of equal width. Show that there are $\frac{1}{2}(c^n + c^{[\frac{n+1}{2}]})$ ways to color it using c colors. [The baton can be rotated through $180°$.]

4. The faces of a regular tetrahedron are to be painted using c different colors. Prove that there are $\frac{1}{12}c^2(c^2+11)$ ways to do it.

5. A necklace has p beads of identical shape and size where p is an odd prime number. Beads of c colors available. How many necklace designs are possible?

6. How many ways are there to place eight identical checkers on an 8×8 chessboard of squares if rotation of the board is allowed?

7. The number of isomorphism types of graph with n vertices is at most $2^{n(n-1)/2}$.

8. Show that there are four isomorphism types of graph with three vertices.

9. Show that there are 34 isomorphism types of graph with five vertices.

10. The number of ways to design a necklace with n beads of c different colors is

$$\frac{1}{2n}\left(\sum_{\substack{i \geq 1 \\ i \mid n}} \phi(i) c^{\frac{n}{i}}\right) + \frac{1}{4}(c^{[\frac{n+1}{2}]} + c^{[\frac{n+2}{2}]}),$$

where ϕ is Euler's function.

Chapter 6
Introduction to rings

A *ring* is a set with two binary operations called *addition* and *multiplication*, subject to a number of natural requirements. Thus from the logical point of view, a ring is a more complex object than a group, which is a set with a single binary operation. Yet some of the most familiar mathematical objects are rings – for example, the sets of integers, real polynomials, continuous functions. For this reason some readers may feel more at home with rings than with groups, despite the increased complexity in logical structure. One motivation for the study of rings is to see how far properties of the ring of integers extend to rings in general.

6.1 Definition and elementary properties of rings

A *ring* is a triple
$$(R, +, \times)$$
where R is a set and $+$ and $\times$ are binary operations on R, called *addition* and *multiplication* such that the following properties hold:

(i) $(R, +)$ is an abelian group;

(ii) $(R, \times)$ is a semigroup;

(iii) the left and right distributive laws hold, i.e.,

$$a(b+c) = ab + ac \quad \text{and} \quad (a+b)c = ac + bc, \quad a, b, c \in R.$$

If in addition the commutative law for multiplication holds,

(iv) $ab = ba$ for all $a, b \in R$,

the ring is said to be *commutative*.

If R contains an element $1_R \neq 0_R$ such that $1_R a = a = a 1_R$ for all $a \in R$, then R is called a *ring with identity* and 1_R is the (clearly unique) *identity element* of R. Care must be taken to distinguish between the *additive identity* (or zero element) 0_R, which exists in any ring R, and the multiplicative identity 1_R in a ring R with identity. These will often be written simply 0 and 1. As with groups, we usually prefer to mention only the underlying set R and speak of "the ring R", rather than the triple $(R, +, \times)$.

Examples of rings. There are many familiar examples of rings at hand.

(i) $\mathbb{Z}$, $\mathbb{Q}$, $\mathbb{R}$, $\mathbb{C}$ are commutative rings with identity where the ring operations are the usual addition and multiplication of arithmetic.

(ii) Let m be a positive integer. Then $\mathbb{Z}_m$, the set of congruence classes modulo m, is a commutative ring with identity where the ring operations are addition and multiplication of congruence classes – see 2.3.

(iii) The set of all continuous real-valued functions defined on the interval $[0, 1]$ is a ring when addition and multiplication are given by $f + g(x) = f(x) + g(x)$ and $fg(x) = f(x)g(x)$. This is a commutative ring in which the identity element is the constant function 1.

(iv) Let R be any ring with identity and define $M_n(R)$ to be the set of all $n \times n$ matrices with entries in R. The usual rules for adding and multiplying matrices are to be used. Then $M_n(R)$ is a ring with identity. It is not hard to see that $M_n(R)$ is commutative if and only if R is commutative and $n = 1$.

Of course the ring axioms must be verified in all these examples, but this is a very straightforward matter.

Rings of polynomials. Now we would like to introduce rings of polynomials, which are probably the most fruitful source of rings. Indeed rings of polynomials occupy a position in ring theory similar to that of permutation groups in group theory.

First we must give a clear definition of a polynomial, not involving vague terms like "indeterminate". In essence a polynomial is just the list of its coefficients, of which only finitely many can be non-zero. We proceed on the basis of this. Let R be a ring with identity. A *polynomial over* R is a sequence of elements of R, one for each natural number,

$$f = (a_0, a_1, a_2, \ldots)$$

such that $a_i = 0$ for all but a finite number of i; the a_i are called the *coefficients* of f. The *zero polynomial* is $(0_R, 0_R, 0_R, \ldots)$. If $f = (a_0, a_1, \ldots)$ is not zero, there is a largest integer i such that $a_i \neq 0$; thus $f = (a_0, a_1, \ldots, a_i, 0, 0, \ldots)$. Then i is called the *degree* of f, in symbols

$$\deg(f).$$

It is convenient to assign to the zero polynomial the degree $-\infty$. A polynomial whose degree is ≤ 0, i.e., one of the form $(a_0, 0, 0, \ldots)$, is called a *constant* polynomial and will be written a_0.

The definitions of addition and multiplication of polynomials are just the familiar rules from high school algebra, but adapted to the current notation. Let $f = (a_0, a_1, \ldots)$ and $g = (b_0, b_1, \ldots)$ be polynomials over R; then we define their sum and product to be

$$f + g = (a_0 + b_0, a_1 + b_1, \ldots, a_i + b_i, \ldots)$$

and

$$fg = \left(a_0b_0, a_0b_1 + a_1b_0, a_0b_2 + a_1b_1 + a_2b_0, \ldots, \sum_{j=0}^{n} a_j b_{n-j}, \ldots\right)$$

Here one needs to pause and convince oneself that these are really polynomials, i.e., that all but a finite number of their coefficients are 0.

(6.1.1) *If f and g are polynomials over a ring with identity, then $f + g$ and fg are polynomials. Also*

$$\deg(f+g) \leq \max\{\deg(f), \deg(g)\} \quad \text{and} \quad \deg(fg) \leq \deg(f) + \deg(g).$$

This follows quickly from the definitions. It is also quite routine to verify that the ring axioms hold for polynomials with these binary operations. Thus we have:

(6.1.2) *If R is a ring with identity, then so is the ring of all polynomials over R.*

Of course, the identity element here is the constant polynomial $(1_R, 0_R, 0_R, \ldots)$.

Next we would like to recover the traditional notation for polynomials "in an indeterminate t". This may be accomplished as follows. Let t denote the polynomial $(0, 1, 0, 0, \ldots)$; then the product rule shows that $t^2 = (0, 0, 1, 0, \ldots)$, $t^3 = (0, 0, 0, 1, 0, \ldots)$ etc. If we define the *multiple* of a polynomial by a ring element r by the natural rule $r(a_0, a_1, \ldots) = (ra_0, ra_1, \ldots)$, then it follows that

$$(a_0, a_1, \ldots, a_n, 0, 0, \ldots) = a_0 + a_1 t + \cdots + a_n t^n.$$

Thus we can return with confidence to the traditional notation for polynomials knowing that it is soundly based. The ring of polynomials in t over R will be written

$$R[t].$$

Polynomial rings in more than one indeterminate are defined recursively by the equation

$$R[t_1, \ldots, t_n] = (R[t_1, \ldots, t_{n-1}])[t_n],$$

where $n > 1$. A typical element of $R[t_1, \ldots, t_n]$ is a multinomial expression

$$\sum_{\ell_1, \ldots, \ell_n} r_{\ell_1 \ldots \ell_n} t_1^{\ell_1} \ldots t_n^{\ell_n}.$$

This is a finite sum with $r_{\ell_1 \ldots \ell_n} \in R$, formed over all non-negative integers ℓ_i.

We list next some elementary and frequently used consequences of the ring axioms.

(6.1.3) *Let R be any ring. Suppose that a, b are elements of R and that n in an integer. Then*

(i) $a0 = 0 = 0a$;

(ii) $a(-b) = (-a)b = -(ab)$;

(iii) $(na)b = n(ab) = a(nb)$.

Proof. (i) By the distributive law $a(0+0) = a0 + a0$. Hence $a0 = a0 + a0$ and so $a0 = 0$ after cancellation. Similarly $0a = 0$.

(ii) $a(-b) + ab = a(-b+b) = a0 = 0$. Thus $a(-b) = -(ab)$. Similarly $(-a)b = -(ab)$.

(iii) First assume that $n \geq 0$; then $(na)b = n(ab)$ by an easy induction on n. Next $(-na)b + nab = (-na + na)b = 0b = 0$, so $(-na)b = -(nab)$. Similarly $a(-nb) = -(nab)$. □

Units in rings. Suppose that R is a ring with identity. An element r is called a *unit* of R if it has a *multiplicative inverse*, i.e., an element $s \in R$ such that $rs = 1 = sr$. Notice that 0 cannot be a unit since $0s = 0 \neq 1$ for all $s \in S$ by (6.1.3). Also, if r is a unit, then it has a *unique* inverse, written r^{-1} – this follows from (3.2.1)(iii).

Now suppose that r_1 and r_2 are two units of R. Then $r_1 r_2$ is also a unit since $(r_1 r_2)^{-1} = r_2^{-1} r_1^{-1}$, as is seen by forming the relevant products. Also of course $(r^{-1})^{-1} = r$, so that r^{-1} is a unit. Since 1 is certainly a unit, we can state:

(6.1.4) *If R is a ring with identity, the set of units of R is a multiplicative group $U(R)$ where the group operation is ring multiplication.*

Here are some simple examples of groups of units.

Example (6.1.1) (i) $U(\mathbb{Z}) = \{\pm 1\}$, a group of order 2.

(ii) $U(\mathbb{Q}) = \mathbb{Q}\setminus\{0\}$, i.e., the multiplicative group of non-zero rational numbers.

(iii) If $m > 0$, then $U(\mathbb{Z}_m)$ is the multiplicative group $\mathbb{Z}_m^*$ of all congruence classes $[i]_m$ where $\gcd(i, m) = 1$. This is an abelian group of order $\phi(m)$.

(iv) $U(\mathbb{R}[t])$ is the group of non-zero constant polynomials. For if $fg = 1$, then f and g must both be constant.

Exercises (6.1)

1. Which of the following are rings?

   (a) The sets of even and odd integers, with the usual arithmetic operations;
   (b) the set of all differentiable functions on [0, 1] where $f + g(x) = f(x) + g(x)$ and $fg(x) = f(x)g(x)$;
   (c) the set of all singular 2×2 real matrices, with the usual matrix operations.

2. Let S be a non-empty set. Define two binary operations on the power set $P(S)$ by $X + Y = (X \cup Y) - (X \cap Y)$ and $X \cdot Y = X \cap Y$. Prove that $(P(S), +, \cdot)$ is a commutative ring with identity. Show also that $X^2 = X$ and $2X = 0_{P(S)}$.

3. A ring R is called *Boolean* if $r^2 = r$ for all $r \in R$, (cf. Exercise 6.1.2). If R is a Boolean ring, prove that $2r = 0$ and that R is necessarily commutative.

4. Let A be an arbitrary (additively written) abelian group. Prove that A is the underlying additive group of some commutative ring.

5. Find the unit groups of the following rings:
 (a) $\{\frac{m}{2^n} \mid m, n \in \mathbb{Z}\}$; (b) $M_n(\mathbb{R})$ with the standard matrix operations; (c) the ring of continuous functions on $[0, 1]$.

6. Prove that the Binomial Theorem is valid in any commutative ring R, i.e., $(a+b)^n = \sum_{i=0}^{n} \binom{n}{i} a^i b^{n-i}$ where $a, b \in R$ and n is a non-negative integer.

7. Let R be a ring with identity. Suppose that a is an element of R with a *unique left inverse* b, i.e., b is the unique element in R such that $ba = 1$. Prove that $ab = 1$, so that a is a unit. [Hint: consider the element $ab - 1 + b$.]

8. Let R be a ring with identity. Explain how to define a *formal power series* over R of the form $\sum_{n=0}^{\infty} a_n t^n$ with $a_n \in R$. Then verify that these form a ring with identity with respect to appropriate sum and product operations. (This is called the *ring of formal power series in t over R*).

6.2 Subrings and ideals

In Chapter 3 the concept of a subgroup of a group was introduced, and already this has proved to be valuable in the study of groups and in applications. We aim to pursue a similar course for rings by introducing subrings.

Let $(R, +, \times)$ be a ring and let S be a subset of the underlying set R. Then S is called a *subring* of R if $(S, +_S, \times_S)$ is a ring where $+_S$ and $\times_S$ denote the binary operations $+$ and $\times$ when restricted to S. In particular S is a subgroup of the additive group $(R, +)$.

From (3.3.3) we deduce a more practical description of a subring.

(6.2.1) *Let S be a subset of a ring R. Then S is a subring of R precisely when S contains 0_R and is closed with respect to addition, multiplication and the formation of negatives, i.e., if $a, b \in S$, then $a + b \in S$, $ab \in S$ and $-a \in S$.*

Example (6.2.1) (i) $\mathbb{Z}, \mathbb{Q}, \mathbb{R}$ are successively larger subrings of the ring of complex numbers $\mathbb{C}$.

(ii) The set of even integers $2\mathbb{Z}$ is a subring of $\mathbb{Z}$. Note that it does not contain the identity element; this is not a requirement for a subring.

(iii) In any ring R there are the *zero subring* 0 and the *improper* subring R itself.

(iv) Let $S = \frac{1}{2}\mathbb{Z}$, i.e., $S = \{\frac{m}{2} \mid m \in \mathbb{Z}\}$. Then S is an additive subgroup of the ring $\mathbb{Q}$, but it is not a subring since $\frac{1}{2} \times \frac{1}{2} = \frac{1}{4} \notin S$. Thus the concept of a subring is more special than that of an additive subgroup.

Ideals. It is reasonable to expect there to be an analogy between groups and rings in which subgroups correspond to subrings. The question then arises: what is to correspond to normal subgroups? This is where ideals enter the picture.

An *ideal* of a ring R is an additive subgroup I such that $ir \in I$ and $ri \in I$ whenever $i \in I$ and $r \in R$. So an ideal is an additive subgroup which is closed with respect to multiplication of its elements by *arbitrary ring elements* on the left and the right. In particular *every ideal is a subring*.

Example (6.2.2) (i) Let R be a commutative ring and let $x \in R$. Define

$$Rx$$

to be the subset $\{rx \mid r \in R\}$. Then Rx is an ideal of R. For in the first place it is a subgroup since $r_1 x + r_2 x = (r_1 + r_2)x$ and $-(rx) = (-r)x$; also $s(rx) = (sr)x$ for all $s \in R$, so Rx is an ideal. An ideal of this type is called a *principal ideal*.

(ii) Every subgroup of $\mathbb{Z}$ has the form $n\mathbb{Z}$ where $n \geq 0$ by (4.1.5). Hence every subgroup of $\mathbb{Z}$ is an ideal.

(iii) On the other hand, $\mathbb{Z}$ is a subring but not an ideal of $\mathbb{Q}$ since $\frac{1}{2}(1) \notin \mathbb{Z}$. So *subrings are not always ideals*. Thus we have a hierarchy of distinct substructures of rings:

$$\text{ideal} \Rightarrow \text{subring} \Rightarrow \text{subgroup}.$$

Homomorphisms of rings. It is still not clear why ideals as defined above are the true analogs of normal subgroups. The decisive test of the appropriateness of our definition will come when homomorphisms of rings have been defined. If we are right, the kernel of a homomorphism will be an ideal.

It is fairly obvious how one should define a *homomorphism from a ring R to a ring S*: this is a function $\theta : R \to S$ which respects the ring operations in the sense that:

$$\theta(a+b) = \theta(a) + \theta(b) \quad \text{and} \quad \theta(ab) = \theta(a)\theta(b)$$

for all $a, b \in R$. So in particular θ is a homomorphism of groups.

If in addition θ is bijective, θ is called an *isomorphism of rings*. If there is an isomorphism from ring R to ring S, then R and S are said to be *isomorphic rings*, in symbols

$$R \simeq S.$$

Example (6.2.3) (i) Let m be a positive integer. Then the function $\theta_m : \mathbb{Z} \to \mathbb{Z}_m$ defined by $\theta_m(x) = [x]_m$ is a ring homomorphism. This is a consequence of the way sums and products of congruence classes were defined.

(ii) The *zero homomorphism* $0 : R \to S$ sends every $r \in R$ to 0_S. Also the *identity isomorphism* from R to R is just the identity function on R.

(iii) As a more interesting example, consider the set R of matrices of the form

$$\begin{bmatrix} a & b \\ -b & a \end{bmatrix}, \quad (a, b \in \mathbb{R}).$$

These are easily seen to form a subring of the matrix ring $M_2(\mathbb{R})$. Now define a function

$$\theta : R \to \mathbb{C}$$

by the rule $\theta\left(\begin{bmatrix} a & b \\ -b & a \end{bmatrix}\right) = a + ib$ where $i = \sqrt{-1}$. Then θ is a ring homomorphism: indeed

$$\begin{bmatrix} a_1 & b_1 \\ -b_1 & a_1 \end{bmatrix} \begin{bmatrix} a_2 & b_2 \\ -b_2 & a_2 \end{bmatrix} = \begin{bmatrix} a_1a_2 - b_1b_2 & a_1b_2 + a_2b_1 \\ -a_1b_2 - a_2b_1 & a_1a_2 - b_1b_2 \end{bmatrix},$$

which is mapped by θ to $(a_1a_2 - b_1b_2) + i(a_1b_2 + a_2b_1)$, i.e., to $(a_1 + ib_1)(a_2 + ib_2)$. An easier calculation shows that θ sends

$$\begin{bmatrix} a_1 & b_1 \\ -b_1 & a_1 \end{bmatrix} + \begin{bmatrix} a_2 & b_2 \\ -b_2 & a_2 \end{bmatrix}$$

to $(a_1 + ib_1) + (a_2 + ib_2)$.

Certainly θ is surjective; it is also injective since $a + ib = 0$ implies that $a = 0 = b$. Therefore θ is an isomorphism and we obtain the interesting fact that $R \simeq \mathbb{C}$. Thus complex numbers can be represented by real 2×2 matrices. In fact this is one way to define complex numbers.

Next let us consider the nature of the kernel and image of a ring homomorphism. The following result should be compared with (4.3.2).

(6.2.2) *If $\theta : R \to S$ is a homomorphism of rings, then $\mathrm{Ker}(\theta)$ is an ideal of R and $\mathrm{Im}(\theta)$ is a subring of S.*

Proof. We know already that $\mathrm{Ker}(\theta)$ and $\mathrm{Im}(\theta)$ are subgroups by (4.3.2). Let $k \in \mathrm{Ker}(\theta)$ and $r \in R$. Then $\theta(kr) = \theta(k)\theta(r) = 0_S$ and $\theta(rk) = \theta(r)\theta(k) = 0_S$ since $\theta(k) = 0_S$. Therefore $\mathrm{Ker}(\theta)$ is an ideal of R. Furthermore $\theta(r_1)\theta(r_2) = \theta(r_1r_2)$, so that $\mathrm{Im}(\theta)$ is a subring of S. □

(6.2.3) *If $\theta : R \to S$ is an isomorphism of rings, then so is $\theta^{-1} : S \to R$.*

Proof. We know from (4.3.3) that θ^{-1} is an isomorphism of groups. It still needs to be shown that $\theta^{-1}(s_1 s_2) = \theta^{-1}(s_1)\theta^{-1}(s_2)$, $(s_i \in S)$. Observe that the image of each side under θ is $s_1 s_2$. Since θ is injective, it follows that $\theta^{-1}(s_1 s_2) = \theta^{-1}(s_1)\theta^{-1}(s_2)$. □

Quotient rings. Reassured that ideals are the natural ring theoretic analog of normal subgroups, we proceed to define the quotient of a ring by an ideal. Suppose that I is an ideal of a ring R. Then I is a subgroup of the additive abelian group R, so $I \triangleleft R$ and we can form the quotient group R/I. This is an additive abelian group whose elements are the cosets of I. To make R/I into a ring, a rule for multiplying cosets must be specified: the obvious one to try is

$$(r_1 + I)(r_2 + I) = r_1 r_2 + I, \quad (r_i \in R).$$

To see that this is well-defined, let $i_1, i_2 \in I$ and note that

$$(r_1 + i_1)(r_2 + i_2) = r_1 r_2 + (r_1 i_2 + i_1 r_2 + i_1 i_2) \in r_1 r_2 + I$$

since I is an ideal. Thus the rule is independent of the choice of representatives r_1 and r_2. A further straightforward check shows that the ring axioms hold; therefore R/I is a ring, the *quotient ring* of I in R. Note also that the assignment $r \mapsto r + I$ is a surjective ring homomorphism from R to R/I with kernel I; this is the *canonical homomorphism*, (cf. 4.3).

As one would expect, there are isomorphism theorems for rings similar to those for groups.

(6.2.4) (First Isomorphism Theorem) *If $\theta : R \to S$ is a homomorphism of rings, then $R/\operatorname{Ker}(\theta) \simeq \operatorname{Im}(\theta)$.*

(6.2.5) (Second Isomorphism Theorem) *If I is an ideal and S a subring of a ring R, then $S + I$ is a subring and $S \cap I$ is an ideal of S. Also $S + I/I \simeq S/S \cap I$.*

(6.2.6) (Third Isomorphism Theorem) *Let I and J be ideals of a ring R with $I \subseteq J$. Then J/I is an ideal of R/I and $(R/I)/(J/I) \simeq R/J$.*

Fortunately we can use the proofs of the isomorphism theorems for groups – see (4.3.4), (4.3.5), (4.3.6). The isomorphisms constructed in these proofs still stand if we allow for the additive notation. One has only to check that they are isomorphisms of rings.

For example, take the case of (6.2.4). From (4.3.4) we know that the assignment $r + \operatorname{Ker}(\theta) \mapsto \theta(r)$ yields a group isomorphism $\alpha : R/\operatorname{Ker}(\theta) \to \operatorname{Im}(\theta)$. Since

$$\alpha((r_1 + \operatorname{Ker}(\theta))(r_2 + \operatorname{Ker}(\theta))) = \alpha(r_1 r_2 + \operatorname{Ker}(\theta)) = \theta(r_1 r_2),$$

which equals

$$\theta(r_1)\theta(r_2) = \alpha(r_1 + \operatorname{Ker}(\theta))\alpha(r_2 + \operatorname{Ker}(\theta)),$$

we conclude that α is an isomorphism of rings: this proves (6.2.4). It is left to the reader to complete the proofs of the other two isomorphism theorems.

(6.2.7) (Correspondence Theorem for subrings and ideals) *Let I be an ideal of a ring R. Then the assignment $S \mapsto S/I$ determines a bijection from the set of subrings of R that contain I to the set of subrings of R/I. Furthermore S/I is an ideal of R/I if and only if S is an ideal of R.*

Proof. The correspondence between subgroups described in (4.2.2) applies here. All one has to do is verify that S is a subring (ideal) if and only if S/I is. Again we leave it to the reader to fill in the details. □

Exercises (6.2)

1. Classify each of the following subsets of a ring R as an additive subgroup, subring or ideal, *as is most appropriate*:
    (a) $\{f \in \mathbb{R}[t] \mid f(a) = 0\}$ where $R = \mathbb{R}[t]$ and $a \in \mathbb{R}$ is fixed;
    (b) the set of twice differentiable functions on $[0, 1]$ which satisfy the differential equation $f'' + f' = 0$: here R is the ring of continuous functions on $[0, 1]$;
    (c) $n\mathbb{Z}$ where $R = \mathbb{Z}$;
    (d) $\frac{1}{2}\mathbb{Z}$ where $R = \mathbb{Q}$.
    (e) the set of $n \times n$ real matrices with zero first row where $R = M_n(\mathbb{R})$.

2. The intersection of a set of subrings (ideals) is a subring (respectively ideal).

3. Use Exercise (6.2.2) to define the subring (respectively ideal) generated by a non-empty subset of a ring.

4. Let X be a non-empty subset of R, a commutative ring with identity. Describe the form of a typical element of: (a) the subring of R generated by X; (b) the ideal of R generated by X.

5. Let R be a ring with identity. If I is an ideal containing a unit, then $I = R$.

6. Let I and J be ideals of a ring R such that $I \cap J = 0$. Prove that $ab = 0$ for all $a \in I, b \in J$.

7. Let $a \in \mathbb{R}$ and define $\theta_a : \mathbb{R}[t] \to \mathbb{R}$ by $\theta_a(f) = f(a)$. Prove that θ_a is a ring homomorphism. Identify $\text{Im}(\theta_a)$ and $\text{Ker}(\theta_a)$.

8. Let $\alpha : R \to S$ be a surjective ring homomorphism and suppose that R has an identity element and S is not a zero ring. Prove that S has an identity element.

6.3 Integral domains, division rings and fields

The purpose of this section is to introduce some special types of ring with desirable properties. Specifically we are interested in rings having a satisfactory theory of division. For this reason we wish to exclude the phenomenon in which the product of two non-zero ring elements is zero.

If R is a ring, a *left zero divisor* is a non-zero element a such that $ab = 0$ for some $b \neq 0$ in R. Of course b is called a *right zero divisor*. Clearly the presence of zero divisors makes it difficult to envision a reasonable theory of division.

Example (6.3.1) Let n be a positive integer. The zero divisors in $\mathbb{Z}_n$ are the congruence classes $[m]$ where m and n are not relatively prime and $1 < m < n$. Thus $\mathbb{Z}_n$ has zero divisors if and only if n is not a prime.

For, if m and n are not relatively prime and $d > 1$ is a common divisor of m and n, then $[m][\frac{n}{d}] = [\frac{m}{d}][n] = [0]$ since $[n] = [0]$, while $[\frac{n}{d}] \neq [0]$; thus $[m]$ is a zero divisor. Conversely, suppose that $[m][\ell] = [0]$ where $[\ell] \neq [0]$. Then $n \mid m\ell$; thus, if m and n are relatively prime, $n \mid \ell$ and $[\ell] = [0]$ by Euclid's Lemma. This contradiction shows that m and n cannot be relatively prime.

Next we introduce an important class of rings with no zero divisors. An *integral domain* (or more briefly a *domain*) is a commutative ring with identity which has no zero divisors. Thus $\mathbb{Z}$ is a domain, and by Example (6.3.1) $\mathbb{Z}_n$ is a domain if and only if n is a prime. Domains can be characterized by a cancellation property.

(6.3.1) *Let R be a commutative ring with identity. Then R is a domain if and only if the cancellation law is valid in R, i.e., $ab = ac$ and $a \neq 0$ always imply that $b = c$.*

Proof. If $ab = ac$ and $b \neq c$, $a \neq 0$, then $a(b - c) = 0$, so that a is a zero divisor and R is not a domain. Conversely, if R is not a domain and $ab = 0$ with $a, b \neq 0$, then $ab = a0$, so the cancellation law fails. □

The next result tells us that when we are working with polynomials, life is much simpler if the coefficient ring is a domain.

(6.3.2) *Let R be an integral domain and let $f, g \in R[t]$. Then*

$$\deg(fg) = \deg(f) + \deg(g).$$

Hence $fg \neq 0$ if $f \neq 0$ and $g \neq 0$, and thus $R[t]$ is an integral domain.

Proof. If $f = 0$, then $fg = 0$ and $\deg(f) = -\infty = \deg(fg)$; hence the formula is valid in this case. Assume that $f \neq 0$ and $g \neq 0$, and let at^m and bt^n be the terms of highest degree in f and g respectively; here $a \neq 0$ and $b \neq 0$. Then $fg = abt^{m+n}$ + terms of lower degree, and $ab \neq 0$ since R is a domain. Therefore $fg \neq 0$ and $\deg(fg) = m + n = \deg(f) + \deg(g)$. □

Recall that a unit in a ring with identity is an element with a multiplicative inverse. A ring with identity in which every non-zero element is a unit is called a *division ring*. Commutative division rings are called a *fields*. Clearly $\mathbb{Q}$, $\mathbb{R}$ and $\mathbb{C}$ are examples of fields, while $\mathbb{Z}$ is not a field. Fields are one of the most frequently used types of ring since the ordinary operations of arithmetic can be performed in a field. Note that *a division ring cannot have zero divisors*: for if $ab = 0$ and $a \neq 0$, then $a^{-1}ab = a^{-1}0 = 0$, so that $b = 0$. Thus two types of ring in which zero divisors cannot occur are integral domains and division rings.

6.3 Integral domains, division rings and fields

The ring of quaternions. The reader may have noticed that the examples of division rings given so far are commutative, i.e., they are fields. We will now describe a famous example of a non-commutative division ring, the *ring of Hamilton's*[1] *quaternions*.

First of all consider the following three 2×2 matrices over $\mathbb{C}$:

$$I = \begin{bmatrix} i & 0 \\ 0 & -i \end{bmatrix}, \quad J = \begin{bmatrix} 0 & 1 \\ -1 & 0 \end{bmatrix}, \quad K = \begin{bmatrix} 0 & i \\ i & 0 \end{bmatrix}$$

where $i = \sqrt{-1}$. These are known in physics as the *Pauli*[2] *Spin Matrices*. Simple matrix computations show that the following relations hold:

$$I^2 = J^2 = K^2 = -1,$$

and

$$IJ = K = -JI, \quad JK = I = -KJ, \quad KI = J = -IK.$$

Here 1 denotes the identity 2×2 matrix and it should be distinguished from I. If a, b, c, d are rational numbers, we may form the matrix

$$a1 + bI + cJ + dK = \begin{bmatrix} a+bi & c+di \\ -c+di & a-bi \end{bmatrix}.$$

This is called a *rational quaternion*. Let R be the set of all rational quaternions. Then R is a subring containing the identity of the matrix ring $M_2(\mathbb{C})$; for

$$(a1 + bI + cJ + dK) + (a'1 + b'I + c'J + d'K)$$
$$= (a+a')1 + (b+b')I + (c+c')J + (d+d')K,$$

while $(a1 + bI + cJ + dK)(a'1 + b'I + c'J + d'K)$ equals

$$(aa' - bb' - cc' - dd')1 - (ab' + a'b + cd' - c'd)I$$
$$+ (ac' + a'c + b'd - bd')J + (ad' + a'd + bc' - b'c)K,$$

as may be seen by multiplying out and using the properties of I, J, K above.

The interesting fact about the ring R is that if $0 \neq Q = a1 + bI + cJ + dK$, then Q is a non-singular matrix. For

$$\det(Q) = \begin{vmatrix} a+bi & c+di \\ -c+di & a-bi \end{vmatrix} = a^2 + b^2 + c^2 + d^2 \neq 0,$$

and so by elementary matrix algebra Q has as its inverse

$$\frac{1}{\det(Q)} \begin{bmatrix} a-bi & -c-di \\ c-di & a+bi \end{bmatrix},$$

which is also a quaternion. This allows us to state:

---
[1]William Rowan Hamilton (1805–1865)
[2]Wolfgang Ernst Pauli (1900–1958)

(6.3.3) *The ring of rational quaternions is a non-commutative division ring.*

Notice that the ring of quaternions is infinite. This is no accident since by a famous theorem of Wedderburn[3] *a finite division ring is a field*. This result will not be proved here. However we will prove the corresponding statement for domains, which is much easier.

(6.3.4) *A finite integral domain is a field.*

Proof. Let R be a finite domain and let $0 \neq r \in R$; we need to show that r is a unit. Consider the function $\alpha : R \to R$ defined by $\alpha(x) = rx$. Now α is injective since $rx = ry$ implies that $x = y$ by (6.3.1). However R is a *finite* set, so it follows that α must also be surjective. Hence $1 = rx$ for some $x \in R$, and $x = r^{-1}$. □

We shall now consider the role of ideals in commutative ring theory. First we show that the presence of proper non-zero ideals is counter-indicative for the existence of units.

(6.3.5) *Let R be a commutative ring with identity. Then the set of non-units of R is equal to the union of all the proper ideals of R.*

Proof. Suppose that r is not a unit of R; then $Rx = \{rx \mid x \in R\}$ is a proper ideal containing r since $1 \notin Rx$. Conversely, if a unit r were to belong to a proper ideal I, then for any x in R we would have $x = (xr^{-1})r \in I$, showing that $I = R$. Thus no unit can belong to a proper ideal. □

Recalling that fields are exactly the commutative rings with identity in which each non-zero element is a unit, we deduce:

Corollary (6.3.6) *A commutative ring with identity is a field if and only if it has no proper non-zero ideals.*

Maximal ideals and prime ideals. Let R be a commutative ring with identity. A *maximal ideal* of R is a proper ideal I such that the only ideals containing I are I itself and R. Thus maximal means *maximal proper*. For example, if p is a prime, $p\mathbb{Z}$ is a maximal ideal of $\mathbb{Z}$: for $|\mathbb{Z}/p\mathbb{Z}| = p$ and the Correspondence Theorem (6.2.7) shows that no ideal can occur strictly between $p\mathbb{Z}$ and $\mathbb{Z}$.

A related concept is that of a prime ideal. If R is a commutative ring with identity, a *prime ideal* of R is a proper ideal with the property: $ab \in I$ implies that $a \in I$ or $b \in I$, where $a, b \in R$.

There are enlightening characterizations of prime and maximal ideals in terms of quotient rings.

---

[3] Joseph Henry Maclagan Wedderburn (1881–1948).

6.3 Integral domains, division rings and fields

(6.3.7) *Let I be a proper ideal of a commutative ring with identity R.*

(i) *I is a prime ideal of R if and only if R/I is an integral domain;*

(ii) *I is a maximal ideal of R if and only if R/I is a field.*

Proof. (i) Let $a, b \in R$; then $ab \in I$ if and only if $(a + I)(b + I) = I = 0_{R/I}$. Thus I is prime exactly when R/I has no zero divisors, i.e., it is a domain.

(ii) By the Correspondence Theorem for ideals, I is maximal in R if and only if R/I has no proper non-zero ideals; by (6.3.6) this property is equivalent to R/I being a field. □

Since every field is a domain, we deduce:

Corollary (6.3.8) *Every maximal ideal of a commutative ring with identity is a prime ideal.*

On the other hand, prime ideals need not be maximal. Indeed, if R is any domain, the zero ideal is certainly prime, but it will not be maximal unless R is a field. More interesting examples of non-maximal prime ideals can be constructed in polynomial rings.

Example (6.3.2) Let $R = \mathbb{Q}[t_1, t_2]$, the ring of polynomials in t_1, t_2 with rational coefficients. Let I be the subset of all polynomials in R which are multiples of t_1. Then I is a prime ideal of R, but it is not maximal.

For consider the function $\alpha : R \to \mathbb{Q}[t_2]$ which carries a polynomial $f(t_1, t_2)$ to $f(0, t_2)$. Then α is a surjective ring homomorphism. Now if $f(0, t_2) = 0$, then f is a multiple of t_1, and hence the kernel is I. By (6.2.4) $R/I \simeq \mathbb{Q}[t_2]$. Since $\mathbb{Q}[t_2]$ is a domain but not a field, it follows from (6.3.7) that I is a prime ideal of R which is not maximal.

The characteristic of an integral domain. Suppose that R is a domain; we wish to consider $S = \langle 1 \rangle$, the additive subgroup of R generated by 1. Suppose for the moment that S is finite, with order n say. Then we claim that n must be a prime. For suppose that $n = n_1 n_2$ where $n_i \in \mathbb{Z}$ and $1 < n_i < n$. Then $0 = n1 = (n_1 n_2)1 = (n_1 1)(n_2 1)$ by (6.1.3). However R is a domain, so $n_1 1 = 0$ or $n_2 1 = 0$, which shows that n divides n_1 or n_2, a contradiction. Therefore n is a prime.

This observation is the essence of:

(6.3.9) *Let R be an integral domain and put $S = \langle 1 \rangle$. Then either S is infinite or else it has prime order p. In the latter event $pa = 0$ for all $a \in R$.*

To prove the last statement, simply note that $pa = (p1_R)a = 0a = 0$.

Again let R be an integral domain. If $\langle 1_R \rangle$ has prime order p, then R is said to have *characteristic p*. The other possibility is that $\langle 1_R \rangle$ is infinite, in which case R is said to have *characteristic* 0. Thus the characteristic of R,

$$\text{char}(R),$$

is either 0 or a prime. For example, $\mathbb{Z}_p$ and $\mathbb{Z}_p[t]$ are domains with characteristic p, while $\mathbb{Q}, \mathbb{R}$ and $\mathbb{R}[t]$ all have characteristic 0.

The field of fractions of an integral domain. Suppose that F is a field and R is a subring of F containing 1_F. Then R is a domain since there cannot be zero divisors in F. Conversely, one may ask if every domain arises in this way as a subring of a field. We shall answer the question positively by showing how to construct the *field of fractions of a domain*. It will be helpful for the reader to keep in mind that the procedure to be described is a generalization of the way in which the rational numbers are constructed from the integers.

Let R be any integral domain. We have first to decide how to define a fraction over R. Consider the set

$$S = \{(a, b) \mid a, b \in R, \ b \neq 0\}.$$

Here a will correspond to the numerator and b to the denominator of the fraction. We introduce a binary relation $\sim$ on S which will allow for cancellation between numerator and denominator:

$$(a_1, b_1) \sim (a_2, b_2) \Leftrightarrow a_1 b_2 = a_2 b_1.$$

Of course this is motivated by a familiar arithmetic rule: $\frac{m_1}{n_1} = \frac{m_2}{n_2}$ if and only if $m_1 n_2 = m_2 n_1$.

Next we verify that $\sim$ is an equivalence relation on S. Only transitivity requires a comment: suppose that $(a_1, b_1) \sim (a_2, b_2)$ and $(a_2, b_2) \sim (a_3, b_3)$; then $a_1 b_2 = a_2 b_1$ and $a_2 b_3 = a_3 b_2$. Multiply the first equation by b_3 and use the second equation to derive $a_1 b_3 b_2 = a_2 b_3 b_1 = a_3 b_2 b_1$. Cancel b_2 to obtain $a_1 b_3 = a_3 b_1$; thus $(a_1, b_1) \sim (a_3, b_3)$. Now define a *fraction* over R to be a $\sim$-equivalence class

$$\frac{a}{b} = [(a, b)]$$

where $a, b \in R, b \neq 0$. Note that $\frac{ac}{bc} = \frac{a}{b}$ since $(a, b) \sim (ac, bc)$; thus cancellation can be performed in a fraction.

Let F denote the set of all fractions over R. We would like to make F into a ring. To this end define addition and multiplication in R by the rules

$$\frac{a}{b} + \frac{a'}{b'} = \frac{ab' + a'b}{bb'} \quad \text{and} \quad \left(\frac{a}{b}\right)\left(\frac{a'}{b'}\right) = \frac{aa'}{bb'}.$$

Here again we have been guided by the arithmetic rules for adding and multiplying fractions. It is necessary to show that these operations are well-defined, i.e., there is no dependence on the chosen representative (a, b) of the equivalent class $\frac{a}{b}$. For example, let $(a, b) \sim (c, d)$ and $(a', b') \sim (c', d')$. Then $(ab' + a'b, bb') \sim (cd' + c'd, dd')$; for

$$(ab' + a'b)dd' = ab'dd' + a'bdd' = bcb'd' + b'c'bd = (cd' + c'd)bb'.$$

The next step is to verify the ring axioms; as an example let us check the validity of the distributive law

$$\left(\frac{a}{b} + \frac{c}{d}\right)\left(\frac{e}{f}\right) = \left(\frac{a}{b}\right)\left(\frac{e}{f}\right) + \left(\frac{c}{d}\right)\left(\frac{e}{f}\right),$$

leaving the reader to verify the other axioms. By definition

$$\left(\frac{a}{b}\right)\left(\frac{e}{f}\right) + \left(\frac{c}{d}\right)\left(\frac{e}{f}\right) = \frac{ae}{bf} + \frac{ce}{df} = \frac{aedf + cebf}{bd\,f^2} = \frac{ade + bce}{bd\,f},$$

which equals

$$\left(\frac{ad + bc}{bd}\right)\left(\frac{e}{f}\right) = \left(\frac{a}{b} + \frac{c}{d}\right)\left(\frac{e}{f}\right),$$

as claimed.

After all the axioms have been checked we will know that F is a ring; note that the zero element of F is $0_F = \frac{0_R}{1_R}$. Clearly F is commutative and it has as its identity element $1_F = \frac{1_R}{1_R}$. Furthermore, if $a, b \neq 0$,

$$\left(\frac{a}{b}\right)\left(\frac{b}{a}\right) = \frac{ab}{ab} = \frac{1_R}{1_R} = 1_F,$$

so that, as expected, the inverse of $\frac{a}{b}$ is $\frac{b}{a}$. Therefore F is a field, the *field of fractions* of the domain R.

In order to relate F to R we introduce a natural function

$$\theta : R \to F$$

defined by $\theta(a) = \frac{a}{1}$. It is quite simple to check that θ is an injective ring homomorphism. Therefore $R \simeq \mathrm{Im}(\theta)$ and of course $\mathrm{Im}(\theta)$ is a subring of F containing 1_F. Thus the original domain is isomorphic with a subring of a field. Our conclusions are summed up in the following result.

(6.3.10) *Let R be an integral domain and F the set of all fractions over R with the addition and multiplication specified above. Then F is a field and the assignment $a \mapsto \frac{a}{1}$ determines is an injective ring homomorphism from R to F.*

Example (6.3.3) (i) When $R = \mathbb{Z}$, the field of fractions is, up to isomorphism, the field of rational numbers $\mathbb{Q}$. This example motivated the general construction.

(ii) Let K be any field and put $R = K[t]$; this is a domain by (6.3.2). The field of fractions F of R is the *field of rational functions* in t over K; these are formally quotients of polynomials in t over K

$$\frac{f}{g}$$

where $f, g \in R$, $g \neq 0$. The notation

$$K\{t\}$$

is often used for the field of rational functions in t over K.

Exercises (6.3)

1. Find all the zero divisors in the following rings: $\mathbb{Z}_6$, $\mathbb{Z}_{15}$, $\mathbb{Z}_2[t]$, $\mathbb{Z}_4[t]$, $M_n(\mathbb{R})$.

2. Let R be a commutative ring with identity such that the degree formula $\deg(fg) = \deg(f) + \deg(g)$ is valid in $R[t]$. Prove that R is a domain.

3. A *left ideal* of a ring R is an additive subgroup L such that $ra \in L$ whenever $r \in R$ and $a \in L$. *Right ideals* are defined similarly. If R is a division ring, prove that the only left ideals and right ideals are 0 and R.

4. Let R be a ring with identity. If R has no left or right ideals except 0 and R, then R is a division ring.

5. Let $\theta : D \to R$ be a non-zero ring homomorphism. If D is a division ring, show that it must be isomorphic with some subring of R.

6. Let I and J be non-zero ideals of a domain. Prove that $I \cap J \neq 0$.

7. Let I be the principal ideal $(\mathbb{Z}[t])t$ of $\mathbb{Z}[t]$. Prove that I is prime but not maximal.

8. The same problem for $I = (\mathbb{Z}[t])(t^2 - 2)$.

9. Let F be a field. If $a, b \in F$ and $a \neq 0$, define a function $\theta_{a,b} : F \to F$ by the rule $\theta_{a,b}(x) = ax + b$. Prove that the set of all $\theta_{a,b}$'s is a group with respect to functional composition.

10. Let F be the field of fractions of a domain R and let $\alpha : R \to F$ be the canonical injective homomorphism $r \mapsto \frac{r}{1}$. Suppose that $\beta : R \to K$ is an injective ring homomorphism into some other field K. Prove that there is an injective homomorphism $\theta : F \to K$ such that $\theta\alpha = \beta$. (So in a sense F is the smallest field containing an isomorphic copy of R.)

Chapter 7
Division in rings

Our aim in this chapter is to construct a theory of division in rings that mirrors the familiar theory of division in the ring of integers $\mathbb{Z}$. To simplify matters let us agree to restrict attention to commutative rings – in non-commutative rings questions of left and right divisibility arise. Also, remembering from 6.3 the unpleasant phenomenon of zero divisors, we shall further restrict ourselves to integral domains. In fact even this class of rings is too wide, although it provides a reasonable target for our theory. For this reason we will introduce some well-behaved types of domains.

7.1 Euclidean domains

Let R be a commutative ring with identity and let $a, b \in R$. Then a is said to *divide* b in R, in symbols
$$a \mid b,$$
if $ac = b$ for some $c \in R$. From the definition there follow very easily some elementary facts about division.

(7.1.1) *Let R be a commutative ring with identity, and let a, b, c, x, y be elements of R. Then*

(i) $a \mid a$ and $a \mid 0$ for all $a \in R$;

(ii) $0 \mid a$ if and only if $a = 0$;

(iii) *if $a \mid b$ and $b \mid c$, then $a \mid c$, (so division is a transitive relation);*

(iv) *if $a \mid b$ and $a \mid c$, then $a \mid bx + cy$ for all x, y;*

(v) *if u is a unit, then $u \mid a$ for all $a \in R$;*

(vi) *if u is a unit, then $a \mid u$ if and only if a is a unit.*

For example, taking the case of (iv), we have $b = ad$ and $c = ae$ for some $d, e \in R$. Then $bx + cy = a(dx + ey)$, so that a divides $bx + cy$. The other proofs are equally simple exercises that are left to the reader.

One situation we must expect to encounter is a pair of ring elements each of which divides the other: such elements are called *associates*.

(7.1.2) *Let R be an integral domain and let $a, b \in R$. Then $a \mid b$ and $b \mid a$ if and only if $b = au$ where u is a unit of R.*

Proof. In the first place, $a \mid au$ and if u is a unit, then $a = (au)u^{-1}$, so $au \mid a$. Conversely, assume that $a \mid b$ and $b \mid a$. If $a = 0$, then $b = 0$, and the statement is certainly true. So let $a \neq 0$. Now $a = bc$ and $b = ad$ for some $c, d \in R$. Therefore $a = bc = adc$, and by the cancellation law (6.3.1), $dc = 1$, so that d is a unit. □

Thus two elements of a domain are associates precisely when one is a unit times the other. For example, two integers a and b are associates if and only if $b = \pm a$.

Irreducible elements. Let R be a commutative ring with identity. An element a of R is called *irreducible* if it is neither 0 nor a unit and if its only divisors are units and associates of a, i.e., the elements that we know must divide a. Thus irreducible elements have as few divisors as possible.

Example (7.1.1) (i) The irreducible elements of $\mathbb{Z}$ are the prime numbers and their negatives.

(ii) A field has no irreducible elements since every non-zero element is a unit.

(iii) If F is a field, the irreducible elements of the polynomial ring $F[t]$ are the so-called *irreducible polynomials*, i.e., the non-constant polynomials which are not expressible as a product of polynomials of lower degree.

Almost every property of division in the integers depends ultimately on the Division Algorithm. Thus it will be difficult to develop a satisfactory theory of division in a ring unless some version of this algorithm is valid for the ring. This motivates us to introduce a special class of domains, the so-called Euclidean domains.

A domain R is called *Euclidean* if there is a function

$$\delta : R - \{0_R\} \to \mathbb{N}$$

with the following properties:

(i) $\delta(a) \leq \delta(ab)$ if $0 \neq a, b \in R$;

(ii) if $a, b \in R$ and $b \neq 0$, there exist $q, r \in R$ such that $a = bq + r$ where either $r = 0$ or $\delta(r) < \delta(b)$.

The standard example of a Euclidean domain is $\mathbb{Z}$ where δ the absolute value function, i.e., $\delta(a) = |a|$. Note that property (i) holds since $|ab| = |a| \cdot |b| \geq |a|$ if $b \neq 0$. Of course (ii) is just the usual statement of the Division Algorithm for $\mathbb{Z}$.

New and important examples of Euclidean domains are given by the next result.

(7.1.3) *If F is a field, the polynomial ring $F[t]$ is a Euclidean domain where the associated function δ is given by $\delta(f) = \deg(f)$.*

7.1 Euclidean domains

Proof. We already know from (6.3.2) that $R = F[t]$ is a domain. Also, by the same result, if $f, g \neq 0$, then $\delta(fg) = \deg(fg) = \deg(f) + \deg(g) \geq \deg(f) = \delta(f)$. Hence property (i) is valid. To prove the validity of (ii), put $S = \{f - gq \mid q \in R\}$. If $0 \in S$, then $f = gq$ for some q, and we may take r to be 0. Assuming that $0 \notin S$, we note that every element of S has degree ≥ 0, so by the Well-Ordering Principle there exists $r \in S$ with smallest degree, say $r = f - gq$ where $q \in R$. Thus $f = gq + r$.

Now suppose that $\deg(r) \geq \deg(g)$. Write $g = at^m + \cdots$ and $r = bt^n + \cdots$ where $m = \deg(g), n = \deg(r), 0 \neq a, b \in F$ and dots indicate terms of lower degree in t. Since $m \leq n$, we may form the polynomial

$$s = r - (a^{-1}bt^{n-m})g \in R.$$

Now the term bt^n cancels in s, so either $s = 0$ or $\deg(s) < n$. But

$$s = f - (q + a^{-1}bt^{n-m})g,$$

so that $s \in S$ and hence $s \neq 0$, which contradicts the choice of r. Therefore $\deg(r) < \deg(g)$, as required. □

A less obvious example of a Euclidean domain is the ring of Gaussian integers. A *Gaussian integer* is a complex number of the form

$$u + iv$$

where $u, v \in \mathbb{Z}$ and of course $i = \sqrt{-1}$. It is easily seen that the Gaussian integers form a subring of $\mathbb{C}$ containing 1, and hence constitute a domain.

Example (7.1.2) The ring R of Gaussian integers is a Euclidean domain.

Proof. In this case we define a function $\delta : R - \{0\} \to \mathbb{N}$ by the rule

$$\delta(u + iv) = |u + iv|^2 = u^2 + v^2.$$

We argue that δ satisfies the two requirements for a Euclidean domain. In the first place, if $0 \neq a, b \in R$, then $\delta(ab) = |ab|^2 = |a|^2|b|^2 \geq |a|^2$ since $|b| \geq 1$.

Verifying the second property is a little harder. First write $ab^{-1} = u' + iv'$ where u', v' are rational numbers. Now choose integers u and v as close as possible to u' and v' respectively; then $|u - u'| \leq \frac{1}{2}$ and $|v - v'| \leq \frac{1}{2}$. Next

$$a = b(u' + iv') = b(u + iv) + b(u'' + iv'')$$

where $u'' = u' - u$ and $v'' = v' - v$. Finally, let $q = u + iv$ and $r = b(u'' + iv'')$. Then $a = bq + r$; also $q \in R$ and hence $r = a - bq \in R$. If $r \neq 0$, then

$$\delta(r) = |b|^2|u'' + iv''|^2 = |b|^2(u''^2 + v''^2) \leq |b|^2\left(\frac{1}{4} + \frac{1}{4}\right) = \frac{1}{2}|b|^2 < |b|^2 = \delta(b)$$

since $|u''| \leq \frac{1}{2}$ and $|v''| \leq \frac{1}{2}$. Therefore $\delta(r) < \delta(b)$ as required. □

Exercises (7.1)

1. Complete the proof of (7.1.1).

2. Identify the irreducible elements in the following rings:
   (a) the ring of rational numbers with odd denominators;
   (b) $\mathbb{Z}[t]$.

3. Let R be a commutative ring with identity. If R has no irreducible elements, show that either R is a field or there exists an infinite strictly increasing chain of principal ideals $I_1 \subset I_2 \subset \cdots$ in R.

4. Let $R = F[[t]]$ be the ring of formal power series in t over a field F, (see Exercise (6.1.8)). Prove that the irreducible elements of R are those of the form tf where $f \in R$ and $f(0) \neq 0$.

5. Let $f = t^5 - 3t^2 + t + 1$, $g = t^2 + t + 1$ belong to $\mathbb{Q}[t]$. Find q, r such that $f = gq + r$ and $\deg(r) \leq 1$.

6. Let R be a Euclidean domain with associated function $\delta : R - \{0\} \to \mathbb{N}$.
   (a) Show that $\delta(a) \geq \delta(1)$ for all $a \neq 0$ in R;
   (b) If a is a unit of R, prove that $\delta(a) = \delta(1)$;
   (c) Conversely, show that if $\delta(a) = \delta(1)$, then a is a unit of R.

7. Prove that $t^3 + t + 1$ is irreducible in $\mathbb{Z}_2[t]$, but $t^3 + t^2 + t + 1$ is reducible.

7.2 Principal ideal domains

Let R be a commutative ring with identity. If $r \in R$, recall from Example (6.2.2) that the subset $Rr = \{rx \mid x \in R\}$ is an ideal of R containing r, called a *principal ideal*. Often one writes simply

$$(r)$$

for Rr. If every ideal of R is principal, then R is a *principal ideal ring*. A domain in which every ideal is principal is called a *principal ideal domain* or PID: these form an extremely important class of domains. For example $\mathbb{Z}$ is a PID since every ideal, being a cyclic subgroup, has the form $\mathbb{Z}n$ where $n \geq 0$

A good source of PID's is indicated by the next result.

(7.2.1) *Every Euclidean domain is a principal ideal domain.*

Proof. Let R be a Euclidean domain with associated function $\delta : R - 0 \to \mathbb{N}$ and let I be an ideal of R; we need to show that I is principal. Now if I is the zero ideal, then $I = (0)$, so I is principal. Assume that $I \neq 0$. We apply the Well-Ordering Law to pick an x in $I - 0$ such that $\delta(x)$ is minimal. Then certainly $(x) \subseteq I$; our claim is that also $I \subseteq (x)$. To substantiate, this let $y \in I$ and write $y = xq + r$ for some $q, r \in R$ where either $r = 0$ or $\delta(r) < \delta(x)$. This is possible since δ is the associated function for the Euclidean domain R. If $r = 0$, then $y = xq \in (x)$. Otherwise $\delta(r) < \delta(x)$; but then $r = y - xq \in I$, which contradicts the choice of x in $I - 0$. Therefore $I = (x)$. □

From (7.1.3) and (7.2.1) we deduce the very important result:

Corollary (7.2.2) *If F is a field, then $F[t]$ is a principal ideal domain.*

Another example of a PID obtained from (7.2.1) and Example (7.1.2) is the ring of Gaussian integers.

Our next objective is to show that PID's have good division properties, even although there may be not be a division algorithm available.

Greatest common divisors. Let a, b be elements in a domain R. A *greatest common divisor* (or gcd) of a and b is a ring element d such that the following hold:

(i) $d \mid a$ and $d \mid b$;

(ii) if $c \mid a$ and $c \mid b$ for some $c \in R$, then $c \mid d$.

The definition here has been carried over directly from the integers.

Notice that if d and d' are two gcd's of a, b, then $d \mid d'$ and $d' \mid d$, so that d and d' are associate. Thus by (7.1.2) $d' = du$ with u a unit of R. Hence gcd's are unique only up to a unit. Of course in the case of $\mathbb{Z}$, where the units are ± 1, we are able to make gcd's unique by insisting that they be positive. But this course of action is not possible in a general domain since there is no concept of positivity.

In general there is no reason to expect that gcd's exist in a domain. However the situation is very satisfactory for PID's.

(7.2.3) *Let a and b be elements of a principal ideal domain R. Then a and b have a greatest common divisor d which has the form $d = ax + by$ with $x, y \in R$.*

Proof. Define $I = \{ax + by \mid x, y \in R\}$ and observe that I is an ideal of R. Hence $I = (d)$ for some $d \in I$, with $d = ax + by$ say. If $c \mid a$ and $c \mid b$, then $c \mid ax + by = d$ by (7.1.1). Also $a \in I = (d)$, so that $d \mid a$, and similarly $d \mid b$. Hence d is a gcd of a and b. □

Elements a and b of a domain R are called *relatively prime* if 1 is a gcd of a and b. Of course this means that $ax + by = 1$ for some $x, y \in R$

(7.2.4) (Euclid's Lemma) *Let a, b, c be elements of a principal ideal domain and assume that $a \mid bc$ where a and b are relatively prime. Then $a \mid c$.*

Corollary (7.2.5) *If R is a principal ideal domain and $p \mid bc$ where $p, b, c \in R$ and p is irreducible, then $p \mid b$ or $p \mid c$.*

The proofs of these results are exactly the same as those given in 2.2 for $\mathbb{Z}$. The reader should supply the details.

Maximal ideals in principal ideal domains. In a PID the maximal ideals and the prime ideals coincide and admit a nice description in terms of irreducible elements.

(7.2.6) *Let I be a non-zero ideal of a principal ideal domain R. Then the following statements about I are equivalent:*

(i) *I is maximal;*

(ii) *I is prime;*

(iii) *$I = (p)$ where p is an irreducible element of R.*

Proof. (i) $\Rightarrow$ (ii). This was proved in (6.3.8).

(ii) $\Rightarrow$ (iii). Assume that I is prime. Since R is a PID, $I = (p)$ for some $p \in R$. Note that p cannot be a unit since $I \neq R$. Suppose that $p = ab$ where neither a nor b is associate to p. Now $ab \in I$ and I is prime, so that $a \in I$ or $b \in I$, i.e., $p \mid a$ or $p \mid b$. Since we also have $a \mid p$ and $b \mid p$, we obtain the contradiction that a or b is associate to p. This shows that p is irreducible.

(iii) $\Rightarrow$ (i). Assume that $I = (p)$ with p irreducible, and let $I \subseteq J \subseteq R$ where J is an ideal of R. Then $J = (x)$ for some $x \in R$, and $p \in (x)$, so that $x \mid p$. This means that either x is a unit or else it is associate to p, i.e., $J = R$ or $J = I$. Therefore I is maximal as claimed. □

Corollary (7.2.7) *Let F be a field. Then the maximal ideals of $F[t]$ are exactly those of the form (f) where f is an irreducible polynomial which is monic, (i.e., its leading coefficient is 1).*

This is because $F[t]$ is a PID by (7.2.2) and the irreducible elements of $F[t]$ are just the irreducible polynomials. The corollary provides us with an important method of constructing a field from an irreducible polynomial $f \in F[t]$ since $F[t]/(f)$ is a field: this will be exploited in 7.4 below.

The ascending chain condition on ideals. We conclude this section by noting a useful property of the ideals of a PID which will be crucial in the following section when we address the question of unique factorization.

A ring R satisfies the *ascending chain condition on ideals* if there exist no infinite strictly ascending chains $I_1 \subset I_2 \subset \ldots$ of ideals of R. Obviously any finite ring satisfies this condition. More interesting for our purposes is:

(7.2.8) *Every principal ideal domain satisfies the ascending chain condition on ideals.*

Proof. Let R be a PID and assume there is an infinite ascending chain of ideals of R, say $I_1 \subset I_2 \subset \cdots$. Put $U = \bigcup_{j=1,2,\ldots} I_j$; then it is easy to verify that U is also an ideal of R. Hence $U = (r)$ for some r since R is a PID. Now $r \in U$, so $r \in I_j$ for some j. Thus $U = (r) \subseteq I_j$, which delivers the contradiction $I_{j+1} = I_j$. Consequently R satisfies the ascending chain condition for ideals. □

For example, $\mathbb{Z}$ and $F[t]$, where F is a field, satisfy the ascending chain condition since these rings are PID's.

Exercises (7.2)

1. Show that $\mathbb{Z}[t]$ is not a PID.

2. Show that $F[t_1, t_2]$ is not a PID for any field F.

3. Let R be a commutative ring with identity. If $R[t]$ is a PID, prove that R must be a field.

4. Prove that Euclid's Lemma is valid for a PID.

5. Let $f = t^3 + t + 1 \in \mathbb{Z}_2[t]$. Show that $\mathbb{Z}_2[t]/(f)$ is field and find its order.

6. Prove that the ring of rational numbers with odd denominators is a PID.

7. Prove that $F[[t]]$, the ring of formal power series in t over a field F, is a PID and describe its ideals.

7.3 Unique factorization in integral domains

In the present section we will look for domains in which there is unique factorization in terms of irreducible elements. Our model here is the Fundamental Theorem of Arithmetic (2.2.7), which asserts that such factorizations exist in $\mathbb{Z}$. First we need to clarify what is meant by uniqueness of factorization.

Let R be a domain and let S denote the set of all irreducible elements in R, which could of course be empty. Observe that "being associate to" is an equivalence relation on S, so that S splits up into equivalence classes. Choosing one element from each equivalence class, we form a subset C of S. (Strictly speaking this procedure involves the Axiom of Choice – see 12.1). Now observe that the set C has the following properties:

(i) every irreducible element of R is associate to some element of C;

(ii) distinct elements of C are not associate.

A subset C with these properties is called a *complete set of irreducibles for R*. We have just proved the following fact.

(7.3.1) *Every integral domain has a (possibly empty) complete set of irreducible elements.*

Our interest in complete sets of irreducibles stems from the observation that if we are to have *unique* factorization in terms of irreducibles, then only irreducibles from a complete set can be used: otherwise there will be different factorizations of the type $ab = (ua)(u^{-1}b)$ where a, b are irreducible and u is a unit.

An integral domain R is called a *unique factorization domain*, or UFD, if there exists a complete set of irreducibles C for R such that each non-zero element a of R has an expression of the form $a = u p_1 p_2 \ldots p_k$ where u is a unit and $p_i \in C$, and furthermore this expression is unique up to order of the factors.

At present the one example of a UFD we know is $\mathbb{Z}$, where we can take C to be the set of all prime numbers. The next theorem provides us with many more examples.

(7.3.2) *Every principal ideal domain is a unique factorization domain.*

Proof. Let R be a PID and let C be any complete set of irreducibles of R. We will prove that there is unique factorization for elements of R in terms of units and irreducibles in C. This will be accomplished in three steps, the first of which establishes the existence of irreducibles when R contains non-zero non-unit elements, i.e., R is not a field.

(i) *If a is a non-zero, non-unit element of R, then a is divisible by at least one irreducible element of R.*

Suppose this is false. Then a itself must be reducible, so $a = a_1 a_1'$ where a_1 and a_1' are non-units, and $(a) \subseteq (a_1)$. Also $(a) \neq (a_1)$; for otherwise $a_1 \in (a)$, so that $a \mid a_1$, as well as $a_1 \mid a$, and by (7.1.2) this implies that a_1' is a unit. Therefore $(a) \subset (a_1)$.

Next a_1 cannot be irreducible since $a_1 \mid a$. Thus $a_1 = a_2 a_2'$ where a_2, a_2' are non-units, and it follows that $(a_1) \subset (a_2)$ by the argument just given. Continuing in this way, we recognize that the procedure cannot terminate: for otherwise we would find an irreducible divisor of a. So there is an infinite strictly ascending chain of ideals $(a) \subset (a_1) \subset (a_2) \subset \cdots$; but this is impossible by (7.2.8).

(ii) *If a is a non-zero, non-unit element of R, then a is a product of irreducibles.*

Again suppose this is false. By (i) there is an irreducible p_1 dividing a, with $a = p_1 a_1$ say. Now a_1 cannot be a unit, so there is an irreducible p_2 dividing a_1, with say $a_1 = p_2 a_2$ and $a = p_1 p_2 a_2$, and so on indefinitely. Notice that $(a) \subset (a_1) \subset (a_2) \subset \cdots$ is a strictly ascending infinite chain of ideals, which again contradicts (7.2.8).

(iii) *If a is a non-zero element of R, then a is the product of a unit and irreducible elements in C.*

This is clear if a is a unit – no irreducibles are needed. Otherwise by (ii) a is a product of irreducibles, each of which is associate to an element of C. The result now follows at once.

The final step in the proof will establish uniqueness.

(iv) Suppose that
$$a = u p_1 \ldots p_k = v q_1 \ldots q_\ell$$
where u, v are units of R and $p_i, q_j \in C$. Argue by induction on k: if $k = 0$, then $a = u$, a unit, so $\ell = 0$ and $u = v$. Now assume that $k > 0$.

Since $p_1 \mid a = v q_1 \ldots q_\ell$, Euclid's Lemma tells us that p_1 must divide one of $q_1, \ldots, q_\ell$. Relabelling the q_j's, we may assume that $p_1 \mid q_1$. Thus p_1 and q_1 are two associate members of C, which can only mean that $p_1 = q_1$. Hence, on cancelling p_1, we obtain $a' = u p_2 \ldots p_k = v q_2 \ldots q_\ell$. By the induction hypothesis $k - 1 = \ell - 1$, so that $k = \ell$ and, after further relabelling, $p_i = q_i$ for $i = 2, 3, \ldots, k$, and finally $u = v$. Therefore uniqueness has been established. □

Corollary (7.3.3) *If F is any field, then $F[t]$ is a UFD.*

This is because $F[t]$ is a PID by (7.2.2). The natural choice for a complete set of irreducibles in $F[t]$ is the set of all monic irreducible polynomials. So we have unique factorization in $F[t]$ in terms of constants and monic irreducible polynomials.

A further example of a UFD is the ring of Gaussian integers,

$$\{a + b\sqrt{-1} \mid a, b \in \mathbb{Z}\}:$$

this a Euclidean domain and hence a PID. However some quite similar domains are not UFD's.

Example (7.3.1) Let R be the subring of $\mathbb{C}$ consisting of all $a+b\sqrt{-3}$ where $a, b \in \mathbb{Z}$. Then R is not a UFD.

First observe that ± 1 are the only units of R. This is because

$$(a + b\sqrt{-3})^{-1} = \frac{1}{a^2 + 3b^2}(a - b\sqrt{-3}) \in R$$

if and only if $\frac{a}{a^2+3b^2}$ and $\frac{b}{a^2+3b^2}$ are integers: this happens only when $b = 0$ and $\frac{1}{a} \in \mathbb{Z}$, i.e., $a = \pm 1$. It follows that no two of the elements $2, 1 + \sqrt{3}, 1 - \sqrt{3}$ are associate.

Next we claim that $2, 1 + \sqrt{3}, 1 - \sqrt{3}$ are irreducible elements of R. Fortunately all three elements can be handled simultaneously. Suppose that

$$(a + \sqrt{-3}b)(c + \sqrt{-3}d) = 1 \pm \sqrt{3} \text{ or } 2$$

where $a, b, c, d \in \mathbb{Z}$. Taking the modulus squared of both sides, we obtain $(a^2 + 3b^2)(c^2 + 3d^2) = 4$ in every case. But this implies that $a^2 = 1$ and $b = 0$ or $c^2 = 1$ and $d = 0$, i.e., $a + \sqrt{-3}b$ or $c + \sqrt{-3}d$ is a unit.

Finally unique factorization fails because

$$4 = 2 \cdot 2 = (1 + \sqrt{-3})(1 - \sqrt{-3})$$

and $2, 1 + \sqrt{-3}, 1 - \sqrt{-3}$ are non-associate irreducibles. It follows that R is not a UFD.

Two useful properties of UFD's are recorded in the next result.

(7.3.4) *Let R be a UFD. Then*

(i) *gcd's exist in R;*

(ii) *Euclid's Lemma holds in R.*

Proof. (i) Let $a = up_1^{e_1} \ldots p_k^{e_k}, b = vp_1^{f_1} \ldots p_k^{f_k}$ where u, v are units of R, $p_1, \ldots, p_k$ are members of a complete set of irreducibles for R, and $e_i, f_i \geq 0$. Define $d = p_1^{g_1} \ldots p_k^{g_k}$ where g_i is the minimum of e_i and f_i. Then it is quickly verified that d is a gcd of a and b. For $d \mid a$ and $d \mid b$, and on the other hand, if $c \mid a$ and $c \mid b$, the unique factorization property shows that c will have the form $wp_1^{h_1} \ldots p_k^{h_k}$ where w is a unit and $0 \leq h_i \leq g_i$. Hence $c \mid d$.

(ii) The proof is left to the reader as an exercise □

Although polynomial rings in more than one variable over a field are not PID's – see Exercise (7.2.2)) – they are in fact UFD's. It is our aim in the remainder of the section to prove this important result.

Primitive polynomials and Gauss's Lemma. Let R be a UFD and let $0 \neq f \in R[t]$. Recalling that gcd's exist in R by (7.3.4), we form the gcd of the coefficients of f; this is called the *content* of f,

$$c(f).$$

Keep in mind that content is unique only up to a unit of R, and equations involving content have to be interpreted in that light. If $c(f) = 1$, i.e., $c(f)$ is a unit, the polynomial f is said to be *primitive*. For example $2 + 4t - 3t^3 \in \mathbb{Z}[t]$ is a primitive polynomial. We will now establish two useful results about the content of polynomials.

(7.3.5) *Let $0 \neq f \in R[t]$ where R is a UFD. Then $f = cf_0$ where $c = c(f)$ and f_0 is primitive in $R[t]$.*

Proof. Write $f = a_0 + a_1 t + \cdots + a_n t^n$; then $c(f) = \gcd\{a_0, \ldots, a_n\} = c$, say. Write $a_i = cb_i$ with $b_i \in R$, and put $f_0 = b_0 + b_1 t + \cdots + b_n t^n \in R[t]$. Thus $f = cf_0$. If $d = \gcd\{b_0, b_1, \ldots, b_n\}$, then $d \mid b_i$ and so $cd \mid cb_i = a_i$. Since c is the gcd of the a_i, it follows that cd divides c and hence that d is a unit and f_0 is primitive. □

(7.3.6) *Let R be a UFD and let f, g be non-zero polynomials over R. Then $c(fg) = c(f)c(g)$. In particular, if f and g are primitive, then so is fg.*

Proof. Consider first the special case where f and g are primitive. If fg is not primitive, then $c(fg)$ is not a unit and it must be divisible by some irreducible element p of R. Write $f = \sum_{i=0}^{m} a_i t^i$, $g = \sum_{j=0}^{n} b_j t^j$ and $fg = \sum_{k=0}^{m+n} c_k t^k$. Since f is primitive, p cannot divide all its coefficients, and so there is an integer $r \geq 0$ such that $p \mid a_0, a_1, \ldots, a_{r-1}$ but $p \nmid a_r$. Similarly there is an $s \geq 0$ such that p divides each of $b_0, b_1, \ldots, b_{s-1}$ but p does not divide b_s. Now consider the coefficient c_{r+s} of t^{r+s} in fg; this equals

$$(a_0 b_{r+s} + a_1 b_{r+s-1} + \cdots + a_{r-1} b_{s+1}) + a_r b_s + (a_{r+1} b_{s-1} + \cdots + a_{r+s} b_0).$$

We know that $p \mid c_{r+s}$; also p divides both the expressions in parentheses in the expression above. It follows that $p \mid a_r b_s$. By Euclid's Lemma for UFD's (see (7.3.4)), we must have $p \mid a_r$ or $p \mid b_s$, both of which are impossible. By this contradiction fg is primitive.

Now we are ready for the general case. Using (7.3.5) we write $f = cf_0$ and $g = dg_0$ where $c = c(f)$ and $d = c(g)$, and the polynomials f_0, g_0 are primitive in $R[t]$. Then $fg = cd(f_0g_0)$ and, as has just been shown, f_0g_0 is primitive. In consequence $c(fg) = cd = c(f)c(g)$. □

The next result can be helpful in deciding whether a polynomial is irreducible.

(7.3.7) (Gauss's Lemma) *Let R be a unique factorization domain and let F denote the field of fractions of R. If $f \in R[t]$, then f is irreducible over R if and only if it is irreducible over F.*

Proof. Of course irreducibility over F certainly implies irreducibility over R. It is the converse implication which needs proof. Assume that f is irreducible over R but reducible over F. Here we can assume that f is primitive on the basis of (7.3.5). Then $f = gh$ where $g, h \in F[t]$ are not constant. Since F is the field of fractions of R (and we can assume that $R \subseteq F$), there exist elements $a, b \neq 0$ in R such that $g_1 = ag \in R[t]$ and $h_1 = bh \in R[t]$. Write $g_1 = c(g_1)g_2$ where $g_2 \in R[t]$ is primitive. Then $ag = c(g_1)g_2$, so we can divide both sides by $\gcd\{a, c(g_1)\}$. On these grounds it is permissible to assume that $c(g_1)$ and a are relatively prime, and that the same holds for $c(h_1)$ and b.

Next we have $(ab)f = (ag)(bh) = g_1h_1$. Taking the content of each side and using (7.3.6), we obtain $ab = c(g_1)c(h_1)$ since f is primitive. But $c(g_1)$ and a are relatively prime, so $a \mid c(h_1)$, and for a similar reason $b \mid c(g_1)$. Therefore we have the factorization $f = (b^{-1}g_1)(a^{-1}h_1)$ and now both factors are polynomials over R. But this contradicts the irreducibility of f over R and so the proof is complete. □

For example, to show that a polynomial over $\mathbb{Z}$ is $\mathbb{Q}$-irreducible, it is enough to show that it is $\mathbb{Z}$-irreducible, usually an easier task.

Polynomial rings in several variables. Let us now use the theory of polynomial content to show that unique factorization occurs in polynomial rings with more than one variable. Here it is important to keep in mind that such rings are not PID's and so are not covered by (7.3.2). The main result is:

(7.3.8) *If R is a unique factorization domain, then so is the polynomial ring $R[t_1, \ldots, t_k]$.*

Proof. In the first place we need only prove this when $k = 1$. For if $k > 1$,

$$R[t_1, \ldots, t_k] = R[t_1, \ldots, t_{k-1}][t_k],$$

and induction on k can be used once the case $k = 1$ is settled. From now on we restrict attention to $S = R[t]$. The first step in the proof is to establish:

(a) *Any non-constant polynomial f in S is expressible as a product of irreducible elements of R and primitive irreducible polynomials over R.*

The key idea in the proof is to introduce the field of fractions F of R, and exploit the fact that $F[t]$ is known to be a PID and hence a UFD. First of all write $f = c(f)f_0$ where $f_0 \in S$ is primitive, using (7.3.5). Here $c(f)$ is either a unit or a product of irreducibles of R. So we can assume that f is primitive. Regarding f as an element of the UFD $F[t]$, we write $f = p_1 p_2 \ldots p_m$ where $p_i \in F[t]$ is irreducible over F. Now find $a_i \neq 0$ in R such that $f_i = a_i p_i \in S$. Writing $c(f_i) = c_i$, we have $f_i = c_i q_i$ where $q_i \in R[t]$ is primitive. Hence $p_i = a_i^{-1} f_i = a_i^{-1} c_i q_i$, and q_i is F-irreducible since p_i is. Thus q_i is certainly R-irreducible.

Combining these expressions for p_i, we find that

$$f = (a_1^{-1} \ldots a_m^{-1} c_1 \ldots c_m) q_1 \ldots q_m,$$

and hence $(a_1 \ldots a_m) f = (c_1 \ldots c_m) q_1 \ldots q_m$. Now take the content of both sides of this equation to get $a_1 \ldots a_m = c_1 \ldots c_m$ up to a unit, since f and the q_i are primitive. Consequently $f = u q_1 \ldots q_m$ for some unit u of R. This is what we had to prove.

(b) The next step is to assemble a complete set of irreducibles for S. First take a complete set of irreducibles C_1 for R. Then consider the set of all primitive irreducible polynomials in S. Now being associate is an equivalence relation on this set, so we can choose an element from each equivalence class. This yields a set of non-associate primitive irreducible polynomials C_2. Every primitive irreducible polynomial in $R[t]$ is associate to some element of C_2. Now put

$$C = C_1 \cup C_2.$$

Two distinct elements of C cannot be associate, so C is a complete set of irreducibles for S. If $0 \neq f \in S$, it follows from (a) that f is expressible as a product of elements of C and a unit of R.

(c) There remains the question of uniqueness. Suppose that

$$f = u a_1 \ldots a_k f_1 \ldots f_r = v b_1 \ldots b_\ell g_1 \ldots g_s$$

where u, v are units, $a_x, b_y \in C_1$, $f_i, g_j \in C_2$. By Gauss's Lemma (7.3.7), the f_i and g_j are F-irreducible. Since $F[t]$ is a UFD and C_2 is a complete set of irreducibles for $F[t]$, we conclude that $r = s$ and $f_i = w_i g_i$, (after possible relabelling), where $w_i \in F$. Write $w_i = c_i d_i^{-1}$ where $c_i, d_i \in R$. Then $d_i f_i = c_i g_i$, and on taking contents we find that $c_i = d_i$ up to a unit. This implies that w_i is a unit of R and so f_i, g_i are associate. Hence $f_i = g_i$.

Cancelling the f_i and g_i, we are left with $u a_1 \ldots a_k = v b_1 \ldots b_k$. But R is a UFD with a complete set of irreducibles C_1, so we have $k = \ell$, $u = v$ and $a_i = b_i$ (after further relabelling). This completes the proof. □

This theorem provides us with some important new examples of UFD's.

Corollary (7.3.9) *The following rings are unique factorization domains: $\mathbb{Z}[t_1, \ldots, t_k]$ and $F[t_1, \ldots, t_k]$ where F is a field.*

Exercises (7.3)

1. Let R be a UFD. Prove that R satisfies the ascending chain condition on *principal* ideals, i.e., there does not exist an infinite strictly ascending chain of principal ideals.

2. If R is a UFD and C is *any* complete set of irreducibles for R, show that there is unique factorization in terms of C.

3. If C_1 and C_2 are two complete sets of irreducibles for a domain R, then $|C_1| = |C_2|$.

4. The domain $\{a + b\sqrt{-5} \mid a, b \in \mathbb{Z}\}$ is not a UFD.

5. Prove that $t^3 + at + 1 \in \mathbb{Z}[t]$ is $\mathbb{Q}$-irreducible if and only if $a \neq 0$ or -2.

6. The ring of rational numbers with odd denominators is a UFD: find a complete set of primes for it.

7. The same question for the power series ring $F[[t]]$ where F is a field.

8. Prove that Euclid's Lemma is valid in any UFD.

7.4 Roots of polynomials and splitting fields

Let R be a commutative ring with identity, let $f = b_0 + b_1 t + \cdots + b_n t^n \in R[t]$ and let $a \in R$. Then the *value* of the polynomial f at a is defined to be

$$f(a) = b_0 + b_1 a + \cdots + b_n a^n \in R.$$

So we have a function $\theta_a : R[t] \to R$ which evaluates polynomials at a, i.e., $\theta_a(f) = f(a)$. Now $f + g(a) = f(a) + g(a)$ and $(fg)(a) = f(a)g(a)$, because the ring elements $f(a)$ and $g(a)$ are added and multiplied by the same rules as polynomials f and g. It follows that $\theta_a : R[t] \to R$ is a ring homomorphism. Its kernel consists of all $f \in R[t]$ such that $f(a) = 0$, i.e., all $f \in R[t]$ which have a as a root.

The following criterion for an element to be a root of a polynomial should be familiar from elementary algebra.

(7.4.1) (The Remainder Theorem) *Let R be an integral domain, let $f \in R[t]$ and let $a \in R$. Then a is a root of f if and only if $t - a$ divides f in the ring $R[t]$.*

Proof. If $t - a$ divides f, then $f = (t - a)g$ where $g \in R[t]$; thus $f(a) = 0$, as may be seen by applying the homomorphism θ_a to the equation, i.e., by evaluating the polynomials at a. Hence a is a root of f. Conversely, assume that $f(a) = 0$ and let F denote the field of fractions of R. Then we can divide f by $t - a$ to get a quotient and remainder in $F[t]$, say $f = (t - a)q + r$ where $q, r \in F[t]$ and $\deg(r) < 1$, i.e., r is constant. Here the Division Algorithm for the ring $F[t]$ has been applied, (see (7.1.3)). However notice that the usual long division process tell us that q and r actually belong to $R[t]$. Finally, apply the evaluation homomorphism θ_a to $f = (t-a)q + r$ to obtain $0 = r$ since r is constant. Therefore $t - a$ divides f. □

Corollary (7.4.2) *The kernel of the evaluation homomorphism θ_a is the principal ideal $(t - a)$.*

This is simply a restatement of (7.4.1)

The multiplicity of a root. Let R be a domain and suppose that $f \in R[t]$ is not constant and has a as a root. Now there is a largest positive integer n such that $(t - a)^n \mid f$ since the degree of a divisor of f cannot exceed $\deg(f)$. Then a is said to be a *root of multiplicity n*. If $n > 1$, then a is called a *multiple root* of f.

(7.4.3) *Let R be a domain and let $0 \neq f \in R[t]$ have degree n. Then the sum of the multiplicities of the roots of f in R is at most n.*

Proof. Let a be a root of f, so that $t - a$ divides f and $f = (t - a)g$ where $g \in R[t]$ has degree $n - 1$. By induction on n the sum of the multiplicities of the roots of g is at most $n - 1$. Now a root of f either equals a or else is a root of g. Consequently the sum of the multiplicities of the roots of f is at most $1 + (n - 1) = n$. □

Example (7.4.1) (i) The polynomial $t^2 + 1 \in \mathbb{Q}[t]$ has no roots in $\mathbb{Q}$. Thus the sum of the multiplicities of the roots can be less than the degree.

(ii) Consider the polynomial $t^4 - 1 \in R[t]$ where R is the ring of rational quaternions (see 6.3). Then f has 8 roots in R, namely $\pm 1, \pm I, \pm J, \pm K$. Therefore (7.4.3) does not hold for non-commutative rings, which is another reason to keep our rings commutative.

Next we state another very well-known theorem.

(7.4.4) (The Fundamental Theorem of Algebra) *Let f be a non-zero polynomial of degree n over the field of complex numbers $\mathbb{C}$. Then the sum of the multiplicities of the roots of f in $\mathbb{C}$ equals n, i.e., f is a product of n linear factors over $\mathbb{C}$.*

The proof of this theorem will be postponed until Chapter 11 – see (11.3.6). Despite the name all the known proofs use some form of analysis.

Derivatives. Derivatives are useful in detecting multiple roots of polynomials. Since we are not dealing with polynomials over $\mathbb{R}$ here, limits cannot be used. For this reason we adopt the following formal definition of the *derivative* f' of $f \in R[t]$ where R is any ring with identity. If $f = a_0 + a_1 t + \cdots + a_n t^n$, then

$$f' = a_1 + 2a_2 t + \cdots + n a_n t^{n-1} \in \mathbb{R}[t].$$

On the basis of this definition one can establish the usual rules governing differentiation.

(7.4.5) *Let $f, g \in R[t]$ and $c \in R$ where R is a commutative ring with identity. Then $(f + g)' = f' + g'$, $(cf)' = cf'$ and $(fg)' = f'g + fg'$.*

Proof. Only the last statement will be proved. Write $f = \sum_{i=0}^{m} a_i t^i$ and $g = \sum_{j=0}^{n} b_j t^j$. Then $fg = \sum_{i=0}^{m+n} \left(\sum_{k=0}^{i} a_k b_{i-k} \right) t^i$, so the coefficient of t^{i-1} in $(fg)'$ is $i \left(\sum_{k=0}^{i} a_k b_{i-k} \right)$. On the other hand, the coefficient of t^{i-1} in $f'g + fg'$ is

$$\sum_{k=0}^{i-1} (k+1) a_{k+1} b_{i-k-1} + \sum_{k=0}^{i-1} (i-k) a_k b_{i-k},$$

which equals

$$i a_i b_0 + \sum_{k=0}^{i-2} (k+1) a_{k+1} b_{i-k-1} + i a_0 b_i + \sum_{k=1}^{i-1} (i-k) a_k b_{i-k}.$$

On adjusting the summation in the second sum, this becomes

$$i a_i b_0 + \sum_{k=0}^{i-2} (k+1) a_{k+1} b_{i-k-1} + \sum_{k=0}^{i-2} (i-k-1) a_{k+1} b_{i-k-1} + i a_0 b_i,$$

which reduces to

$$i \left(a_0 b_i + \sum_{k=0}^{i-2} a_{k+1} b_{i-k-1} + a_i b_0 \right) = i \left(\sum_{k=0}^{i} a_k b_{i-k} \right).$$

It follows that $(fg)' = f'g + fg'$. □

Corollary (7.4.6) $(f^\ell)' = \ell f^{\ell-1} f'$ *where ℓ is a positive integer.*

This is proved by induction on ℓ using (7.4.5).

A criterion for a multiple root can now be given.

(7.4.7) *Let R be a domain and let a be a root of $f \in R[t]$ in R. Then a is a multiple root if and only if $t - a$ divides both f and f'.*

Proof. Let ℓ be the multiplicity of the root a. Then $\ell \geq 1$ and $f = (t-a)^\ell g$ where $t - a \nmid g \in R[t]$. Hence $f' = \ell(t-a)^{\ell-1} g + (t-a)^\ell g'$ by (7.4.5) and (7.4.6). If a is a multiple root of f, then $\ell \geq 2$ and $f'(a) = 0$; by (7.4.1) $t - a$ divides f', as well as f.

Conversely, suppose that $t - a \mid f' = \ell(t-a)^{\ell-1} g + (t-a)^\ell g'$. If $\ell = 1$, then $t - a$ divides $g + (t-a) g'$, which implies that $t - a$ divides g, a contradiction. Thus $\ell > 1$ and a is a multiple root. □

130   7 Division in rings

Example (7.4.2) Let F be a field whose characteristic does not divide the positive integer n. Then $t^n - 1$ has no multiple roots in F.

For if $f = t^n - 1$, then $f' = nt^{n-1} \neq 0$ since $\text{char}(F) \nmid n$. Hence $t^n - 1$ and nt^{n-1} are relatively prime, so that f and f' have no non-constant common roots. Therefore f has no multiple roots.

Splitting fields. From now on we will consider roots of polynomials over a field F. If $f \in F[t]$ is not constant, then we know that f has at most $\deg(f)$ roots in F by (7.4.3). Now f need not have any roots in F, as the example of $t^2 + 1$ in $\mathbb{R}[t]$ shows: on the other hand, $t^2 + 1$ has two roots in the larger field $\mathbb{C}$. The question to be addressed is this: can we construct a field K, larger than F in some sense, in which f will have exactly $\deg(f)$ roots, i.e., over which f splits into a product of linear factors? A smallest such field is called a *splitting field* of f. In the case of the polynomial $t^2 + 1 \in \mathbb{R}[t]$, the situation is quite clear; its splitting field is $\mathbb{C}$ since $t^2 + 1 = (t + i)(t - i)$ where $i = \sqrt{-1}$. However for a general field F we do not have a convenient larger field like $\mathbb{C}$ at hand. Thus splitting fields will have to be constructed from scratch.

We begin by reformulating the definition of a splitting field. Let F be a field and f a non-constant polynomial over F. A *splitting field* for f over F is a field K containing an isomorphic copy F_1 of F such that the polynomial in $F_1[t]$ corresponding to f can be expressed in the form $a(t - c_1) \ldots (t - c_n)$ where $a, c_i \in K$ and K is a smallest field containing F_1 and $a, c_1, \ldots, c_n$. There is nothing lost in assuming that $F \subseteq K$ since F can be replaced by the isomorphic field F_1. Thus F is a *subfield* of K, i.e., a subring containing the identity element which is closed under forming inverses of non-zero elements.

Our first concern must be to demonstrate that splitting fields actually exist.

(7.4.8) *If f is any non-constant polynomial over a field F, then f has a splitting field over F.*

Proof. We shall argue by induction on $n = \deg(f)$; note that we may assume $n > 1$ since otherwise F itself is a splitting field for f. Assume the result is true for all polynomials of degree less than n. Consider first the case where f is reducible, so $f = gh$ where g, h in $F[t]$ both have degree less than n. By induction hypothesis g has a splitting field over F, say K_1, which we may suppose contains F as a subfield. For the same reason h has a splitting field over K_1, say K, with $K_1 \subseteq K$. Then clearly f is a product of linear factors over K. If f were such a product over some subfield S of K containing F, then we would obtain first that $K_1 \subseteq S$ and then $K \subseteq S$. Hence $K = S$ and K is a splitting field of f.

So we can assume f is irreducible. Then by (7.2.6) the ideal (f) is a maximal in $F[t]$, and consequently the quotient ring

$$K_1 = F[t]/(f)$$

7.4 Roots of polynomials and splitting fields

is a field. Next the assignment $a \mapsto a + (f)$, where $a \in F$, is an injective ring homomorphism from F to K_1. Thus its image is a subfield F_1 of K_1 and $F \simeq F_1$. So we may regard f as a polynomial over F_1.

The critical observation to make is that f has a root in K_1, namely $a_1 = t+(f)$; for $f(a_1) = f(t)+(f) = (f) = 0_{K_1}$. By the Remainder Theorem (7.4.1) $f = (t-a_1)g$ where $g \in K_1[t]$, and of course $\deg(g) = n - 1$. By induction hypothesis g has a splitting field K containing K_1. Since $a_1 \in K_1 \subseteq K$, we see that K is a splitting field for f: for any subfield of K containing F and the roots of f must contain K_1 since each element of K_1 has the form $h + (f) = h(a_1)$ for some $h \in F[t]$. □

Example (7.4.3) Let $f = t^3 - 2 \in \mathbb{Q}[t]$. The roots of f are $2^{1/3}, c2^{1/3}, c^2 2^{1/3}$ where $c = e^{2\pi i/3}$, a complex cube root of unity. Then f has as its splitting field the smallest subfield of $\mathbb{C}$ containing $\mathbb{Q}, 2^{1/3}$ and c.

The next example illustrates how one can construct finite fields from irreducible polynomials.

Example (7.4.4) Show that $f = t^3 + 2t + 1 \in \mathbb{Z}_3[t]$ is irreducible and use it to construct a field of order 27. Prove that this is a splitting field of f.

First note that the only way a cubic polynomial can be reducible is if it has a linear factor, i.e., it has a root in the field. But we easily verify that f has no roots in $\mathbb{Z}_3 = \{0, 1, 2\}$ since $f(0) = f(1) = f(2) = 1$. (Here we have written i for the congruence class $[i]$). It follows that f is irreducible and

$$K = \mathbb{Z}_3[t]/(f)$$

is a field.

If $g \in \mathbb{Z}_2[t]$, then by the Division Algorithm $g = fq + r$ where $q, r \in \mathbb{Z}_3[t]$ and $0 \leq \deg r < 3$. Hence $g + (f) = r + (f)$. This shows that every element of K has the form $a_0 + a_1 t + a_2 t^2 + (f)$ where $a_i \in \mathbb{Z}_3$. Thus $|K| \leq 3^3 = 27$. On the other hand, all such elements are distinct. Indeed, if $r + (f) = s + (f)$ with r and s both of degree < 3, then $f \mid r - s$, so that $r = s$. Therefore $|K| = 27$ and we have constructed a field of order 27.

From the proof of (7.4.8) we see that f has the root $a = t + (f)$ in K. To prove that K is actually a splitting field note that f has two further roots in K, namely $a + 1$ and $a + 2$. Thus $f = (t - a)(t - a - 1)(t - a - 2)$.

Further discussion of fields is postponed until Chapter 10. However we have seen enough to realize that irreducible polynomials play a vital role in the theory of fields. Thus a practical criterion for irreducibility is sure to be useful. Probably the best known test for irreducibility is:

(7.4.9) (Eisenstein's[1] Criterion) *Let R be a unique factorization domain and let $f = a_0 + a_1 t + \cdots + a_n t^n$ be a polynomial over R. Suppose that there is an irreducible element p of R such that $p \mid a_0, p \mid a_1, \ldots, p \mid a_{n-1}$, but $p \nmid a_n$ and $p^2 \nmid a_0$. Then f is irreducible over R.*

Proof. Suppose that f is irreducible and

$$f = (b_0 + b_1 t + \cdots + b_r t^r)(c_0 + c_1 t + \cdots + c_s t^s)$$

where $b_i, c_j \in R$, $r, s < n$, and of course $r + s = n$. Then

$$a_i = b_0 c_i + b_1 c_{i-1} + \cdots + b_i c_0.$$

Now by hypothesis $p \mid a_0 = b_0 c_0$ but $p^2 \nmid a_0$; thus p must divide exactly one of b_0 and c_0, say $p \mid b_0$ and $p \nmid c_0$. Also p cannot divide every b_i since otherwise it would divide $a_n = b_0 c_n + \cdots + b_n c_0$. Therefore there is a smallest positive integer k such that $p \nmid b_k$. Now p divides each of $b_0, b_1, \ldots, b_{k-1}$, and also $p \mid a_k$ since $k \leq r < n$. Since $a_k = (b_0 c_k + \cdots + b_{k-1} c_1) + b_k c_0$, it follows that $p \mid b_k c_0$. By Euclid's Lemma, which is valid in any UFD by (7.3.4), $p \mid b_k$ or $p \mid c_0$, both of which are forbidden. □

Eisenstein's Criterion is often applied in conjunction with Gauss's Lemma (7.3.7) to give a test for irreducibility over the field of fractions of a domain.

Example (7.4.5) Prove that $t^5 - 9t + 3$ is irreducible over $\mathbb{Q}$.

First of all $f = t^5 - 9t + 3$ is irreducible over $\mathbb{Z}$ by Eisenstein's Criterion with $p = 3$. Then Gauss's Lemma shows that f is irreducible over $\mathbb{Q}$.

Example (7.4.6) Show that if p is a prime, the polynomial $f = 1 + t + t^2 + \cdots + t^{p-1}$ is irreducible over $\mathbb{Q}$.

By Gauss's Lemma it suffices to prove that f is $\mathbb{Z}$-irreducible. Now (7.4.9) does not immediately apply to f, so we resort to a trick. Consider the polynomial

$$g = f(t+1) = 1 + (t+1) + \cdots + (t+1)^{p-1} = \frac{(t+1)^p - 1}{t},$$

by the formula for the sum of a geometric series. Expanding $(t+1)^p$ by the Binomial Theorem we arrive at

$$g = t^{p-1} + \binom{p}{p-1} t^{p-2} + \cdots + \binom{p}{2} t + \binom{p}{1}.$$

Now $p \mid \binom{p}{i}$ if $0 < i < p$ by (2.3.3). Therefore g is irreducible over $\mathbb{Z}$ by Eisenstein. But clearly this implies that f is irreducible over $\mathbb{Z}$. (This polynomial f is called the *cyclotomic polynomial of order p*).

---
[1] Ferdinand Gotthold Max Eisenstein (1823–1852).

Exercises (7.4)

1. Let $f \in F[t]$ have degree ≤ 3 where F is a field. Show that f is reducible over F if and only if it has a root in F.

2. Find the multiplicity of the root 2 of the polynomial $t^3 + 2t^2 + t + 2$ in $\mathbb{Z}_5[t]$.

3. List all irreducible polynomials of degree at most 3 in $\mathbb{Z}_2[t]$.

4. Use $t^3 + t + 1 \in \mathbb{Z}_5[t]$ to construct a field of order 125.

5. Let $f = 1 + t + t^2 + t^3 + t^4 \in \mathbb{Q}[t]$.

   (a) Prove that $K = \mathbb{Q}[t]/(f)$ is a field.
   (b) Show that every element of K can be uniquely written in the form $a_0 + a_1 x + a_2 x^2 + a_3 x^3$ where $x = t + (f)$ and $a_i \in \mathbb{Q}$.
   (c) Prove that K is a splitting field of f. [Hint: note that $x^5 = 1$ and check that x^2, x^3, x^4 are roots of f].
   (d) Compute $(1 + x^2)^3$ in K.
   (e) Compute $(1 + x)^{-1}$ in K.

6. Show that $t^6 + 6t^5 + 4t^4 + 2t + 2$ is irreducible over $\mathbb{Q}$.

7. Show that $t^6 + 12t^5 + 49t^4 + 96t^3 + 99t^2 + 54t + 15$ is irreducible over $\mathbb{Q}$. [Hint: use a suitable change of variable].

8. Let $F = \mathbb{Z}_p\{t_1\}$ and $R = F[t]$ where t and t_1 are distinct indeterminates. Show that $t^n - t_1^2 t + t_1 \in R$ is irreducible over F.

9. Find a polynomial of degree 4 in $\mathbb{Z}[t]$ which has $\sqrt{3} - \sqrt{2}$ as a root and is irreducible over $\mathbb{Q}$.

10. Prove that $1 + t + t^2 + \cdots + t^{n-1}$ is reducible over any field if n is composite.

11. Show that $\mathbb{Q}[t]$ contains an irreducible polynomial of each degree $n \geq 1$.

12. Let R be a commutative ring with identity containing a zero divisor c. Show that the polynomial $ct \in R[t]$ has at least two roots in R, so that (7.4.3) fails for R.

Chapter 8
Vector spaces

We have already encountered two of the three most commonly used algebraic structures, namely groups and rings. The third type of structure is a vector space. Vector spaces appear throughout mathematics and they also turn up in many applied areas, for example, in quantum theory and coding theory.

8.1 Vector spaces and subspaces

Let F be a field. A *vector space over F* is a triple consisting of a set V with a binary operation $+$ called *addition*, and an action of F on V called *scalar multiplication*, i.e., a function from $F \times V$ to V written $(a, v) \mapsto av$ ($a \in F, u \in V$): the following axioms are to hold for all $u, v \in V$ and $a \in F$.

(a) $(V, +)$ is an abelian group;

(b) $a(u + v) = au + av$;

(c) $(a + b)v = av + bv$;

(d) $(ab)v = a(bv)$;

(e) $1_F v = v$.

Notice that (d) and (e) assert that the multiplicative group of F *acts* on the set V in the sense of 5.1. Elements of V are called *vectors* and elements of F *scalars*. When there is no chance of confusion, it is usual to refer to V as the vector space.

First of all we record two elementary consequences of the axioms.

(8.1.1) *Let v be a vector in a vector space V over a field F, and let $a \in F$. Then:*

(i) $0_F v = 0_V$ *and* $a 0_V = 0_V$;

(ii) $(-1_F)v = -v$.

Proof. (i) Put $a = 0_F = b$ in axiom (c) to get $0_F v = 0_F v + 0_F v$; then $0_F v = 0$ by the cancellation law for the group $(V, +)$. Similarly, setting $u = 0_V = v$ in (b) yields $a 0_V = 0_V$.

(ii) Using axioms (c) and (e) and property (i), we obtain

$$v + (-1_F)v = (1_F + (-1_F))v = 0_F v = 0_V.$$

Therefore $(-1_F)v$ equals $-v$. □

Examples of vector spaces. (i) *Vector spaces of matrices.* Let F be a field and define

$$M_{m,n}(F)$$

to be the set of all $m \times n$ matrices over F. This is already an abelian group with respect to ordinary matrix addition. There is also a natural scalar multiplication here: if $A = [a_{ij}] \in M_{m,n}(F)$ and $f \in F$, then fA is the matrix which has fa_{ij} as its (i, j) entry. That the vector space axioms hold is guaranteed by elementary results from matrix algebra. Two special cases of interest are the vector spaces

$$F^m = M_{m,1}(F) \quad \text{and} \quad F_n = M_{1,n}(F).$$

Thus F^m is the vector space of *m-column vectors* over F, while F_n is the vector space of *n-row vectors* over F.

The space $\mathbb{R}^n$ is called *Euclidean n-space*. For $n \leq 3$ there is a well-known geometric interpretation of $\mathbb{R}^n$. Consider for example $\mathbb{R}^3$. A vector

$$\begin{bmatrix} a \\ b \\ c \end{bmatrix}$$

in $\mathbb{R}^3$ is represented by a line segment drawn from an arbitrary initial point (p, q, r) to the point $(p+a, q+b, r+c)$. With this interpretation of vectors, the rule of addition in $\mathbb{R}^3$ is equivalent to the well-known *triangle rule* for adding line segments u and v; this is illustrated in the diagram below.

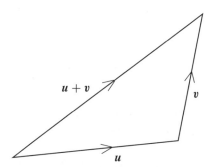

A detailed account of the geometric interpretations of Euclidean 2-space and 3-space may be found in any text on linear algebra – see for example [9].

(ii) *Vector spaces of polynomials.* The set $F[t]$ of all polynomials in t over a field F is a vector space with respect to the natural addition and scalar multiplication of polynomials.

(iii) *Fields as vector spaces.* Suppose that F is a *subfield* of a field K, i.e., F is a subring containing 1 which is a field with respect to the operations of K. Then we can regard K as a vector space over F, using the field operations as vector space operations. At first sight this example may seem confusing since elements of F are simultaneously vectors and scalars. However this point of view will be particularly valuable when we come to investigate the structure of fields in Chapter 10.

Subspaces. In analogy with subgroups of groups and subrings of rings, it is natural to introduce the concept of a subspace of a vector space. Let V be a vector space over a field F, and let S be a subset of V. Then S is called a *subspace* of V if, when we restrict the vector space operations of V to S, we obtain a vector space over F. Taking into account our analysis of the subgroup concept in Chapter 3 – see(3.3.3) – we conclude that *a subspace is simply a subset of V containing 0_V which is closed with respect to addition and multiplication by scalars.*

Obvious examples of subspaces of V are 0_V, the *zero subspace* which contains just the zero vector, and V itself. A more interesting source of examples is given in:

Example (8.1.1) Let A be an $m \times n$ matrix over a field F and define S to be the subset of all X in F^n such that $AX = 0$. Then S is a subspace of F^n.

The verification of the closure properties is very easy. The subspace S is called the *null space* of the matrix A.

Linear combinations of vectors. Suppose that V is a vector space over a field F, and that $v_1, v_2, \ldots, v_k$ are vectors in V. A *linear combination* of these vectors is a vector of the form
$$a_1 v_1 + a_2 v_2 + \cdots + a_k v_k$$
where $a_1, a_2, \ldots, a_k \in F$. If X is any non-empty set of vectors in V, write
$$F \langle X \rangle$$
for the set of all linear combinations of vectors in the set X. It is a fundamental fact that this is always a subspace.

(8.1.2) *Let X be a non-empty subset of a vector space V over a field F. Then $F \langle X \rangle$ is the smallest subspace of V that contains X.*

Proof. In the first place it is easy to verify that $F \langle X \rangle$ is closed with respect to addition and scalar multiplication; of course it also contains the zero vector 0_V. Therefore $F \langle X \rangle$ is a subspace. Also it contains X since $x = 1_F x \in F \langle X \rangle$ for all $x \in X$. Finally, any subspace that contains X automatically contains every linear combination of vectors in X, i.e., it must contain $F \langle X \rangle$ as a subset. □

The subspace $F \langle X \rangle$ is called the subspace *generated* (or *spanned*) by X. If $V = F \langle X \rangle$, then X is said to *generate* the vector space V. If V can be generated by some finite set of vectors, we say that V is a *finitely generated vector space*. What this means is that every vector in V can be expressed as a linear combination of vectors in some finite set.

Example (8.1.2) F^n is a finitely generated vector space.

To see why consider the so-called *elementary vectors* in F^n:

$$E_1 = \begin{bmatrix} 1 \\ 0 \\ 0 \\ \vdots \\ 0 \end{bmatrix}, E_2 = \begin{bmatrix} 0 \\ 1 \\ 0 \\ \vdots \\ 0 \end{bmatrix}, \ldots, E_n = \begin{bmatrix} 0 \\ 0 \\ \vdots \\ 0 \\ 1 \end{bmatrix}.$$

A general vector in F^n,

$$\begin{bmatrix} a_1 \\ a_2 \\ \vdots \\ a_n \end{bmatrix},$$

can be written $a_1 E_1 + a_2 E_2 + \cdots + a_n E_n$. Therefore $F^n = F \langle E_1, E_2, \ldots, E_n \rangle$ and F^n is finitely generated.

On the other hand, infinitely, i.e., non-finitely, generated vector spaces are not hard to find.

Example (8.1.3) The vector space $F[t]$ is infinitely generated.

Indeed suppose that $F[t]$ could be generated by finitely many polynomials p_1, $p_2, \ldots, p_k$ and let m be the maximum degree of the p_i's. Then clearly t^{m+1} cannot be expressed as a linear combination of $p_1, \ldots, p_k$, so we have reached a contradiction.

Exercises (8.1)

1. Which of the following are vector spaces? The operations of addition and scalar multiplication are the natural ones.

    (a) the set of all real 2×2 matrices with determinant 0;

    (b) the set of all solutions $y(x)$ of the homogeneous linear differential equation $a_n(x)y^{(n)} + a_{n-1}(x)y^{(n-1)} + \cdots + a_1(x)y' + a_0(x)y = 0$, where the $a_i(x)$ are given real-valued functions of x;

    (c) the set of all solution X of the matrix equation $AX = B$.

2. In the following cases say whether S is a subspace of the vector space V.

    (a) $V = \mathbb{R}^2$, $S = $ all $\begin{bmatrix} a^2 \\ a \end{bmatrix}$, $a \in \mathbb{R}$;

    (b) V is the vector space of all continuous functions on the interval $[0, 1]$ and S consists of all infinitely differentiable functions in V;

    (c) $V = F[t]$, $S = \{f \in V \mid f(a) = 0\}$ where a is a fixed element of F.

3. Verify that the rule for adding the vectors in $\mathbb{R}^3$ corresponds to the triangle rule for addition of line segments.

4. Does $\begin{bmatrix} 4 & 3 \\ 1 & -2 \end{bmatrix}$ belong to the subspace of $M_2(\mathbb{R})$ generated by the matrices $\begin{bmatrix} 3 & 4 \\ 1 & 2 \end{bmatrix}$, $\begin{bmatrix} 0 & 2 \\ -\frac{1}{3} & 4 \end{bmatrix}, \begin{bmatrix} 0 & 2 \\ 0 & 1 \end{bmatrix}$?

5. Let V be a vector space over a finite field. Prove that V is finitely generated if and only if it is finite.

8.2 Linear independence, basis and dimension

A concept of critical importance in vector space theory is linear independence. Let V be a vector space over a field F and let X be a non-empty subset of V. Then X is called *linearly dependent* if there exist distinct vectors $x_1, x_2, \ldots, x_k$ in X and scalars $a_1, a_2, \ldots, a_k \in F$, *not all zero*, such that

$$a_1 x_1 + a_2 x_2 + \cdots + a_k x_k = 0.$$

This amounts to saying that one of the x_i can be expressed as a linear combination of the others. For if, say, $a_i \neq 0$, we can solve for x_i, obtaining

$$x_i = \sum_{\substack{j=1 \\ j \neq i}}^{k} (-a_i^{-1}) a_j v_j.$$

A subset which is not linearly dependent is said to be *linearly independent*. For example, the elementary vectors $E_1, E_2, \ldots, E_n$ form a linearly independent subset of F^n for any field F.

Homogeneous linear systems. To make any significant progress with linear independence, a basic result about systems of linear equations is needed. Let F be a field and consider a system of m linear equations over F

$$\begin{cases} a_{11}x_1 + \cdots + a_{1n}x_n = 0 \\ a_{21}x_1 + \cdots + a_{2n}x_n = 0 \\ \vdots \\ a_{m1}x_1 + \cdots + a_{mn}x_n = 0 \end{cases}$$

Here $a_{ij} \in F$ and $x_1, \ldots, x_n$ are the unknowns. This system has the trivial solution $x_1 = x_2 = \cdots = x_n = 0$. The interesting question is whether there are any non-trivial solutions. A detailed account of the theory of systems of linear equations can be found in any book on linear algebra, for example [9]. For our purposes the following result will be sufficient.

8.2 Linear independence, basis and dimension

(8.2.1) *If the number of equations m is less than the number of unknowns n, then the homogeneous linear system has a non-trivial solution.*

Proof. We adopt a method of systematic elimination which is often called *Gaussian elimination*. We may assume that $a_{11} \neq 0$ – if this is not true, replace equation 1 by the first equation in which x_1 appears. Since equation 1 can be multiplied by a_{11}^{-1}, we may also assume that $a_{11} = 1$. Then, by subtracting multiples of equation 1 from equations 2 through m, eliminate x_1 from these equations.

Next find the first of equations 2 through m which contains an unknown with smallest subscript > 1, say x_{i_2}. Move this equation up to second position. Now make the coefficient of x_{i_2} equal to 1 and subtract multiples of equation 2 from equations 3 through m. Repeat this procedure until the remaining equations involve no more unknowns, i.e., they are of the form $0 = 0$. Say this happens after r steps. At this point the matrix of coefficients is in *row echelon form* and has r linearly independent rows.

Unknowns other than $x_1 = x_{i_1}, x_{i_2}, \ldots, x_{i_r}$ can be given arbitrary values. The non-trivial equations may then be used to solve back for $x_{i_1}, \ldots, x_{i_r}$. Since $r \leq m < n$, at least one unknown can be given an arbitrary value, so the linear system has a non-trivial solution. □

This result will now be used to establish the fundamental theorem about linear dependence in vector spaces.

(8.2.2) *Let $v_1, v_2, \ldots, v_k$ be vectors in a vector space V over a field F. Then any subset of $F \langle v_1, v_2, \ldots, v_k \rangle$ containing $k + 1$ or more vectors is linearly dependent.*

Proof. Let $u_1, u_2, \ldots, u_{k+1} \in S = F \langle v_1, \ldots, v_k \rangle$. It is enough to show that $\{u_1, u_2, \ldots, u_{k+1}\}$ is a linearly dependent set. This amounts to finding field elements $a_1, a_2, \ldots, a_{k+1}$, not all zero, such that $a_1 u_1 + \cdots + a_{k+1} u_{k+1} = 0$.

Since $u_i \in S$, there is an expression $u_i = d_{1i} v_1 + \cdots + d_{ki} v_k$ where $d_{ji} \in F$. On substituting for the u_i, we obtain

$$a_1 u_1 + \cdots + a_{k+1} u_{k+1} = \sum_{i=1}^{k+1} a_i \left(\sum_{j=1}^{k} d_{ji} v_j \right) = \sum_{j=1}^{k} \left(\sum_{i=1}^{k+1} a_i d_{ji} \right) v_j.$$

Therefore $a_1 u_1 + \cdots + a_{k+1} u_{k+1}$ will equal 0 if the a_i satisfy the equations

$$\sum_{i=1}^{k+1} a_i d_{ji} = 0, \quad j = 1, \ldots, k.$$

But this is a system of k linear homogeneous equations in the $k + 1$ unknowns a_i. By (8.2.1) there is a non-trivial solution $a_1, \ldots, a_{k+1}$. Therefore $\{u_1, \ldots, u_{k+1}\}$ is linearly dependent, as claimed. □

Corollary (8.2.3) *If a vector space V can be generated by k elements, then every subset of V with $k+1$ or more elements is linearly dependent.*

Bases. A *basis* of a vector space V is a non-empty subset X such that:

(i) X is linearly independent;

(ii) X generates V.

Notice that these are contrasting properties since (i) implies that X is not too large and (ii) that X is not too small.

For example, the elementary vectors $E_1, E_2, \ldots, E_n$ form a basis of the vector space F^n – called the *standard basis* – for any field F. More generally a basis of $M_{m,n}(F)$ is obtained by taking the $m \times n$ matrices over F with just one non-zero entry which is required to be 1_F.

A important property of bases is unique expressibility.

(8.2.4) *If $\{v_1, v_2, \ldots, v_n\}$ is a basis of a vector space V over a field F, then every vector v in V is uniquely expressible in the form $v = a_1 v_1 + \cdots + a_n v_n$ with $a_i \in F$.*

Proof. In the first place such expressions for v exist. If v in V had two such expressions $v = \sum_{i=1}^{n} a_i v_i = \sum_{i=1}^{n} b_i v_i$, then we would have $\sum_{i=1}^{n} (a_i - b_i) v_i = 0$, from which it follows that $a_i = b_i$ by linear independence of the v_i. $\square$

This result shows that a basis may be used to introduce coordinates in a vector space. Suppose that V is a vector space over field F and that $\mathcal{B} = \{v_1, v_2, \ldots, v_n\}$ is a basis of V with its elements written in a specific order, i.e., an *ordered basis*. Then by (8.2.4) each $v \in V$ has a unique expression $v = \sum_{i=1}^{n} c_i v_i$ with $c_i \in F$. So v is determined by the column vector in F^n whose entries are $c_1, c_2, \ldots, c_n$; this is called the *coordinate column vector* of v with respect to $\mathcal{B}$ and it is written

$$[v]_{\mathcal{B}}.$$

Coordinate vectors provide a concrete representation of the vectors of an abstract vector space.

The existence of bases. There is nothing in the definition of a basis to make us certain that bases exist. Our first task will be to show that this is true for finitely generated non-zero vector spaces. Notice that the zero space does not have a basis since it has no linearly independent subsets.

(8.2.5) *Let V be a finitely generated vector space and suppose that X_0 is a linearly independent subset of V. Then X_0 is contained in a basis of V.*

Proof. Suppose that V can be generated by m vectors. Then by (8.2.3) no linearly independent subset of V can contain more than m vectors. It follows that X_0 is contained in a largest linearly independent subset X; for otherwise it would be possible to form ever larger finite linearly independent subsets containing X_0.

We complete the proof by showing that X generates V. If this is false, there is a vector u in $V - F\langle X\rangle$. Then $u \notin X$, so $X \neq X \cup \{u\}$ and $X \cup \{u\}$ must be linearly independent by maximality of X. Writing $X = \{v_1, \ldots, v_k\}$, we deduce that there is a relation of the type

$$a_1 v_1 + \cdots + a_k v_k + bu = 0,$$

where $a_1, \ldots, a_k, b \in F$ and not all of these scalars are 0. Now b cannot equal 0: for otherwise $a_1 v_1 + \cdots + a_k v_k = 0$ and $a_1 = \cdots = a_k = 0$ since $v_1, \ldots, v_k$ are known to be linearly independent. Therefore $u = -b^{-1} a_1 v_1 - \cdots - b^{-1} a_k v_k \in F\langle X\rangle$, which is a contradiction. □

Corollary (8.2.6) *Every finitely generated non-zero vector space V has a basis.*

Indeed, since $V \neq 0$, we can choose a non-zero vector v from V and apply (8.2.5) with $X_0 = \{v\}$.

In fact every infinitely generated vector space has a basis, but advanced methods are needed to prove this – see (12.1.1) below.

Dimension. A vector space usually has many bases, but it is a fact that all of them have the same number of elements.

(8.2.7) *Let V be a finitely generated non-zero vector space. Then any two bases of V have the same number of elements.*

Proof. In the first place a basis of V is necessarily finite by (8.2.3). Next let $\{u_1, \ldots, u_m\}$ and $\{v_1, \ldots, v_n\}$ be two bases. Then $V = \langle v_1, \ldots, v_n\rangle$ and so by (8.2.2) there cannot be a linearly independent subset of V with more than n elements. Therefore $m \leq n$. By the same reasoning $n \leq m$, so we obtain $m = n$, as required. □

This result enables us to define the *dimension*

$$\dim(V)$$

of a finitely generated vector space V. If $V = 0$, define $\dim(V)$ to be 0, and if $V \neq 0$, let $\dim(V)$ be the number of elements in a basis of V. By (8.2.7) this definition is unambiguous. In the future we shall speak of *finite dimensional vector spaces* instead of finitely generated ones.

(8.2.8) *Let $X_1, X_2, \ldots, X_k$ be vectors in F^n, F a field. Let $A = [X_1, X_2, \ldots, X_k]$ be the $n \times k$ matrix which has the X_i as columns. Then $\dim(F\langle X_1, \ldots, X_k\rangle) = r$ where r is the rank of the matrix A, (i.e., the number of 1's in the normal form of A).*

Proof. We shall need to use some elementary facts about matrices here. In the first place, $S = F \langle X_1, \ldots, X_k \rangle$ is the *column space* of A, and it is unaffected when column operations are applied to A. Now by applying column operations to A, just as we did for row operations during Gaussian elimination in the proof of (8.2.1), we can replace A by a matrix with the same column space S which has the so-called *column echelon form* with r non-zero columns. Here r is the rank of A. These r columns are linearly independent, so they form a basis of S (if $r > 0$). Hence $\dim(S) = r$. □

Next we consider the relation between the dimension of a vector space and that of a subspace.

(8.2.9) *If V is a vector space with finite dimension n and U is a subspace of V, then $\dim(U) \leq \dim(V)$. Furthermore $\dim(U) = \dim(V)$ if and only if $U = V$.*

Proof. If $U = 0$, then $\dim(U) = 0 \leq \dim(V)$. Assume that $U \neq 0$ and let X be a basis of U. By (8.2.5) X is contained in a basis Y of V. Hence $\dim(U) = |X| \leq |Y| = \dim(V)$.

Finally, suppose that $0 < \dim(U) = \dim(V)$, but $U \neq V$. Then $U \neq 0$. Using the notation of the previous paragraph, we see that Y must have at least one more element than X, so $\dim(U) < \dim(V)$, a contradiction. □

The next result can simplify the task of showing that a subset of a finite dimensional vector space is a basis.

(8.2.10) *Let V be a finite dimensional vector space with dimension n and let X be a subset of V with n elements. Then the following statements about X are equivalent:*

(a) *X is a basis of V;*

(b) *X is linearly independent;*

(c) *X generates V.*

Proof. Of course (a) implies (b) and (c). Assume that (b) holds. Then X is a basis of the subspace it generates, namely $F \langle X \rangle$; hence $\dim(F \langle X \rangle) = n = \dim(V)$ and (8.2.9) shows that $F \langle X \rangle = V$. Thus (b) implies (a).

Finally, assume that (c) holds. If X is not a basis of V, it must be linearly dependent, so one of its elements can be written as a linear combination of the others. Hence $n = \dim(V) = \dim(F \langle X \rangle) < |X| = n$, a contradiction. □

Change of basis. As previously remarked, vector spaces usually have many bases, and a vector may be represented with respect to each basis by a coordinate column vector. A natural question is: how are these coordinate vectors related?

Let $\mathcal{B} = \{v_1, \ldots, v_n\}$ and $\mathcal{B}' = \{v'_1, \ldots, v'_n\}$ be two ordered bases of a finite dimensional vector space V over a field F. Then each v'_i can be expressed as a linear

8.2 Linear independence, basis and dimension

combination of $v_1, \ldots, v_n$, say

$$v'_i = \sum_{j=1}^{n} s_{ji} v_j,$$

where $s_{ji} \in F$. The change of basis $\mathcal{B}' \to \mathcal{B}$ is described by the *transition matrix* $S = [s_{ij}]$. Observe that S is $n \times n$ and its i-th column is the coordinate vector $[v'_i]_{\mathcal{B}}$.

To understand how S determines the change of basis $\mathcal{B}' \to \mathcal{B}$, choose an arbitrary vector v from V and write $v = \sum_{i=1}^{n} c'_i v'_i$ where $c'_1, \ldots, c'_n$ are the entries of the coordinate vector $[v]_{\mathcal{B}'}$. Replace v'_i by $\sum_{j=1}^{n} s_{ji} v_j$ to get

$$v = \sum_{i=1}^{n} c'_i \left(\sum_{j=1}^{n} s_{ji} v_j \right) = \sum_{j=1}^{n} \left(\sum_{i=1}^{n} s_{ji} c'_i \right) v_j.$$

Therefore the entries of the coordinate vector $[v]_{\mathcal{B}}$ are $\sum_{i=1}^{n} s_{ji} c'_i$ for $j = 1, 2, \ldots, n$. This shows that

$$[v]_{\mathcal{B}} = S[v]_{\mathcal{B}'},$$

i.e., left multiplication by the transition matrix S transforms coordinate vectors with respect to $\mathcal{B}'$ into those with respect to $\mathcal{B}$.

Notice that the transition matrix S must be non-singular. For otherwise, by standard matrix theory, there would exist a non-zero $X \in F^n$ such that $SX = 0$; however, if $u \in V$ is defined by $[u]_{\mathcal{B}'} = X$, then $[u]_{\mathcal{B}} = SX = 0$, which can only mean that $u = 0$ and $X = 0$. From $[v]_{\mathcal{B}} = S[v]_{\mathcal{B}'}$ we deduce that $S^{-1}[v]_{\mathcal{B}} = [v]_{\mathcal{B}'}$. Thus S^{-1} is the transition matrix for the change of basis $\mathcal{B} \to \mathcal{B}'$. We sum up these conclusions in:

(8.2.11) *Let $\mathcal{B}$ and $\mathcal{B}'$ be two ordered bases of an n-dimensional vector space V. Define S to be the $n \times n$ matrix whose i-th column is the coordinate vector of the i-th vector of $\mathcal{B}'$ with respect to $\mathcal{B}$. Then S is non-singular: also $[v]_{\mathcal{B}} = S[v]_{\mathcal{B}'}$ and $[v]_{\mathcal{B}'} = S^{-1}[v]_{\mathcal{B}}$ for all v in V.*

Example (8.2.1) Let V be the vector space of all real polynomials in t with degree at most 2. Then $\mathcal{B} = \{1, t, t^2\}$ is clearly a basis of V, and so is $\mathcal{B}' = \{1+t, 2t, 4t^2 - 2\}$ since this set can be verified to be linearly independent. Write down the coordinate vectors of $1+t, 2t, 4t^2 - 2$ with respect to $\mathcal{B}$ as columns of the matrix

$$S = \begin{bmatrix} 1 & 0 & -2 \\ 1 & 2 & 0 \\ 0 & 0 & 4 \end{bmatrix}.$$

This is the transition matrix for the change of basis $\mathcal{B}' \to \mathcal{B}$. The transition matrix for $\mathcal{B} \to \mathcal{B}'$ is

$$S^{-1} = \begin{bmatrix} 1 & 0 & \frac{1}{2} \\ -\frac{1}{2} & \frac{1}{2} & -\frac{1}{4} \\ 0 & 0 & \frac{1}{4} \end{bmatrix}.$$

For example, to express $f = a + bt + ct^2$ in terms of the basis $\mathcal{B}'$, we compute

$$[f]_{\mathcal{B}'} = S^{-1}[f]_{\mathcal{B}} = S^{-1}\begin{bmatrix} a \\ b \\ c \end{bmatrix} = \begin{bmatrix} a + \frac{c}{2} \\ -\frac{1}{2}a + \frac{1}{2}b - \frac{1}{4}c \\ \frac{1}{4}c \end{bmatrix}.$$

Thus $f = (a + \frac{c}{2})(1 + t) + (-\frac{1}{2}a + \frac{1}{2}b - \frac{1}{4}c)(2t) + \frac{1}{4}c(4t^2 - 2)$, which is clearly correct.

Dimension of the sum and intersection of subspaces. Since a vector space V is an additively written abelian group, one can form the sum of two subspaces U, W; thus $U + W = \{u + w \mid u \in U, w \in W\}$. It is easily verified that $U + W$ is a subspace of V. There is a useful formula connecting the dimensions of the subspaces $U + W$ and $U \cap W$.

(8.2.12) *Let U and W be subspaces of a finite dimensional vector space V. Then*

$$\dim(U + W) + \dim(U \cap W) = \dim(U) + \dim(W).$$

Proof. If $U = 0$, then $U + W = W$ and $U \cap W = 0$; in this case the formula is certainly true. Thus we can assume that $U \neq 0$ and $W \neq 0$.

Choose a basis for $U \cap W$, say $z_1, \ldots, z_r$, if $U \cap W \neq 0$; should $U \cap W$ be 0, just ignore the z_i. By (8.2.5) we can extend $\{z_1, \ldots, z_r\}$ to bases of U and of W, say

$$\{z_1, \ldots, z_r, u_{r+1}, \ldots, u_m\} \quad \text{and} \quad \{z_1, \ldots, z_r, w_{r+1}, \ldots, w_n\}.$$

Now the vectors $z_1, \ldots, z_r, u_{r+1}, \ldots, u_n v_{r+1}, \ldots, v_n$ surely generate $U + W$: for any vector of $U + W$ is expressible in terms of them. We claim that these elements are linearly independent.

To establish the claim suppose there is a linear relation

$$\sum_{i=1}^{r} e_i z_i + \sum_{j=r+1}^{m} c_j u_j + \sum_{k=r+1}^{n} d_k w_k = 0$$

where e_i, c_j, d_k are scalars. Then

$$\sum_{k=r+1}^{n} d_k w_k = \sum_{i=1}^{r}(-e_i)z_i + \sum_{j=r+1}^{m}(-c_j)u_j,$$

which belongs to U and to W, and so to $U \cap W$. Hence $\sum_{k=r+1}^{n} d_k w_k$ is a linear combination of the z_i. But $z_1, \ldots, z_r, w_{r+1}, \ldots, w_n$ are linearly independent, which means that $d_k = 0$ for all k. Our linear relation now reduces to

$$\sum_{i=1}^{r} e_i z_i + \sum_{j=r+1}^{m} c_j u_j = 0.$$

8.2 Linear independence, basis and dimension   145

But $z_1, \ldots, z_r, u_{r+1}, \ldots, u_m$ are linearly independent. Hence $c_j = 0$ and $e_i = 0$, which establishes the claim of linear independence.

Finally, $\dim(U + W)$ equals the number of the vectors $z_1, \ldots, z_r, u_{r+1}, \ldots, u_m, v_{r+1}, \ldots, v_n$, which is

$$r + (m - r) + (n - r) = m + n - r = \dim(U) + \dim(W) - \dim(U \cap W),$$

and the required formula follows. □

Direct sums of vector spaces. Since a vector space V is an additive abelian group, we can form the direct sum of subspaces $U_1, \ldots, U_k$ of V – see 4.2. Thus

$$U = U_1 \oplus \cdots \oplus U_k$$

where $U = \{u_1 + \cdots + u_k \mid u_i \in U_i\}$ and $U_i \cap \sum_{j \neq i} U_j = 0$. Clearly U is a subspace of V. Note that by (8.2.12) and induction on k

$$\dim(U_1 \oplus \cdots \oplus U_k) = \dim(U_1) + \cdots + \dim(U_k).$$

Next let $\{v_1, \ldots, v_n\}$ be a basis of V. Then clearly $V = F \langle v_1 \rangle \oplus \cdots \oplus F \langle v_n \rangle$, so that we have established:

(8.2.13) *An n-dimensional vector space is the direct sum of n 1-dimensional subspaces.*

This is also true when $n = 0$ if the direct sum is interpreted as 0.

Quotient spaces. Suppose that V is a vector space over a field F and U is a subspace of V. Since V is an abelian group and U is a subgroup, the quotient

$$V/U = \{v + U \mid v \in V\}$$

is already defined as an abelian group. Now make V/U into a vector space over F by defining scalar multiplication in the natural way:

$$a(v + U) = av + U \quad (a \in F).$$

This is evidently a well-defined operation. After an easy check of the vector space axioms, we conclude that V/U is a vector space over F, the *quotient space* of U in V. The dimension of a quotient space is easily computed.

(8.2.14) *Let U be a subspace of a finite dimensional space V. Then*

$$\dim(V/U) = \dim(V) - \dim(U).$$

Proof. If $U = 0$, the statement is obviously true because $V/0 \simeq V$. Assuming $U \neq 0$, we choose a basis $\{v_1, \ldots, v_m\}$ of U and extend it to a basis of V, say $\{v_1, \ldots, v_m, v_{m+1}, \ldots, v_n\}$. We will argue that $\{v_{m+1} + U, \ldots, v_n + U\}$ is a basis of V/U.

Assume that $\sum_{i=m+1}^{n} a_i(v_i + U) = U$ where $a_i \in F$. Then $\sum_{i=m+1}^{n} a_i v_i \in U$, so this element is a linear combination of $v_1, \ldots, v_m$. It follows by linear independence that each $a_i = 0$, and thus $\{v_{m+1}+U, \ldots, v_n+U\}$ is linearly independent. Next, if $v \in V$, write $v = \sum_{i=1}^{n} a_i v_i$, with scalars a_i, and observe that $v+U = \sum_{i=m+1}^{n} a_i(v_i+U)$ since $v_1, \ldots, v_m \in U$. It follows that $v_{m+1} + U, \ldots, v_n + U$ form a basis of V/U and $\dim(V/U) = n - m = \dim(V) - \dim(U)$, as required. □

Two applications. To conclude this section let us show that the mere existence of a basis in a finite dimensional vector space is enough to prove two important results about abelian groups and finite fields.

Let p be a prime. An additively written abelian group A is called an *elementary abelian p-group* if $pa = 0$ for all a in A. For example, the Klein 4-group is an elementary abelian 2-group. The structure of finite elementary abelian p-groups is given in the next result.

(8.2.15) *Let A be a finite abelian group. Then A is an elementary abelian p-group if and only if A is a direct sum of copies of $\mathbb{Z}_p$.*

Proof. The essential idea of the proof is to view A as a vector space over the field $\mathbb{Z}_p$. Here the scalar multiplication is the natural one, namely $(i + p\mathbb{Z})a = ia$ where $i \in \mathbb{Z}, a \in A$. One has to verify that this operation is well-defined, which is true since $(i + pm)a = ia + mpa = ia$ for all $a \in A$. Since A is finite, it is a finite dimensional vector space over $\mathbb{Z}_p$. By (8.2.13) $A = A_1 \oplus A_2 \oplus \cdots \oplus A_n$ where each A_i is a 1-dimensional subspace, i.e., $|A_i| = p$ and $A_i \simeq \mathbb{Z}_p$. Conversely, any direct sum of copies of $\mathbb{Z}_p$ certainly satisfies $pa = 0$ for every element a and so is elementary abelian p. □

The second result to be proved states that the number of elements in a finite field is always a prime power. This is in marked contrast to the behavior of groups and rings, of which examples exist with any finite order.

(8.2.16) *Let F be a finite field. Then $|F|$ is a power of a prime.*

Proof. By (6.3.9) the field F has characteristic a prime p and $pa = 0$ for all $a \in F$. Thus, as an additive group, F is elementary abelian p. It now follows from (8.2.15) that $|F|$ is a power of p. □

Exercises (8.2)

1. Show that $X_1 = \begin{bmatrix} 4 \\ 2 \\ 1 \end{bmatrix}$, $X_2 = \begin{bmatrix} -5 \\ 2 \\ -3 \end{bmatrix}$, $X_3 = \begin{bmatrix} 1 \\ 3 \\ 0 \end{bmatrix}$ form a basis of $\mathbb{R}^3$, and express the elementary vectors E_1, E_2, E_3 in terms of X_1, X_2, X_3.

2. Find a basis for the null space of the matrix $A = \begin{bmatrix} 2 & 3 & 1 & 1 \\ -3 & 1 & 4 & -7 \\ 1 & 2 & 1 & 0 \end{bmatrix}$.

3. Find the dimension of the vector space $M_{m,n}(F)$ where F is a arbitrary field.

4. Let $v_1, v_2, \ldots, v_n$ be vectors in a vector space V. Assume that each element of V is *uniquely* expressible as a linear combination of $v_1, v_2, \ldots, v_n$. Prove that the v_i's form a basis of V.

5. Let $\mathcal{B} = \{E_1, E_2, E_3\}$ be the standard ordered basis of $\mathbb{R}^3$ and let

$$\mathcal{B}' = \left\{ \begin{bmatrix} 2 \\ 0 \\ 0 \end{bmatrix}, \begin{bmatrix} -1 \\ 2 \\ 0 \end{bmatrix}, \begin{bmatrix} 1 \\ 1 \\ 1 \end{bmatrix} \right\}.$$

Show that $\mathcal{B}'$ is a basis of $\mathbb{R}^3$ and find the transition matrices for the changes of bases $\mathcal{B} \to \mathcal{B}'$ and $\mathcal{B}' \to \mathcal{B}$.

6. Let V be a vector space of dimension n and let i be an integer such that $0 \leq i \leq n$. Show that V has at least one subspace of dimension i.

7. The same as Exercise 6 with "subspace" replaced by "quotient space".

8. Let U be a subspace of a finite dimensional vector space V. Prove that there is a subspace W such that $V = U \oplus W$. Show also that $W \simeq V/U$.

9. Let V be a vector space of dimension $2n$ and assume that U and W are subspaces of dimensions n and $n+1$ respectively. Prove that $U \cap W \neq 0$.

10. Let vectors $v_1, v_2, \ldots, v_m$ generate a vector space V. Prove that some subset of $\{v_1, v_2, \ldots, v_m\}$ is a basis of V.

8.3 Linear mappings

Just as there are homomorphisms of groups and of rings, there are homomorphisms of vector spaces. Traditionally these are called *linear mappings* or *transformations*. Let V and W be two vector spaces over the same field F. Then a function

$$\alpha : V \to W$$

is called a *linear mapping* from V to W if the following rules are valid for all $v_1, v_2 \in V$ and $a \in F$:

(i) $\alpha(v_1 + v_2) = \alpha(v_1) + \alpha(v_2)$;

(ii) $\alpha(av_1) = a\alpha(v_1)$.

If α is also bijective, it is called an *isomorphism of vector spaces*. Should there exist an isomorphism between vector spaces V and W over a field F, we will write $V \stackrel{F}{\simeq} W$ or $V \simeq W$. Notice that a linear mapping is automatically a homomorphism of additive groups by (i). So all results established for group homomorphisms may be carried over to linear mappings. A linear mapping $\alpha : V \to V$ is called a *linear operator* on V.

Example (8.3.1) Let A be an $m \times n$ matrix over a field F and define a function $\alpha : F^n \to F^m$ by the rule $\alpha(X) = AX$ where $X \in F^n$. Then simple properties of matrices reveal that α is a linear mapping.

Example (8.3.2) Let V be an n-dimensional vector space over a field F and let $\mathcal{B} = \{v_1, v_2, \ldots, v_n\}$ be an ordered basis of V. Recall that to each vector v in V there corresponds a unique coordinate vector $[v]_\mathcal{B}$ with respect to $\mathcal{B}$. Use this correspondence to define a function $\alpha : V \to F^n$ by $\alpha(v) = [v]_\mathcal{B}$. By simple calculations we see that $[u + v]_\mathcal{B} = [u]_\mathcal{B} + [v]_\mathcal{B}$ and $[av]_\mathcal{B} = a[v]_\mathcal{B}$ where $u, v \in V, a \in F$. Hence α is a linear mapping. Clearly $[v]_\mathcal{B} = 0$ implies that $v = 0$; thus α is injective, and it is obviously surjective. Our conclusion is that α is an isomorphism and $V \stackrel{F}{\simeq} F^n$.

We state this result formally as:

(8.3.1) *If V is a vector space with dimension n over a field F, then $V \stackrel{F}{\simeq} F^n$. Thus any two vector spaces with the same finite dimension over F are isomorphic.*

(The second statement follows from the fact that isomorphism of vector spaces is an equivalence relation).

An important way of defining a linear mapping is by specifying its effect on a basis.

(8.3.2) *Let $\{v_1, \ldots, v_n\}$ be a basis of a vector space V over a field F and let $w_1, \ldots, w_n$ be any n vectors in another F-vector space W. Then there is a unique linear mapping $\alpha : V \to W$ such that $\alpha(v_i) = w_i$ for $i = 1, 2, \ldots, n$.*

Proof. Let $v \in V$ and write $v = \sum_{i=1}^{n} a_i v_i$, where a_i in F are unique. Define a function $\alpha : V \to W$ by the rule

$$\alpha(v) = \sum_{i=1}^{n} a_i w_i.$$

Then an easy check shows that α is a linear mapping, and of course $\alpha(v_i) = w_i$. If $\alpha' : V \to W$ is another such linear mapping, then $\alpha' = \alpha$; for $\alpha'(v) = \sum_{i=1}^{n} a_i \alpha'(v_i) = \sum_{i=1}^{n} a_i w_i = \alpha(v)$. □

Our experience with groups and rings suggests it may be worthwhile to examine the kernel and image of a linear mapping.

(8.3.3) *Let $\alpha : V \to W$ be a linear mapping. Then $\mathrm{Ker}(\alpha)$ and $\mathrm{Im}(\alpha)$ are subspaces of V and W respectively.*

Proof. Since α is a group homomorphism, it follows from (4.3.2) that $\mathrm{Ker}(\alpha)$ and $\mathrm{Im}(\alpha)$ are additive subgroups. We leave the reader to complete the proof by showing that these subgroups are also closed under scalar multiplication. □

Just as for groups and rings, there are *Isomorphism Theorems* for vector spaces.

(8.3.4) *If $\alpha : V \to W$ is a linear mapping between vector spaces over a field F, then $V/\mathrm{Ker}(\alpha) \stackrel{F}{\simeq} \mathrm{Im}(\alpha)$.*

In the next two results U and W are subspaces of a vector space V over a field F.

(8.3.5) $(U + W)/W \stackrel{F}{\simeq} U/(U \cap W)$.

(8.3.6) *If $U \subseteq W$, then $(V/U)/(W/U) \stackrel{F}{\simeq} V/W$.*

Since the Isomorphism Theorems for groups are applicable, all one has to prove here is that the functions introduced in the proofs of (4.3.4), (4.3.5) and (4.3.6) are linear mappings, i.e., they act appropriately on scalar multiples.

For example, in (8.3.4) the function in question is $\theta : V/\mathrm{Ker}(\alpha) \to \mathrm{Im}(\alpha)$ where $\theta(v + \mathrm{Ker}(\alpha)) = \alpha(v)$. Then

$$\theta(a(v + \mathrm{Ker}(\alpha))) = \theta(av + \mathrm{Ker}(\alpha)) = \alpha(av) = a\alpha(v) = a\theta(v + \mathrm{Ker}(\alpha)),$$

and it follows that θ is a linear mapping.

There is an important dimension formula connecting kernel and image.

(8.3.7) *If $\alpha : V \to W$ is a linear mapping between finite dimensional vector spaces, then $\dim(\mathrm{Ker}(\alpha)) + \dim(\mathrm{Im}(\alpha)) = \dim(V)$.*

This follows directly from (8.3.4) and (8.2.14). An immediate application is to the null space of a matrix.

Corollary (8.3.8) *Let A be an $m \times n$ matrix with rank r over a field F. Then the dimension of the null space of A is $n - r$.*

Proof. Let α be the linear mapping from F^n to F^m defined in Example (8.3.1) by $\alpha(X) = AX$. Now $\mathrm{Ker}(\alpha)$ is the null space of A, and it is readily seen that $\mathrm{Im}(\alpha)$ is the column space. By (8.2.8) $\dim(\mathrm{Im}(\alpha)) = r$, the rank of A, and by (8.3.7) $\dim(\mathrm{Ker}(\alpha)) = n - r$, so the result follows. □

As another application of (8.3.7) we give a different proof of the dimension formula for sum and intersection of subspaces – see (8.2.12).

(8.3.9) *If U and W are subspaces of a finite dimensional vector space, then*
$$\dim(U + W) + \dim(U \cap W) = \dim(U) + \dim(W).$$

Proof. By (8.3.5) $(U+W)/W \simeq U/(U \cap W)$. Hence, taking dimensions and applying (8.2.14), we find that $\dim(U + W) - \dim(W) = \dim(U) - \dim(U \cap W)$. □

Vector spaces of linear mappings. It is useful to endow sets of linear mappings with the structure of a vector space. Suppose that V and W are two vector spaces over the same field F. We will write
$$L(V, W)$$
for the set of all linear mappings from V to W. Define addition and scalar multiplication in $L(V, W)$ by the natural rules
$$\alpha + \beta(v) = \alpha(v) + \beta(v), \quad a \cdot \alpha(v) = a(\alpha(v)),$$
where $\alpha, \beta \in L(V, W)$, $v \in V$, $a \in F$. It is simple to verify that $\alpha + \beta$ and $a \cdot \alpha$ are linear mappings. The basic result about $L(V, W)$ is:

(8.3.10) *Let V and W be vector spaces over a field F. Then:*

(i) *$L(V, W)$ is a vector space over F;*

(ii) *if V and W are finite dimensional, then $L(V, W)$ has finite dimension equal to $\dim(V) \cdot \dim(W)$.*

Proof. We omit the routine proof of (i) and concentrate and (ii). Let $\{v_1, \ldots, v_m\}$ and $\{w_1, \ldots, w_n\}$ be bases of V and W respectively. By (8.3.2), for $i = 1, 2, \ldots, m$ and $j = 1, 2, \ldots, n$, there is a unique linear mapping $\alpha_{ij} : V \to W$ such that
$$\alpha_{ij}(v_k) = \begin{cases} w_j & \text{if } k = i \\ 0 & \text{if } k \neq i. \end{cases}$$

Thus α_{ij} sends basis element v_i to basis element w_j and all other v_k's to 0. First we show that the α_{ij} are linearly independent in the vector space $L(V, W)$.

Let $a_{ij} \in F$; then by definition of α_{ij} we have for each k
$$\left(\sum_{i=1}^{m} \sum_{j=1}^{n} a_{ij} \alpha_{ij} \right)(v_k) = \sum_{j=1}^{n} \sum_{i=1}^{m} a_{ij}(\alpha_{ij}(v_k)) = \sum_{j=1}^{n} a_{kj} w_j. \quad (*)$$

Therefore $\sum_{i=1}^{m} \sum_{j=1}^{n} a_{ij} \alpha_{ij} = 0$ if and only if $a_{kj} = 0$ for all j, k. It follows that the α_{ij} are linearly independent.

Finally, we claim that α_{ij} actually generate $L(V, W)$. To prove this let $\alpha \in L(V, W)$ and write $\alpha(v_k) = \sum_{j=1}^{n} a_{kj} w_j$ where $a_{kj} \in F$. Then from the equation $(*)$ above we see that $\alpha = \sum_{i=1}^{m} \sum_{j=1}^{n} a_{ij} \alpha_{ij}$. Therefore the α_{ij}'s form a basis of $L(V, W)$ and $\dim(L(V, W)) = mn = \dim(V) \cdot \dim(W)$. □

8.3 Linear mappings

The dual space. If V is a vector space over a field F, the vector space

$$V^* = L(V, F)$$

is called the *dual space* of V; here F is to be regarded as a 1-dimensional vector space. Elements of V^* are linear mappings from V to F: these are called *linear functionals* on V.

Example (8.3.3) Let $Y \in F^n$ and define $\alpha : F^n \to F$ by the rule $\alpha(X) = Y^T X$. Then α is a linear functional on F^n.

If V has finite dimension n, then

$$\dim(V^*) = \dim(L(V, F)) = \dim(V)$$

by (8.3.10). Thus V, V^* and the *double dual* $V^{**} = (V^*)^*$ all have the same dimension. Therefore these vector spaces are isomorphic by (8.3.1).

In fact there is a *canonical* linear mapping $\theta : V \to V^{**}$. Let $v \in V$ and define $\theta(v) \in V^{**}$ by the rule

$$\theta(v)(\alpha) = \alpha(v)$$

where $\alpha \in V^*$. So $\theta(v)$ simply evaluates each linear functional on V at v. Concerning the function θ, we will prove:

(8.3.11) *The function* $\theta : V \to V^{**}$ *is an injective linear mapping. If V has finite dimension, then θ is an isomorphism.*

Proof. In the first place $\theta(v) \in V^{**}$: for

$$\theta(v)(\alpha + \beta) = \alpha + \beta(v) = \alpha(v) + \beta(v) = \theta(v)(\alpha) + \theta(v)(\beta).$$

Also $\theta(v)(a \cdot \alpha) = a \cdot \alpha(v) = a(\alpha(v)) = a(\theta(v)(\alpha))$.

Next we have

$$\theta(v_1 + v_2)(\alpha) = \alpha(v_1 + v_2) = \alpha(v_1) + \alpha(v_2)$$
$$= \theta(v_1)(\alpha) + \theta(v_2)(\alpha) = (\theta(v_1) + \theta(v_2))(\alpha).$$

Hence $\theta(v_1 + v_2) = \theta(v_1) + \theta(v_2)$. We leave the reader to verify in a similar way that $\theta(a \cdot v) = a(\theta(v))$. So far it what we have shown that θ is a linear mapping from V to V^{**}.

Now suppose that $\theta(v) = 0$. Then $0 = \theta(v)(\alpha) = \alpha(v)$ for *all* $\alpha \in V^*$. This can only mean that $v = 0$: for if $v \neq 0$, we could include v in a basis of V and construct a linear functional α such that $\alpha(v) = 1_F$, by (8.3.2). It follows that θ is injective.

Finally, assume that V has finite dimension; then $\dim(V) = \dim(V^*) = \dim(V^{**})$. Also $\dim(V^{**}) = \dim(\mathrm{Im}(\theta))$ since θ is injective. It follows from (8.2.9) that $\mathrm{Im}(\theta) = V^{**}$, so that θ^* is an isomorphism. □

8 Vector spaces

Representing linear mappings by matrices. A linear mapping between finite dimensional vector spaces can be described by matrix multiplication, which provides us with a concrete way of representing linear mappings.

Let V and W be vector spaces over the same field F with finite dimensions m and n respectively. Choose ordered bases for V and W, say $\mathcal{B} = \{v_1, v_2, \ldots, v_m\}$ and $\mathcal{C} = \{w_1, w_2, \ldots, w_n\}$ respectively. Now let $\alpha \in L(V, W)$; then

$$\alpha(v_i) = \sum_{j=1}^{n} a_{ji} w_j$$

for some $a_{ji} \in F$. This enables us to form the $n \times m$ matrix over F

$$A = [a_{ij}],$$

which is to represent α. Notice that the i-th column of A is precisely the coordinate column vector of $\alpha(v_i)$ with respect to the basis $\mathcal{C}$. Thus we have a function

$$\theta : L(V, W) \to M_{n,m}(F)$$

defined by the rule that column i of $\theta(\alpha)$ is $[\alpha(v_i)]_\mathcal{C}$

To understand how the matrix $A = \theta(\alpha)$ reproduces the effect of α on an arbitrary vector $v = \sum_{i=1}^{m} b_i v_i$ of V, we compute

$$\alpha(v) = \sum_{i=1}^{m} b_i(\alpha(v_i)) = \sum_{i=1}^{m} b_i \left(\sum_{j=1}^{n} a_{ji} w_j \right) = \sum_{j=1}^{n} \left(\sum_{i=1}^{m} a_{ji} b_i \right) w_j.$$

Hence the coordinate column vector of $\alpha(v)$ with respect to $\mathcal{C}$ has entries $\sum_{i=1}^{m} a_{ji} b_i$, for $j = 1, \ldots, n$, i.e., it is just $A \begin{bmatrix} b_1 \\ \vdots \\ b_m \end{bmatrix} = A[v]_\mathcal{B}$. Therefore we have the basic formula

$$[\alpha(v)]_\mathcal{C} = A[v]_\mathcal{B} = \theta(\alpha)[v]_\mathcal{B}.$$

Concerning the function θ we shall prove:

(8.3.12) *The function $\theta : L(V, W) \to M_{n,m}(F)$ is an isomorphism of vector spaces.*

Proof. We show first that θ is a linear mapping. Let $\alpha, \beta \in L(V, W)$ and $v \in V$. Then the basic formula above shows that

$$\theta(\alpha + \beta)[v]_\mathcal{B} = [\alpha + \beta(v)]_\mathcal{C} = [\alpha(v) + \beta(v)]_\mathcal{C} = [\alpha(v)]_\mathcal{C} + [\beta(v)]_\mathcal{C},$$

which equals

$$\theta(\alpha)[v]_\mathcal{B} + \theta(\beta)[v]_\mathcal{B} = (\theta(\alpha) + \theta(\beta))[v]_\mathcal{B}.$$

Hence $\theta(\alpha + \beta) = \theta(\alpha) + \theta(\beta)$, and in a similar fashion it may be shown that $\theta(a \cdot \alpha) = a(\theta(\alpha))$, where $a \in F$.

Next, if $\theta(\alpha) = 0$, then $[\alpha(v)]_{\mathcal{C}} = 0$ for all $v \in V$, so $\alpha(v) = 0$ and $\alpha = 0$. Hence θ is injective. Both the vector spaces concerned have dimension mn, so $\text{Im}(\theta) = M_{m,n}(F)$ and θ is an isomorphism. □

Example (8.3.4) Consider the dual space $V^* = L(V, F)$, where V is an n-dimensional vector space over a field F. Choose an ordered basis $\mathcal{B}$ of V and use the basis $\{1_F\}$ for V. Then a linear functional $\alpha \in V^*$ is represented by a row vector, i.e., by X^T where $X \in F^n$, according to the rule $\alpha(v) = X^T[v]_{\mathcal{B}}$. So the effect of a linear functional is produced by left multiplication of coordinate vectors by a row vector, (cf. Example (8.3.3)).

The effect of a change of basis. We have seen that any linear mapping between finite dimensional vector spaces can be represented by multiplication by a matrix. However the matrix depends on the choice of ordered bases of the vector spaces. The precise nature of this dependence will now be investigated.

Let $\mathcal{B}$ and $\mathcal{C}$ be ordered bases of finite dimensional vector spaces V and W over a field F, and let $\alpha : V \to W$ be a linear mapping. Then α is represented by a matrix A over F where $[\alpha(v)]_{\mathcal{C}} = A[v]_{\mathcal{B}}$. Now suppose now that two different ordered bases $\mathcal{B}'$ and $\mathcal{C}'$ are chosen for V and W. Then α will be represented by another matrix A'. The question is: how are A and A' related?

To answer the question we introduce the transition matrices S and T for the respective changes of bases $\mathcal{B} \to \mathcal{B}'$ and $\mathcal{C} \to \mathcal{C}'$ (see (8.2.11)). Thus for any $v \in V$ and $w \in W$ we have

$$[v]_{\mathcal{B}'} = S[v]_{\mathcal{B}} \quad \text{and} \quad [w]_{\mathcal{C}'} = T[w]_{\mathcal{C}}.$$

Therefore

$$[\alpha(v)]_{\mathcal{C}'} = T[\alpha(v)]_{\mathcal{C}} = TA[v]_{\mathcal{B}} = TAS^{-1}[v]_{\mathcal{B}'},$$

and it follows that $A' = TAS^{-1}$. We record this conclusion in:

(8.3.13) *Let V and W be non-zero finite dimensional vector spaces over the same field. Let $\mathcal{B}$, $\mathcal{B}'$ be two ordered bases of V and $\mathcal{C}$, $\mathcal{C}'$ two ordered bases of W. Suppose further that S and T are the transition matrices for the changes of bases $\mathcal{B} \to \mathcal{B}'$ and $\mathcal{C} \to \mathcal{C}'$ respectively. If the linear mapping $\alpha : V \to W$ is represented by matrices A and A' with respect to the respective pairs of bases $(\mathcal{B}, \mathcal{C})$ and $(\mathcal{B}', \mathcal{C}')$, then $A' = TAS^{-1}$.*

The case where α is a linear operator on V is especially important. Here $V = W$ and we can take $\mathcal{B} = \mathcal{C}$ and $\mathcal{B}' = \mathcal{C}'$. Thus $S = T$ and $A' = SAS^{-1}$, i.e., A and A' are *similar matrices*. Consequently, *the matrices which can represent a given linear operator are all similar*.

The algebra of linear operators. Let V be a vector space over a field F and suppose also that V is a ring with respect to some multiplication operation. Then V is called an *F-algebra* if, in addition to the vector space and ring axioms, we have the law

$$a(uv) = (au)v = u(av)$$

for all $a \in F$, $u, v \in V$. For example, the set of all $n \times n$ matrices $M_n(F)$ is an F-algebra with respect to the usual matrix operations.

Now let V be any vector space over a field F; we shall write

$$L(V)$$

for the vector space $L(V, V)$ of all linear operators on V. Our aim is to make $L(V)$ into an F-algebra. Now there is a natural product operation on $L(V)$, namely functional composition. Indeed, if $\alpha_1, \alpha_2 \in L(V)$, then $\alpha_1 \alpha_2 \in L(V)$ by an easy check. We claim that with this product operation $L(V)$ becomes an F-algebra.

The first step is to verify that all the ring axioms hold for $L(V)$. This is fairly routine. For example,

$$\alpha_1(\alpha_2 + \alpha_3)(v) = \alpha_1(\alpha_2(v) + \alpha_3(v)) = \alpha_1\alpha_2(v) + \alpha_1\alpha_3(v) = (\alpha_1\alpha_2 + \alpha_1\alpha_3)(v).$$

Hence $\alpha_1(\alpha_2 + \alpha_3) = \alpha_1\alpha_2 + \alpha_1\alpha_3$. Once the ring axioms have been dealt with, we have to check that $a(\alpha_1\alpha_2) = (a\alpha_1)\alpha_2 = \alpha_1(a\alpha_2)$. This too is not hard; indeed all three mappings send v to $a(\alpha_1(\alpha_2(v)))$. Therefore $L(V)$ is an F-algebra.

A function $\alpha : A_1 \to A_2$ between two F-algebras is called an *algebra isomorphism* if it is bijective and it is both a linear mapping of vector spaces and a homomorphism of rings.

(8.3.14) *Let V be a vector space with finite dimension n over a field F. Then $L(V)$ and $M_n(F)$ are isomorphic as F-algebras.*

Proof. Choose an ordered basis $\mathcal{B}$ of V and let $\Phi : L(V) \to M_n(F)$ be the function which associates with a linear operator α the $n \times n$ matrix that represents α with respect to $\mathcal{B}$. Thus $[\alpha(v)]_\mathcal{B} = \Phi(\alpha)[v]_\mathcal{B}$ for all $v \in V$. Clearly Φ is bijective, so to prove that it is an F-algebra isomorphism we need to establish

$$\Phi(\alpha + \beta) = \Phi(\alpha) + \Phi(\beta), \quad \Phi(a \cdot \alpha) = a \cdot \Phi(\alpha) \quad \text{and} \quad \Phi(\alpha\beta) = \Phi(\alpha)\Phi(\beta).$$

For example, take the third statement. If $v \in V$, then

$$[\alpha\beta(v)]_\mathcal{B} = \Phi(\alpha)[\beta(v)]_\mathcal{B} = \Phi(\alpha)(\Phi(\beta)[v]_\mathcal{B}) = \Phi(\alpha)\Phi(\beta)[v]_\mathcal{B}.$$

Therefore $\Phi(\alpha\beta) = \Phi(\alpha)\phi(\beta)$. □

Thus (8.3.14) tells us is in a precise way that linear operators on an n-dimensional vector space over F behave in very much the same manner as $n \times n$ matrices over F.

Exercises (8.3)

1. Which of the following functions are linear mappings?
   (a) $\alpha : \mathbb{R}_3 \to \mathbb{R}$ where $\alpha([x_1 \ x_2 \ x_3]) = \sqrt{x_1^2 + x_2^2 + x_3^2}$;
   (b) $\alpha : M_{m,n}(F) \to M_{n,m}(F)$ where $\alpha(A)$ is the transpose of A;
   (c) $\alpha : M_n(F) \to F$ where $\alpha(A) = \det(A)$.

2. A linear mapping $\alpha : \mathbb{R}^4 \to \mathbb{R}^3$ is defined by
$$\alpha([x_1 \ x_2 \ x_3 \ x_4]^T) = [x_1 - x_2 + x_3 - x_4, \ 2x_1 + x_2 - x_3, \ x_2 - x_3 + x_4]^T.$$
Find the matrix which represents α when the standard bases of $\mathbb{R}^4$ and $\mathbb{R}^3$ are used.

3. Answer Exercise (8.3.2) when the ordered basis $\{[1 \ 1 \ 1]^T, [0 \ 1 \ 1]^T, [0 \ 0 \ 1]^T\}$ of $\mathbb{R}^3$ is used, together with the standard basis of $\mathbb{R}^4$.

4. Find bases for the kernel and image of the following linear mappings:
   (a) $\alpha : F^4 \to F$ where α maps a column vector to the sum of its entries;
   (b) $\alpha : \mathbb{R}[t] \to \mathbb{R}[t]$ where $\alpha(f) = f'$, the derivative of f;
   (c) $\alpha : \mathbb{R}^2 \to \mathbb{R}^2$ where $\alpha([x \ y]^T) = [2x + 3y \ 4x + 6y]^T$.

5. A linear mapping $\alpha : V \to W$ is injective if and only if α maps linearly independent subsets of V to linearly independent subsets of W.

6. A linear mapping $\alpha : V \to W$ is surjective if and only if α maps generating sets of V to generating sets of W.

7. Let U and W be subspaces of a finite dimensional vector space V. Prove that there is a linear operator α on V such that $\text{Ker}(\alpha) = U$ and $\text{Im}(\alpha) = W$ if and only if $\dim(U) + \dim(W) = \dim(V)$.

8. Suppose that $\alpha : V \to W$ is a linear mapping. Explain how to define a corresponding linear mapping $\alpha^* : W^* \to V^*$. Then prove that $(\alpha\beta)^* = \beta^*\alpha^*$.

9. Let $U \xrightarrow{\alpha} V \xrightarrow{\beta} W \to 0$ be an *exact sequence* of vector spaces and linear mappings. (This means that $\text{Im}(\alpha) = \text{Ker}(\beta)$ and $\text{Im}(\beta) = \text{Ker}(W \to 0) = W$, i.e., β is surjective). Prove that the corresponding sequence of dual spaces $0 \to W^* \xrightarrow{\beta^*} V^* \xrightarrow{\alpha^*} U^*$ is exact, (i.e., β^* is injective and $\text{Im}(\beta^*) = \text{Ker}(\alpha^*)$).

8.4 Orthogonality in vector spaces

The aim of this section is to show how to introduce a concept of orthogonality in real vector spaces, the idea being to generalize the geometrical notion of perpendicularity in the Euclidean spaces $\mathbb{R}^2$ and $\mathbb{R}^3$. The essential tool is the concept of an inner product on a real vector space.

8 Vector spaces

Let V be a vector space over the field of real numbers $\mathbb{R}$. An *inner product* on V is a function from $V \times V$ to $\mathbb{R}$, assigning to each pair of vectors u, v a real number

$$\langle u, v \rangle,$$

called the *inner product* of u and v: the following properties are required to hold for all $u, v, w \in V$ and $a, b \in F$:

(a) $\langle v, v \rangle \geq 0$ and $\langle v, v \rangle = 0$ if and only if $v = 0$;
(b) $\langle u, v \rangle = \langle v, u \rangle$;
(c) $\langle au + bv, w \rangle = a \langle u, w \rangle + b \langle v, w \rangle$.

Two vectors u and v are said to be *orthogonal* if $\langle u, v \rangle = 0$. The non-negative real number $\sqrt{\langle v, v \rangle}$ is called the *norm* of the vector v and is written

$$\|v\|.$$

The pair $(V, \langle \; \rangle)$ constitutes what is called an *inner product space*.

Example (8.4.1) The *standard inner product* on $\mathbb{R}^n$ is defined by

$$\langle X, Y \rangle = X^T Y.$$

Elementary matrix algebra shows that the axioms for an inner product are valid here. Thus $(\mathbb{R}^n, \langle \; \rangle)$ is an inner product space. Notice that $|X| = \sqrt{x_1^2 + \cdots + x_n^2}$ where $X = [x_1, \ldots, x_n]$.

When $n \leq 3$, orthogonality as just defined coincides with the geometrical notion of two lines being perpendicular. Recall that a vector in $\mathbb{R}^3$ can be represented by a line segment. Take two vectors X and Y in $\mathbb{R}^3$ and represent them by line segments AB and AC with common initial point A. Denote the angle between the line segments by θ where $0 \leq \theta \leq \pi$. In the first place, if X has entries x_1, x_2, x_3, then $\|X\| = \sqrt{x_1^2 + x_2^2 + x_3^2}$ is the length of the line segment representing X. We further claim that

$$\langle X, Y \rangle = \|X\| \cdot \|Y\| \cos \theta.$$

To see this apply the cosine rule to the triangle ABC, keeping in mind the triangle rule for addition of vectors in $\mathbb{R}^3$:

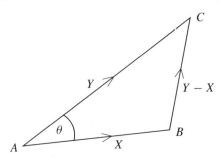

This gives
$$BC^2 = AB^2 + AC^2 - 2(AB \cdot AC)\cos\theta.$$
In vector notation this becomes
$$\|Y - X\|^2 = \|X\|^2 + \|Y\|^2 - 2\|X\| \cdot \|Y\|\cos\theta.$$
Now $\|X\|^2 = x_1^2 + x_2^2 + x_3^2$ and $\|Y\|^2 = y_1^2 + y_2^2 + y_3^2$, while
$$\|Y - X\|^2 = (y_1 - x_1)^2 + (y_2 - x_2)^2 + (y_3 - x_3)^2.$$
Substitute for $\|X\|^2$, $\|Y\|^2$ and $\|Y - X\|^2$ in the equation above and then solve for $\|X\| \cdot \|Y\|\cos\theta$. After some simplification we obtain
$$\|X\| \cdot \|Y\|\cos\theta = x_1 y_1 + x_2 y_2 + x_3 y_3 = X^T Y = \langle X, Y\rangle,$$
as claimed.

Now assume that the vectors X and Y are non-zero. Then $\langle X, Y\rangle = 0$ if and only if $\cos\theta = 0$, i.e., $\theta = \pi/2$. Thus our concept of orthogonality coincides with perpendicularity in $\mathbb{R}^3$: of course the argument is the same for $\mathbb{R}^2$.

The following is a very different example of an inner product.

Example (8.4.2) Let $C[a, b]$ be the vector space of all continuous functions on the interval $[a, b]$ and define
$$\langle f, g\rangle = \int_a^b fg\,dt$$
for f and g in $C[a, b]$. Then $\langle\ \rangle$ is an inner product on $C[a, b]$, as basic properties of definite integrals show.

A fundamental inequality is established next.

(8.4.1) (The Cauchy–Schwartz[1] Inequality) *If u and v are vectors in an inner product space, then $|\langle u, v\rangle| \leq \|u\| \cdot \|v\|$.*

Proof. For any real number t we have
$$\langle u - tv, u - tv\rangle = \langle u, u\rangle - 2t\langle u, v\rangle + t^2\langle v, v\rangle,$$
by the defining properties of the inner product. Hence, setting $a = \langle u, u\rangle$, $b = \langle u, v\rangle$ and $c = \langle w, w\rangle$, we have
$$at^2 - 2bt + c = \|u - tv\|^2 \geq 0.$$
Now the condition for the quadratic expression $at^2 - 2bt + c$ to be non-negative for all values of t is that $b^2 \leq ac$, as may be seen by completing the square. Hence, taking square roots, we obtain $|b| \leq \sqrt{ac}$ or $|\langle u, v\rangle| \leq \|u\| \cdot \|v\|$. □

---
[1] Hermann Amandus Schwartz (1843–1921).

The inequality of (8.4.1) allows us to introduce the concept of angle in an arbitrary inner product space $(V, \langle \ \rangle)$. Let $0 \neq u, v \in V$; since $\langle u, v \rangle / \|u\| \cdot \|v\|$ lies in the interval $[-1, 1]$, there is a unique angle θ in the range $0 \leq \theta \leq \pi$ satisfying

$$\cos \theta = \frac{\langle u, v \rangle}{\|u\| \cdot \|v\|}.$$

Call θ the *angle between the vectors u and v*. In the case of Euclidean 3-space this coincides with the usual geometrical notion of the angle between two line segments.

Another important inequality valid in any inner product space is:

(8.4.2) (The Triangle Inequality) *If u and v are vectors in an inner product space, then $\|u + v\| \leq \|u\| + \|v\|$.*

Proof. From the properties of inner products we have

$$\|u + v\|^2 = \langle u + v, u + v \rangle = \|u\|^2 + 2\langle u, v \rangle + \|v\|^2,$$

and by (8.4.1) $\langle u, v \rangle \leq \|u\| \cdot \|v\|$. Therefore

$$\|u + v\|^2 \leq \|u\|^2 + 2\|u\| \cdot \|v\| + \|v\|^2 = (\|u\| + \|v\|)^2,$$

whence the inequality follows. □

Notice that for $\mathbb{R}^3$ the assertion of (8.4.2) is that the sum of the lengths of two sides of a triangle cannot exceed the length of the third side; hence the terminology "Triangle Inequality".

It was seen that in $\mathbb{R}^3$ the norm of a vector is just the length of any representing line segment. One can introduce a concept of length in an abstract real vector space by postulating the existence of a norm function.

Let V be a real vector space. A *norm* on V is a function from V to $\mathbb{R}$, written $v \mapsto \|v\|$, such that for all $u, v \in V$ and $a \in F$:

(a) $\|v\| \geq 0$ and $\|v\| = 0$ if and only if $v = 0$;

(b) $\|av\| = |a| \cdot \|v\|$;

(c) the triangle inequality holds, $\|u + v\| \leq \|u\| + \|v\|$.

The pair $(V, \| \ \|)$ is then called a *normed linear space*.

If $(V, \langle \ \rangle)$ is an inner product space and we define $\|v\|$ to be $\sqrt{\langle v, v \rangle}$, then $(V, \| \ \|)$ is a normed linear space. For (a) holds by definition and (c) by (8.4.2): also $\|av\| = \sqrt{\langle av, av \rangle} = \sqrt{a^2 \langle v, v \rangle} = |a| \|v\|$. Thus every inner product space is a normed linear space.

8.4 Orthogonality in vector spaces

Orthogonal complements. Suppose that S is a subspace of an inner product space V. Then the *orthogonal complement* of S is the subset

$$S^\perp$$

of all vectors in V that are orthogonal to every vector in S. The most basic facts about $S^\perp$ are contained in:

(8.4.3) *Let S be a subspace of an inner product space V. Then:*

(a) $S^\perp$ *is a subspace of V;*

(b) $S \cap S^\perp = 0$;

(c) *a vector v belongs to $S^\perp$ if and only if it is orthogonal to every vector in a set of generators of S.*

Proof. We leave the proofs of (a) and (b) an exercise. To prove (c) suppose that $S = F\langle X \rangle$ and assume that $v \in V$ is orthogonal to every vector in X. If $s \in S$, then $s = a_1 x_1 + \cdots + a_r x_r$ where $x_i \in X$ and the a_i are scalars. Then $\langle v, s \rangle = \sum_{i=1}^{r} a_i \langle v, x_i \rangle = 0$ since each $\langle v, x_i \rangle$ is 0. Hence $v \in S^\perp$. The converse is obvious. □

Example (8.4.3) Consider the Euclidean inner product space $\mathbb{R}^n$ with $\langle X, Y \rangle = X^T Y$. Choose a real $m \times n$ matrix A and let S denote the column space of the transpose A^T, i.e., the subspace of $\mathbb{R}^n$ generated by the columns of A^T. Then by (8.4.3) a vector X belongs to $S^\perp$ if and only if $\langle C, X \rangle = C^T X = 0$ for every column C of A^T. Since C^T is a row of A, this amounts to requiring that $AX = 0$. In short $S^\perp$ is precisely the null space of A.

The main result about orthogonal complements is:

(8.4.4) *Let S be a subspace of a finite dimensional real inner product space V. Then $V = S \oplus S^\perp$ and $\dim(V) = \dim(S) + \dim(S^\perp)$.*

Proof. If $S = 0$, then clearly $S^\perp = V$ and it is obvious that $V = S \oplus S^\perp$. So assume that $S \neq 0$. Choose a basis $\{v_1, v_2, \ldots, v_m\}$ of S and extend it to a basis of V, say $\{v_1, \ldots, v_m, v_{m+1}, \ldots, v_n\}$. An arbitrary vector $v \in V$ can be written in the form $v = \sum_{j=1}^{n} a_j v_j$, with scalars a_j. Now according to (8.4.3) $v \in S^\perp$ if and only if v is orthogonal to each of the vectors $v_1, \ldots, v_m$: the conditions for this to hold are that

$$\langle v_i, v \rangle = \sum_{j=1}^{n} \langle v_i, v_j \rangle a_j = 0$$

for $i = 1, 2, \ldots, m$. This is a system of m homogeneous linear equations in the n unknowns $a_1, \ldots, a_n$ with coefficient matrix $A = [\langle v_i, v_j \rangle]$. Thus $\dim(S^\perp)$ equals the dimension of the solution space, i.e., the null space of A. But we know from (8.3.8) that this dimension is $n - r$ where r is the rank of A. Now it is clear that $r \leq m$ since A is $m \times n$. We shall argue that in fact $r = m$. If this were false, the m rows of A would be linearly dependent, and there would exist scalars $d_1, d_2, \ldots, d_m$, not all 0, such that

$$0 = \sum_{i=1}^{m} d_i \langle v_i, v_j \rangle = \left\langle \left(\sum_{i=1}^{m} d_i v_i \right), v_j \right\rangle$$

for $j = 1, 2, \ldots, n$. But this in turn would mean that the vector $\sum_{i=1}^{m} d_i v_i$ is orthogonal to every basis vector of V, so $\sum_{i=1}^{m} d_i v_i$ is orthogonal to itself and hence must be 0. Finally $d_i = 0$ for all i because the v_i are linearly independent; thus a contradiction has been reached

The argument given shows that $\dim(S^\perp) = n - r = n - m = n - \dim(S)$. Hence

$$\dim(V) = \dim(S) + \dim(S^\perp) = \dim(S \oplus S^\perp).$$

From (8.2.9) it follows that $V = S \oplus S^\perp$. □

Corollary (8.4.5) *If S is a subspace of a finite dimensional real inner product space, V, then $(S^\perp)^\perp = S$.*

Proof. Clearly $S \subseteq (S^\perp)^\perp$. Also by (8.4.4)

$$\dim(S^\perp)^\perp = \dim(V) - \dim(S^\perp) = \dim(V) - (\dim(V) - \dim(S)) = \dim(S).$$

Thus $(S^\perp)^\perp = S$ by (8.2.9) once again. □

Orthonormal bases. A non-empty subset S of an inner product space is called *orthonormal* if each pair of vectors in S is orthogonal and each vector in S has norm 1. For example, the standard basis $\{E_1, E_2, \ldots, E_n\}$ in Euclidean n-space is an orthonormal basis.

As an application of (8.4.4) we prove:

(8.4.6) *Every non-zero finite dimensional real inner product space V has an orthonormal basis.*

Proof. Choose $u \neq 0$ in V and put $v = \frac{1}{\|u\|} u$. Thus $\|v\| = 1$. Let $S = F\langle v \rangle$ and apply (8.4.4) to get $V = S \oplus S^\perp$. Now $\dim(S^\perp) = \dim(V) - \dim(S) = \dim(V) - 1$. By induction on the dimension of V either the subspace $S^\perp$ is 0 or it has an orthonormal basis. Adjoin v_1 to this basis to get an orthonormal basis of V. □

8.4 Orthogonality in vector spaces

The idea behind this proof may be used to give an algorithm which, when a basis for S is given, constructs an orthonormal basis: this is the well-known Gram[2]–Schmidt[3] Process – see Exercise (8.4.10) below.

Exercises (8.4)

1. Which of the following are inner product spaces?
    (a) $\mathbb{R}^n$ where $\langle X, Y \rangle = -X^T Y$;
    (b) $\mathbb{R}^n$ where $\langle X, Y \rangle = 2X^T Y$;
    (c) $C[a, b]$, the vector space of continuous functions on the interval $[a, b]$, where $\langle f, g \rangle = \int_a^b (f + g) dx$.

2. Let $w \in C[a, b]$ be positive valued and define $\langle f, g \rangle = \int_a^b f(x)w(x)g(x)dx$ for $f, g \in C[a, b]$. Prove that $\langle \ \rangle$ is an inner product on $C[a, b]$. What does the Cauchy–Schwartz Inequality tell us in this case?

3. Let $\langle \ \rangle$ be an inner product on a finite dimensional inner product space V. Choose an ordered basis $\mathcal{B} = \{v_1, \ldots, v_n\}$ of V, define a_{ij} to be $\langle v_i, v_j \rangle$ and let A be the matrix $[a_{ij}]$.
    (a) Prove that $\langle u, v \rangle = [u]_\mathcal{B}^T A[v]_\mathcal{B}$ for all vectors u, v in V.
    (b) Show that the matrix A is symmetric and *positive definite*, i.e., $X^T A X > 0$ for all $X \neq 0$.

4. Let A be any real symmetric, positive definite $n \times n$ matrix. Define $\langle X, Y \rangle$ to be $X^T A Y$. Show that $\langle \ \rangle$ is an inner product on $\mathbb{R}^n$.

5. Let V be the vector space of real polynomials in t with degree at most 2. An inner product on V is defined by $\langle f, g \rangle = \int_0^1 f(t)g(t)dt$. If S is the subspace of V generated by $1 - t^2$ and $1 - t + t^2$, find a basis for $S^\perp$.

6. Let S and T be subspaces of a finite dimensional inner product space V. Prove that:
    (a) $(S + T)^\perp = S^\perp \cap T^\perp$;
    (b) $S^\perp = T^\perp$ implies that $S = T$;
    (c) $(S \cap T)^\perp = S^\perp + T^\perp$.

7. Suppose that $\{v_1, v_2, \ldots, v_n\}$ is an orthonormal basis of an inner product space V. If v is any vector in V, then $v = \sum_{i=1}^n \langle v, v_i \rangle v_i$ and $\|v\|^2 = \sum_{i=1}^n \langle v, v_i \rangle^2$.

8. (*Complex inner products.*) Let V be a vector space over the field of complex numbers $\mathbb{C}$. A *complex inner product* on V is a function $\langle \ \rangle$ from $V \times V$ to $\mathbb{C}$ such that
    (a) $\langle v, v \rangle$ is real and non-negative with $\langle v, v \rangle = 0$ only if $v = 0$;
    (b) $\langle u, v \rangle = \overline{\langle v, u \rangle}$;
    (c) $\langle au + bv, w \rangle = \overline{a}\langle u, w \rangle + \overline{b}\langle v, w \rangle$.

---
[2] Jorgen Pedersen Gram (1850–1916).
[3] Erhart Schmidt (1876–1959).

(Here a "bar" denotes the complex conjugate). If $X, Y \in \mathbb{C}^n$, define $\langle X, Y \rangle = \overline{(X^T)}Y$. Show that $\langle \ \rangle$ is a complex inner product on $\mathbb{C}^n$.

9. Let V be a complex inner product space, i.e., a vector space over $\mathbb{C}$ with a complex inner product.

   (a) Prove that $\langle u+av, u+av \rangle = \|u\|^2 + 2\,\text{Re}(a\langle u, v\rangle) + |a|^2 \|v\|^2$ where $u, v \in V$ and $a \in \mathbb{C}$.

   (b) Prove that the Cauchy–Schwartz Inequality is valid in V. [Hint: assume $v \neq 0$ and take a in (a) to be $-\overline{\langle u, v \rangle}/\|v\|^2$].

   (c) Prove that the Triangle Inequality is valid in V.

10. (*The Gram–Schmidt Process*). Let $\{u_1, u_2, \ldots, u_n\}$ be a basis of a real inner product space V. Define vectors $v_1, v_2, \ldots, v_n$ by the rules

$$v_1 = \frac{1}{\|u_1\|} u_1 \quad \text{and} \quad v_{i+1} = \frac{1}{\|u_{i+1} - p_i\|}(u_{i+1} - p_i),$$

where $p_i = \langle u_{i+1}, v_1 \rangle v_1 + \cdots + \langle u_{i+1}, v_i \rangle v_i$. Show that $\{v_1, v_2, \ldots, v_n\}$ is an orthonormal basis of V. [Hint: Show that $u_{i+1} - p_i$ is orthogonal to each of $v_1, v_2, \ldots, v_{i-1}$].

11. Apply the Gram–Schmidt Process to find an orthonormal basis of the vector space of all real polynomials in t of degree ≤ 2 where $\langle f, g \rangle = \int_{-1}^{+1} f(t)g(t)dt$, starting with the basis $\{1, t, t^2\}$.

Chapter 9
The structure of groups

In this chapter we shall pursue the study of groups at a deeper level. A common method of investigation in algebra is to try to break up a complex structure into simpler substructures. The hope is that, by repeated application of this procedure, one will eventually arrive at structures that are easy to understand. It may then be possible, in some sense, to synthesise these substructures to reconstruct the original one. While it is rare for the procedure just described to be brought to such a perfect state of completion, the analytic-synthetic method can yield valuable information and suggest new concepts. We shall consider how far it can be used in group theory.

9.1 The Jordan–Hölder Theorem

The basic concept here is that of a *series in a group* G. By this is meant a chain of subgroups $S = \{G_i \mid i = 0, 1, \ldots, n\}$ leading from the identity subgroup up to G, with each term normal in its successor, i.e.,

$$1 = G_0 \triangleleft G_1 \triangleleft \cdots \triangleleft G_n = G.$$

The G_i are the *terms* of the series and the quotient groups G_{i+1}/G_i are the *factors*. The *length* of the series is defined to be the number of non-trivial factors. Note that G_i may not be normal in G since normality is not a transitive relation – see Exercise (4.2.6). A subgroup H which appears in some series is called a *subnormal subgroup* of G. Clearly this is equivalent to requiring that there be a chain of subgroups leading from H to G, i.e., $H = H_0 \triangleleft H_1 \triangleleft \cdots \triangleleft H_m = G$.

A partial order on the set of series in a group G can be introduced in the following way. A series S is called a *refinement* of a series T if every term of T is also a term of S. If S has at least one term which is not a term of T, then S is a *proper refinement* of T. It is easy to see that the relation of being a refinement is a partial order on the set of series in G.

Example (9.1.1) The symmetric group S_4 has the series $1 \triangleleft V \triangleleft A_4 \triangleleft S_4$, where as usual V is the Klein 4-group. This series is a refinement of the series $1 \triangleleft A_4 \triangleleft S_4$.

Isomorphic series. Two series S and T in a group G are called *isomorphic* if there is a bijection from the set of non-trivial factors of S to the set of non-trivial factors of T such that corresponding factors are isomorphic groups. Thus isomorphic series must have the same length, but the isomorphic factors may occur at different points in the series.

Example (9.1.2) In $\mathbb{Z}_6$ the series $0 \triangleleft \langle [2] \rangle \triangleleft \mathbb{Z}_6$ and $0 \triangleleft \langle [3] \rangle \triangleleft \mathbb{Z}_6$ are isomorphic since $\langle [2] \rangle \simeq \mathbb{Z}_3 \simeq \mathbb{Z}_6/\langle [3] \rangle$ and $\langle [3] \rangle \simeq \mathbb{Z}_2 \simeq \mathbb{Z}_6/\langle [2] \rangle$.

The foundation for the entire theory of series in groups is the following technical result. It should be viewed as a generalization of the Second Isomorphism Theorem – see (4.3.5).

(9.1.1) (Zassenhaus's[1] Lemma) *Let A_1, A_2, B_1, B_2 be subgroups of a group G such that $A_1 \triangleleft A_2$ and $B_1 \triangleleft B_2$. Define $D_{ij} = A_i \cap B_j$, $(i, j = 1, 2)$. Then $A_1 D_{21} \triangleleft A_1 D_{22}$ and $B_1 D_{12} \triangleleft B_1 D_{22}$. Furthermore $A_1 D_{22}/A_1 D_{21} \simeq B_1 D_{22}/B_1 D_{12}$.*

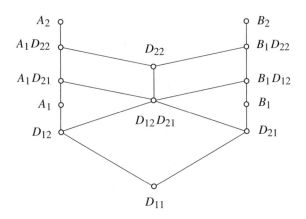

Proof. The Hasse diagram above displays all the relevant subgroups. From $B_1 \triangleleft B_2$ we obtain $D_{21} \triangleleft D_{22}$. Since $A_1 \triangleleft A_2$, it follows that $A_1 D_{21} \triangleleft A_1 D_{22}$, on applying the canonical homomorphism $A_2 \to A_2/A_1$. Similarly $B_1 D_{12} \triangleleft B_1 D_{22}$. Now use the Second Isomorphism Theorem (4.3.5) with $H = D_{22}$ and $N = A_1 D_{21}$; thus $HN/N \simeq H/H \cap N$. But $HN = A_1 D_{22}$ and $H \cap N = D_{22} \cap (A_1 D_{21}) = D_{12} D_{21}$ by the Modular Law – see (4.1.11). The conclusion is that $A_1 D_{22}/A_1 D_{21} \simeq D_{22}/D_{12} D_{21}$. By the same argument $B_1 D_{22}/B_1 D_{12} \simeq D_{22}/D_{12} D_{21}$, from which the result follows. □

The main use of Zassenhaus's Lemma is to prove a theorem about refinements: its statement is remarkably simple.

(9.1.2) (Schreier's[2] Refinement Theorem). *Any two series in a group have isomorphic refinements.*

---

[1] Hans Zassenhaus (1912–1991).
[2] Otto Schreier (1901–1929).

Proof. Let $1 = H_0 \triangleleft H_1 \triangleleft \cdots \triangleleft H_l = G$ and $1 = K_0 \triangleleft K_1 \triangleleft \cdots \triangleleft K_m = G$ be two series in a group G. Define subgroups $H_{ij} = H_i(H_{i+1} \cap K_j)$ for $0 \le i \le l-1$, $0 \le j \le m$ and $K_{ij} = K_j(H_i \cap K_{j+1})$ for $0 \le i \le l, 0 \le j \le m-1$. Now apply (9.1.1) with $A_1 = H_i$, $A_2 = H_{i+1}$, $B_1 = K_j$ and $B_2 = K_{j+1}$; then the conclusion is that $H_{ij} \triangleleft H_{ij+1}$ and $K_{ij} \triangleleft K_{i+1j}$, and also that $H_{ij+1}/H_{ij} \simeq K_{i+1j}/K_{ij}$. Therefore the series $\{H_{ij} \mid i = 0, 1, \ldots, l-1, j = 0, 1, \ldots m\}$ and $\{K_{ij} \mid i = 0, 1, \ldots, l, j = 0, 1, \ldots, m-1\}$ are isomorphic refinements of $\{H_i \mid i = 0, 1, \ldots, l\}$ and $\{K_j \mid j = 0, 1, \ldots, m\}$ respectively. □

Composition series. A series which has no proper refinements is called a *composition series* and its factors are *composition factors*. If G is a finite group, we can start with any series, for example $1 \triangleleft G$, and keep refining it until we reach a composition series. Thus *every finite group has a composition series*. However not every infinite group has a composition series, as is shown in (9.1.6) below.

A composition series can be recognized from the structure of its factors.

(9.1.3) *A series is a composition series if and only if all its factors are simple groups.*

Proof. Let X/Y be a factor of a series in a group G. If X/Y is not simple, there is a subgroup W such that $Y < W < X$ and $W \triangleleft X$; here the Correspondence Theorem (4.2.2) has been invoked. Adjoining W to the given series, we obtain a new series which is a proper refinement of it, with terms $Y \triangleleft W \triangleleft X$ in place of $Y \triangleleft X$.

Conversely, if a series in G has a proper refinement, there must be two consecutive terms $Y \triangleleft X$ of the series with additional terms of the refined series between Y and X. Hence there is a subgroup W such that $Y < W < X$ and $W \triangleleft X$. But then W/Y is a proper non-trivial subgroup of X/Y and the latter cannot be a simple group. So the result is proved. □

The main result about composition series is a celebrated theorem associated with the names of two prominent 19th Century algebraists, Camille Jordan (1838–1922) and Otto Hölder (1859–1937).

(9.1.4) (The Jordan–Hölder Theorem) *Let S be a composition series in a group G and suppose that T is any series in G. Then T has a refinement which is isomorphic with S.*

The most important case is when T is itself a composition series and the conclusion is that T is isomorphic with S. Thus we obtain:

Corollary (9.1.5) *Any two composition series in a group are isomorphic.*

Proof of (9.1.4). By the Refinement Theorem (9.1.2), the series S and T have isomorphic refinements. But S is a composition series, so it must be isomorphic with a refinement of T. □

Example (9.1.3) Consider the symmetric group S_4. This has a series

$$1 \triangleleft C \triangleleft V \triangleleft A_4 \triangleleft S_4$$

where $|C| = 2$ and V is the Klein 4-group. Now C, V/C and S_4/A_4 all have order 2, while A_4/V has order 3, so all factors of the series are simple. By (9.1.3) the series is a composition series, with composition factors $\mathbb{Z}_2, \mathbb{Z}_2, \mathbb{Z}_3, \mathbb{Z}_2$.

The next result shows that not every group has a composition series.

(9.1.6) *An abelian group A has a composition series if and only if it is finite.*

Proof. Only necessity is in doubt, so suppose that A has a composition series. Then each factor of the series is simple and also abelian. So by (4.1.9) each factor has prime order and A is therefore A finite. □

Example (9.1.4) Let n be an integer greater than 1. The group $\mathbb{Z}_n$ has a composition series with factors of prime order. Since the product of the orders of the composition factors is equal to n, the group order, it follows that n is a product of primes, which is the first part of the of the Fundamental Theorem of Arithmetic. In fact we can also get the uniqueness part.

Suppose that $n = p_1 p_2 \ldots p_k$ is an expression for n as a product of primes. Define H_i to be the subgroup of $\mathbb{Z}_n$ generated by the congruence class $[p_{i+1} p_{i+2} \ldots p_k]$ where $0 \leq i < k$ and let $H_k = \mathbb{Z}_n$. Then $0 = H_0 \triangleleft H_1 \triangleleft \cdots \triangleleft H_{k-1} \triangleleft H_k = \mathbb{Z}_n$ is a series in $\mathbb{Z}_n$. Now clearly $|H_i| = p_1 p_2 \ldots p_i$ and thus $|H_{i+1}/H_i| = p_{i+1}$. So we have constructed a composition series in $\mathbb{Z}_n$ with factors of orders $p_1, p_2, \ldots, p_k$.

Suppose next that $n = q_1 q_2 \ldots q_l$ is another expression for n as product of primes. Then there is a corresponding composition series with factors of orders $q_1, q_2, \ldots, q_l$. But the Jordan–Hölder Theorem tells us that these composition series are isomorphic. Consequently the q_j's must be the p_i's in another order. In short we have recovered the Fundamental Theorem of Arithmetic from the Jordan–Hölder Theorem.

Some simple groups. Our investigations so far show that in a certain sense every finite group decomposes into a number of simple groups, namely its composition factors. Now the only simple groups we currently know are the groups of prime order and the alternating group A_5 – see (5.3.10). It is now definitely time to expand this list, which we do by proving:

(9.1.7) *The alternating group A_n is simple if and only if $n \neq 1, 2$ or 4.*

The proof uses a result about 3-cycles, which will be established first.

(9.1.8) *If $n \geq 3$, the alternating group A_n is generated by 3-cycles.*

Proof. First of all note that 3-cycles are even and hence belong to A_n. Next each element of A_n is the product of an even number of transpositions by (3.1.4). Finally, the equations $(a\ c)(a\ b) = (a\ b\ c)$ and $(a\ b)(c\ d) = (a\ d\ b)(a\ d\ c)$ demonstrate that every element of A_n is a product of 3-cycles. □

Proof of (9.1.7). In the first place A_4 has a normal subgroup of order 4, so it cannot be simple. Also A_1 and A_2 have order 1 and are therefore excluded. However $|A_3| = 3$ and thus A_3 is simple. So we can assume that $n \geq 5$ and aim to show that A_n is simple. If this is false, there is a proper non-trivial normal subgroup, say N. The proof analyzes the possible forms of elements of N.

Assume first that N contains a 3-cycle $(a\ b\ c)$. If $(a'\ b'\ c')$ is another 3-cycle and π in S_n sends a, b, c to a', b', c' respectively, then $\pi(a\ b\ c)\pi^{-1} = (a'\ b'\ c')$. If π is even, it follows that $(a'\ b'\ c') \in N$. If however π is odd, we can replace it by the even permutation $\pi \circ (e\ f)$ where e, f are different from a', b', c' – notice that this uses $n \geq 5$. We will still have $\pi(a\ b\ c)\pi^{-1} = (a'\ b'\ c')$. Consequently N contains all 3-cycles and by (9.1.8) $N = A_n$, a contradiction. Hence N cannot contain a 3-cycle.

Assume next that N contains a permutation π whose disjoint cycle decomposition involves a cycle of length at least 4, say

$$\pi = (a_1\ a_2\ a_3\ a_4\ \ldots)\ \ldots$$

where the final dots indicate the possible presence of further disjoint cycles. Now N also contains the conjugate of π

$$\pi' = (a_1\ a_2\ a_3)\pi(a_1\ a_2\ a_3)^{-1} = (a_2\ a_3\ a_1\ a_4\ \ldots)\ \ldots.$$

Therefore N contains $\pi'\pi^{-1} = (a_1\ a_2\ a_4)$. Note that the other cycles cancel here. Since this conclusion is untenable, elements in N must have disjoint cycle decompositions involving cycles of length at most 3. Furthermore, such elements cannot involve just one 3-cycle, otherwise by squaring we would obtain a 3-cycle in N.

Assume next that N contains a permutation with at least two 3-cycles, say $\pi = (a\ b\ c)(a'\ b'\ c')\ \ldots$. Then N will also contain the conjugate

$$\pi' = (a'\ b'\ c)\pi(a'\ b'\ c)^{-1} = (a\ b\ a')(c\ c'\ b')\ \ldots,$$

and hence N contains $\pi\pi' = (a\ c\ a'\ b\ b')\ \ldots$, which has been seen to be impossible. Therefore each element of N must be the product of an *even* number of disjoint transpositions.

If $\pi = (a\ b)(a'\ b') \in N$, then N will contain $\pi' = (a\ c\ b)\pi(a\ c\ b)^{-1} = (a\ c)(a'\ b')$ for any c unaffected by π. But then N will contain $\pi\pi' = (a\ c\ b)$, which is false. Consequently, if $1 \neq \pi \in N$, then $\pi = (a_1\ b_1)(a_2\ b_2)(a_3\ b_3)(a_4\ b_4)\ldots$, with at least four transpositions. But then N will also contain

$$\pi' = (a_3\ b_2)(a_2\ b_1)\pi(a_2\ b_1)(a_3\ b_2) = (a_1\ a_2)(a_3\ b_1)(b_2\ b_3)(a_4\ b_4)\ \ldots.$$

Finally, N contains $\pi\pi' = (a_1\ b_2\ a_3)(a_2\ b_1\ b_3)$, our final contradiction. □

Thus A_5 and A_6 are simple groups of order $\frac{1}{2}(5!) = 60$ and $\frac{1}{2}(6!) = 360$ respectively. There are of course infinitely many such examples. We will now use the simplicity of A_n to determine the composition series of S_n.

Example (9.1.5) If $n \geq 5$, then $1 \lhd A_n \lhd S_n$ is the unique composition series of S_n.

In the first place this is a composition series since A_n and $S_n/A_n \simeq \mathbb{Z}_2$ are simple groups. Suppose that N is a proper non-trivial normal subgroup of S_n. We shall argue that $N = A_n$, which will settle the matter. Note first that $N \cap A_n \lhd A_n$, so that either $N \cap A_n = 1$ or $A_n \leq N$ since A_n is simple. Now $|S_n : A_n| = 2$, so if $A_n \leq N$, then $N = A_n$. Suppose therefore that $N \cap A_n = 1$. Then $NA_n = S_n$ and $|N| = |NA_n/A_n| = |S_n/A_n| = 2$, which implies that $|N| = 2$. Thus N contains a single non-identity element, necessarily an odd permutation. Since N must contain all conjugates of this permutation, it belongs to the center of S_n; however $Z(S_n) = 1$ by Exercise (4.2.10) and a contradiction ensues.

Projective linear groups. We mention in passing another infinite family of finite simple groups. Let F be any field. It is not difficult to prove by direct matrix calculations that the center of the general linear group $\mathrm{GL}_n(F)$ is just the subgroup of all scalar matrices $f 1_n$ where $f \in F$ – see Exercise (4.2.12). The *projective general linear group of degree n over F* is defined to be

$$\mathrm{PGL}_n(F) = \mathrm{GL}_n(F)/Z(\mathrm{GL}_n(F)).$$

Recall that $\mathrm{SL}_n(F)$ is the special linear group consisting of all matrices in $\mathrm{GL}_n(F)$ with determinant 1. The center of $\mathrm{SL}_n(F)$ can be shown to be $Z(\mathrm{GL}_n(F)) \cap \mathrm{SL}_n(F)$. Therefore by the Second Isomorphism Theorem

$$\mathrm{SL}_n(F)Z(\mathrm{GL}_n(F))/Z(\mathrm{GL}_n(F)) \simeq \mathrm{SL}_n(F)/Z(\mathrm{SL}_n(F)).$$

This group is called the *projective special linear group*

$$\mathrm{PSL}_n(F).$$

The projective special linear groups are usually simple; in fact the following result holds true.

(9.1.9) *Let F be a field and let $n > 1$. Then $\mathrm{PSL}_n(F)$ is simple if and only if $n \geq 3$ or $n = 2$ and F has more than three elements.*

This result can be proved by direct, if tedious, matrix calculations, (see [10]). When F is a finite field of order q, which by (8.2.16) is necessarily a prime power, it is easy to compute the orders of the projective groups. It is usual to write

$$\mathrm{GL}_n(q), \; \mathrm{PGL}_n(q), \ldots$$

instead of $\mathrm{GL}_n(F)$, $\mathrm{PGL}_n(F)$, etc., where F is a field of order q. (By (10.3.5) below there is just one field of order q). Now $|Z(\mathrm{GL}_n(F)| = |F^*| = q - 1$, where $F^* =$

$U(F) = F - 0$, and also $|Z(\mathrm{SL}_n(F))| = \gcd\{n, q-1\}$: for the last statement one needs to know that F^* is a cyclic group – for a proof see (10.3.6) below. The orders of the projective groups can now be read off. A simple count of the non-singular $n \times n$ matrices over F reveals that $\mathrm{GL}_n(q)$ has order

$$(q^n - 1)(q^n - q) \ldots (q^n - q^{n-1}),$$

while $|\mathrm{SL}_n(q)| = |\mathrm{GL}_n(q)|/(q-1)$. Thus we have formulas for the orders of the projective groups.

(9.1.10) (i) $|\mathrm{PGL}_n(q)| = |\mathrm{GL}_n(q)|/(q-1)$;
(ii) $|\mathrm{PSL}_n(q)| = |\mathrm{SL}_n(q)|/\gcd\{n, q-1\}$.

For example, $\mathrm{PSL}_2(5)$ is a simple group of order 60. In fact there is only one simple group of this order – see Exercise (9.2.17) – so $\mathrm{PSL}_2(5)$ must be isomorphic with A_5. But $\mathrm{PSL}_2(7)$ of order 168 and $\mathrm{PSL}_2(8)$ of order 504 are simple groups that are not of alternating type.

The projective groups and projective space. We shall indicate briefly how the projective groups arise in geometry. Let V be an $(n+1)$-dimensional vector space over a field F and let V^* be the set of all non-zero vectors in V. An equivalence relation $\sim$ on V^* is introduced by the following rule: $u \sim v$ if and only if $u = fv$ for some $f \neq 0$ in F. Let $[v]$ denote the equivalence class of the vector v: this is just the set of non-zero multiples of v. Then the set

$$\tilde{V}$$

of all $[v]$ with $v \in V^*$, is called an *n-dimensional projective space* over F.

Now let α be a bijective linear operator on V. Then there is an induced mapping $\tilde{\alpha} : \tilde{V} \to \tilde{V}$ defined by the rule

$$\tilde{\alpha}([v]) = [\alpha(v)].$$

Here $\tilde{\alpha}$ is called a *collineation* on $\tilde{V}$. It is not hard to see that the collineations on $\tilde{V}$ form a group $\mathrm{PGL}(V)$ with respect to functional composition.

It is straightforward to check that the assignment $\alpha \mapsto \tilde{\alpha}$ gives rise to a surjective group homomorphism from $\mathrm{GL}(V)$ to $\mathrm{PGL}(\tilde{V})$ with kernel equal to the subgroup of all scalar linear operators. Therefore $\mathrm{PGL}(\tilde{V}) \simeq \mathrm{PGL}_n(F)$, while $\mathrm{PSL}_n(F)$ corresponds to the subgroup of collineations with determinant equal to 1.

The classification of finite simple groups. The projective special linear groups form just one of a number of infinite families of finite simple groups known collectively as *the simple groups of Lie type*. They arise as groups of automorphisms of simple Lie algebras. In addition to the alternating groups and the groups of Lie type, there are 26 isolated simple groups, the so-called *sporadic simple groups*. The smallest of these, the *Mathieu*[3] *group* M_{11}, has order 7920, while the largest one, the so-called *Monster*, has order

$$2^{46} \cdot 3^{20} \cdot 5^9 \cdot 7^6 \cdot 11^2 \cdot 13^3 \cdot 17 \cdot 19 \cdot 23 \cdot 29 \cdot 31 \cdot 41 \cdot 47 \cdot 59 \cdot 71,$$

or approximately 8.08×10^{53}.

It is now widely accepted that the alternating groups, the simple groups of Lie type and the sporadic simple groups account for all the finite non-abelian simple groups. A complete proof of this profound result has yet to appear, but a multi-volume work devoted to this is in preparation. The proof of the classification of finite simple groups is a synthesis of the work of many mathematicians and is by any standard one of the great scientific achievements of all time.

To conclude the section let us assess how far we have come in trying to understand the structure of finite groups. If our aim is to construct all finite groups, the Jordan–Hölder Theorem shows that two steps are necessary:

(i) find all finite simple groups;

(ii) construct all possible group extensions of a given finite group N by a finite simple group S.

Here (ii) means that we have to construct all groups G with a normal subgroup M such that $M \simeq N$ and $G/M \simeq S$. Suppose we accept that step (i) has been accomplished. In fact a formal description of the extensions arising in (ii) is possible. However it should be emphasized that the problem of deciding when two of the constructed groups are isomorphic is likely to be intractable. Thus the practicality of the scheme is questionable. This does not mean that the enterprise was not worthwhile since a vast amount of knowledge about finite groups has been accumulated during the course of the program.

Exercises (9.1)

1. Show that isomorphic groups have the same composition factors.

2. Find two non-isomorphic groups with the same composition factors.

3. Show that S_3 has a unique composition series, while S_4 has exactly three composition series.

4. Let G be a finite group and let $N \triangleleft G$. How are the composition factors of G related to those of N and G/N?

5. Suppose that G is a group generated by normal subgroups $N_1, N_2, \ldots, N_k$ each of which is simple. Prove that G is the direct product of certain of the N_i. [Hint: choose r maximal subject to the existence of normal subgroups $N_{i_1}, \ldots, N_{i_r}$ which generate their direct product. Then show that the direct product equals G].

---

[3]Émile Léonard Mathieu (1835–1890).

6. Let G be as in the previous exercise. If $N \triangleleft G$, show that N is a direct factor of G. [Hint: write $G = N_1 \times N_2 \times \cdots \times N_s$. Choose r maximal subject to $N, N_{i_1}, \ldots, N_{i_r}$ generating their direct product; then prove that this direct product equals G].

7. Let G be a group with a series in which each factor is either infinite cyclic or finite. Prove that any other series of this type has the same number of infinite factors, but not necessarily the same number of finite ones.

8. Suppose that G is a group with a composition series. Prove that G satisfies the *ascending and descending chain conditions for subnormal subgroups*, i.e., there cannot exist an infinite ascending chain $H_1 < H_2 < H_3 < \cdots$ or an infinite descending chain $H_1 > H_2 > H_3 > \cdots$ where the H_i are subnormal subgroups of G.

9. Prove that a group which satisfies both the ascending and descending chain conditions on subnormal subgroups has a composition series. [Hint: start by choosing a minimal non-trivial subnormal subgroup of G].

9.2 Solvable and nilpotent groups

In this section we are going to discuss certain types of group which are wide generalizations of abelian groups, but which retain vestiges of commutativity. The basic concept is that of a *solvable group*, which is defined to be a group with a series all of whose factors are abelian. The terminology derives from the classical problem of solving algebraic equations by radicals, which is discussed in detail in Chapter 11. The length of a shortest series with abelian factors is called the *derived length* of the solvable group. Thus abelian groups are the solvable groups with derived length at most 1. Solvable group with derived length 2 or less are often called *metabelian groups*.

Finite solvable groups are easily characterized in terms of their composition factors.

(9.2.1) *A finite group is solvable if and only if all its composition factors have prime order. In particular a simple group is solvable if and only if it has prime order.*

Proof. Let G be a finite solvable group, so that G has a series S with abelian factors. Refine S to a composition series of G. The factors of this series are simple and they are also abelian since they are essentially quotients of abelian groups. By (4.1.9) a simple abelian group has prime order. Hence composition factors of G have prime orders. The converse is an immediate consequence of the definition of solvability. □

Solvability is well-behaved with respect to the formation of subgroups, quotient groups and extensions.

(9.2.2) (i) *If G is a solvable group, then every subgroup and every quotient group of G is solvable.*

(ii) *Let G be a group with a normal subgroup N such that N and G/N are solvable. Then G is solvable.*

Proof. (i) Let $1 = G_0 \triangleleft G_1 \triangleleft \cdots \triangleleft G_n = G$ be a series in G with abelian factors and let H be a subgroup of G. Then $1 = G_0 \cap H \triangleleft G_1 \cap H \triangleleft \cdots \triangleleft G_n \cap H = H$ is a series in H; furthermore

$$G_{i+1} \cap H / G_i \cap H = G_{i+1} \cap H / (G_{i+1} \cap H) \cap G_i \simeq (G_{i+1} \cap H) G_i / G_i \leq G_{i+1}/G_i$$

by the Second Isomorphism Theorem (4.3.5). Hence $G_{i+1} \cap H / G_i \cap H$ is abelian and H is a solvable group.

Next let $N \triangleleft G$. Then G/N has the series

$$1 = G_0 N/N \triangleleft G_1 N/N \triangleleft \cdots \triangleleft G_n N/N = G/N.$$

By the Second and Third Isomorphism Theorems

$$(G_{i+1} N/N)/(G_i N/N) \simeq G_{i+1} N / G_i N$$
$$= G_{i+1}(G_i N)/G_i N \simeq G_{i+1}/G_{i+1} \cap (G_i N).$$

On applying the Modular Law (4.1.11), we find that $G_{i+1} \cap (G_i N) = G_i(G_{i+1} \cap N)$ and it follows that $G_{i+1}/G_{i+1} \cap (G_i N)$ equals

$$G_{i+1}/G_i(G_{i+1} \cap N) \simeq (G_{i+1}/G_i)/(G_i(G_{i+1} \cap N)/G_i).$$

Consequently G/N has series with abelian factors and hence is solvable.

(ii) The proof left as an exercise for the reader. □

The derived series. Recall from 4.2 that $[x, y] = xyx^{-1}y^{-1}$ is the *commutator* of the group elements x and y. The *derived subgroup* G' of a group G is the subgroup generated by all the commutators:

$$G' = \langle [x, y] \mid x, y \in G \rangle.$$

The *derived chain* $G^{(i)}$, $i = 0, 1, 2, \ldots$, is the descending sequence of subgroups formed by repeatedly taking derived subgroups,

$$G^{(0)} = G, \quad G^{(i+1)} = (G^{(i)})'.$$

Note that $G^{(i)} \triangleleft G$ and $G^{(i)}/G^{(i+1)}$ is an abelian group. The really important properties of the derived chain are that in a solvable group it reaches the identity subgroup and of all series with abelian factors it has shortest length.

(9.2.3) *Let $1 = G_0 \triangleleft G_1 \triangleleft \cdots \triangleleft G_k = G$ be a series with abelian factors in a solvable group G. Then $G^{(i)} \leq G_{k-i}$ for $0 \leq i \leq k$. In particular $G^{(k)} = 1$, so that the length of the derived chain equals the derived length of G.*

Proof. The containment is certainly true for $i = 0$. Assume that it is true for i. Since G_{k-i}/G_{k-i-1} is abelian, $G^{(i+1)} = (G^{(i)})' \leq (G_{k-i})' \leq G_{k-i-1}$, as required. Setting $i = k$, we get $G^{(k)} = 1$. □

9.2 Solvable and nilpotent groups

Notice the consequence: *a solvable group has a normal series with abelian factors*, for example the derived series.

It is sometimes possible to deduce the solvability of a group from arithmetic properties of its order. Examples of group orders for which this can be done are given in the next result.

(9.2.4) *Let p, q, r be primes. Then every group with order of the form p^m, p^2q^2, $p^m q$ or pqr is solvable.*

Proof. First observe that it is enough to show that there are no simple groups with these orders. This is because in the general case this result can then be applied to the composition factors to show they are of prime order.

(i) Let G be a group of order $p^m \neq 1$. Then $Z(G) \neq 1$ by (5.3.6) and $Z(G) \triangleleft G$, so $G = Z(G)$ and G is abelian. Thus $|G| = p$.

(ii) Next consider the case where $|G| = p^m q$. We can of course assume that $p \neq q$. Then $n_p \equiv 1 \pmod{p}$ and $n_p \mid q$, so that $n_p = q$. (For $n_p = 1$ would mean that there is a normal Sylow p-subgroup.)

Now choose two distinct Sylow subgroups P_1 and P_2 such that their intersection $I = P_1 \cap P_2$ has largest order. First of all suppose that $I = 1$. Then distinct Sylow subgroups intersect in 1, which makes it easy to count the number of non-trivial elements with order a power of p; indeed this number is $q(p^m - 1)$ since there are q Sylow p-subgroups. This leaves $p^m q - q(p^m - 1) = q$ elements of order prime to p. These elements must form a unique Sylow q-subgroup, which is therefore normal in G, contradicting the simplicity of the group G. It follows that $I \neq 1$.

By Exercise (5.3.11) or (9.2.7) below, we have $I < N_i = N_{P_i}(I)$ for $i = 1, 2$. Thus $I \triangleleft J = \langle N_1, N_2 \rangle$. Suppose for the moment that J is a p-group. Then J must be contained in some Sylow subgroup P_3 of G. But $P_1 \cap P_3 \geq P_1 \cap J > I$ since $N_1 \leq P_1 \cap J$, which contradicts the maximality of the intersection I. Hence J is not a p-group.

By Lagrange's Theorem $|J|$ divides $|G| = p^m q$ and it is not a power of p, from which it follows that q must divide $|J|$. Let Q be a Sylow q-subgroup of J. Then by (4.1.12)

$$|P_1 Q| = \frac{|P_1| \cdot |Q|}{|P_1 \cap Q|} = \frac{p^m q}{1} = |G|,$$

and thus $G = P_1 Q$. Now let $g \in G$ and write $g = ab$ where $a \in P_1$, $b \in Q$. Then $bIb^{-1} = I$ since $I \triangleleft J$, and hence $gIg^{-1} = a(bIb^{-1})a^{-1} = aIa^{-1} \leq P_1 < G$. But this means that $\bar{I} = \langle gIg^{-1} \mid g \in G \rangle \leq P_1 < G$ and $1 \neq \bar{I} \triangleleft G$, a final contradiction.

The remaining cases are left to the reader as exercises with appropriate hints – see Exercise (9.2.5) and (9.2.6). □

We mention two very important arithmetic criteria for a finite group to be solvable. The first states that *a group of order $p^m q^n$ is solvable if p and q are primes*. This is the celebrated *Burnside p-q Theorem*. It is best proved using group characters and is beyond the scope of this book. An even more difficult result is the *Feit–Thompson Theorem*, which asserts that *every group of odd order is solvable*. The original proof was over 250 pages long. What these results indicate is that there are many finite solvable groups: in fact finite simple groups should be regarded as a rarity.

Nilpotent groups. Nilpotent groups form an important subclass of the class of solvable groups. A group G is said to be *nilpotent* if it has a *central series*, i.e., a series of *normal* subgroups $1 = G_0 \triangleleft G_1 \triangleleft G_2 \triangleleft \cdots \triangleleft G_n = G$ such that G_{i+1}/G_i is contained in the center of G/G_i for all i. The length of a shortest central series is called the *nilpotent class* of G. Abelian groups are just the nilpotent groups with class ≤ 1. Notice that every nilpotent group is solvable, but S_3 is a solvable group that is not nilpotent since its center is trivial.

One great source of nilpotent groups is the groups of prime power order.

(9.2.5) *Let G be a group of order p^m where p is a prime. Then G is nilpotent, and if $m > 1$, the nilpotent class of G is at most $m - 1$.*

Proof. Define a chain of subgroups by repeatedly forming centers. Thus $Z_0 = 1$ and $Z_{i+1}/Z_i = Z(G/Z_i)$. By (5.3.6), if $Z_i \neq G$, then $Z_i < Z_{i+1}$. Since G is finite, there must be a smallest integer n such that $Z_n = G$, and clearly $n \leq m$. Suppose that $n = m$. Then $|Z_{m-2}| \geq p^{m-2}$ and so G/Z_{m-2} has order at most $p^m/p^{m-2} = p^2$, which implies that G/Z_{m-2} is abelian by (5.3.7). This yields the contradiction $Z_{m-1} = G$, so $n \leq m - 1$. □

The foregoing proof suggests a general construction, that of the *upper central chain* of a group G. This is the chain of subgroups defined by repeatedly forming centers,
$$Z_0(G) = 1, \quad Z_{i+1}(G)/Z_i(G) = Z(G/Z_i(G)).$$
Thus $1 = Z_0 \leq Z_1 \leq \ldots$ and $Z_i \triangleleft G$. If G is finite, this chain will certainly terminate, although it may it not reach G. The significance of the upper central chain for nilpotency is shown by the next result.

(9.2.6) *Let $1 = G_0 \triangleleft G_1 \triangleleft \cdots \triangleleft G_k = G$ be a central series in a nilpotent group G. Then $G_i \leq Z_i(G)$ for $0 \leq i \leq k$. In particular the length of the upper central chain equals the nilpotent class of G.*

Proof. We show that $G_i \leq Z_i(G)$ by induction on i; this is certainly true for $i = 0$. If it is true for i, then, since $G_{i+1}/G_i \leq Z(G/G_i)$, we have
$$G_{i+1}Z_i(G)/Z_i(G) \leq Z(G/Z_i(G)) = Z_{i+1}(G)/Z_i(G).$$
Thus $G_{i+1} \leq Z_{i+1}(G)$ and this holds for all i. Consequently $G = G_k \leq Z_k(G)$ and $G = Z_k(G)$. □

9.2 Solvable and nilpotent groups

Example (9.2.1) Let p be a prime and let $n > 1$. Denote by $U_n(p)$ the group of all $n \times n$ upper unitriangular matrices over the field $\mathbb{Z}_p$, i.e., matrices which have 1's on the diagonal and 0's below it. Counting the matrices of this type, we find that $|U_n(p)| = p^{n-1} \cdot p^{n-2} \ldots p \cdot 1 = p^{n(n-1)/2}$. So $U_n(p)$ is a nilpotent group, and in fact its class is $n - 1$ – cf. Exercise (9.2.11).

Characterizations of finite nilpotent groups. There are several different descriptions of finite nilpotent groups which shed light on the nature of nilpotency.

(9.2.7) *Let G be a finite group. Then the following statements are equivalent:*

(i) *G is nilpotent;*

(ii) *every subgroup of G is subnormal;*

(iii) *every proper subgroup of G is smaller than its normalizer;*

(iv) *G is the direct product of its Sylow subgroups.*

Proof. (i) implies (ii). Let $1 = G_0 \triangleleft G_1 \triangleleft \cdots \triangleleft G_n = G$ be a central series and let H be a subgroup of G. Then $G_{i+1}/G_i \leq Z(G/G_i)$, so $HG_i/G_i \triangleleft HG_{i+1}/G_i$. Hence there is a chain of subgroups $H = HG_0 \triangleleft HG_1 \triangleleft \cdots \triangleleft HG_n = G$ and H is subnormal in G.

(ii) implies (iii). Let $H < G$; then there is a chain $H = H_0 \triangleleft H_1 \triangleleft \cdots \triangleleft H_m = G$. Now there is a least $i > 0$ such that $H \neq H_i$, and then $H = H_{i-1} \triangleleft H_i$. Therefore $H_i \leq N_G(H)$ and $N_G(H) \neq H$.

(iii) implies (iv). Let P be a Sylow p-subgroup of G. If P is not normal in G, then $N_G(P) < G$, and hence $N_G(P)$ is smaller than its normalizer. But this contradicts Exercise (5.3.12). Therefore $P \triangleleft G$ and P must be the unique Sylow p-subgroup, which will be written G_p.

Next we need to observe that each element g of G is a product of elements of prime power order. For if g has order $n = p_1^{m_1} \ldots p_k^{m_k}$, with the p_i distinct primes, write $l_i = n/p_i^{m_i}$; the integers l_i are relatively prime, so $1 = l_1 r_1 + \cdots + l_k r_k$ for certain integers r_i. Clearly g^{l_i} has order $p_i^{m_i}$ and $g = (g^{l_1})^{r_1} \ldots (g^{l_k})^{r_k}$. It follows that G is generated by its Sylow subgroups G_p. Also $G_p \cap \langle G_q \mid q \neq p \rangle = 1$; for $|G_p|$ is a power of p and so is relatively prime to $|\langle G_q \mid q \neq p \rangle|$. Therefore G is the direct product of the G_p's.

(iv) implies (i). This follows quickly from the fact that a finite p-group is nilpotent. (The unique Sylow p-subgroup G_p is called the *p-component* of the nilpotent group G). □

The Frattini[4] subgroup. A very intriguing subgroup of a group G is its *Frattini subgroup* $\varphi(G)$, which is defined to be the intersection of all the maximal subgroups

---
[4]Giovanni Frattini (1852–1925).

of G. Here a *maximal subgroup* is a proper subgroup which is not contained in any larger proper subgroup. If there are no maximal subgroups in G, then $\varphi(G)$ is taken to be G. Note that $\varphi(G)$ is always normal in G. For example, S_3 has one maximal subgroup of order 3 and three of order 2: these intersect in 1, so that $\varphi(S_3) = 1$.

There is another, very different, way of describing the Frattini subgroup which uses the notion of a non-generator. An element g of a group G is called a *non-generator* if $G = \langle g, X \rangle$ always implies that $G = \langle X \rangle$ where X is a subset of G. Thus a non-generator can be omitted from any generating set for G.

(9.2.8) *Let G be a finite group. Then $\varphi(G)$ is precisely the set of all non-generators of G.*

Proof. Let g be a non-generator of G and assume that g is not in $\varphi(G)$. Then there is at least one maximal subgroup of G which does not contain g, say M. Thus M is definitely smaller than $\langle g, M \rangle$, which implies that $G = \langle g, M \rangle$ since M is maximal. Hence $G = M$ by the non-generator property, which is a contradiction since by definition maximal subgroups are proper.

Conversely, let $g \in \varphi(G)$ and suppose that $G = \langle g, X \rangle$, but $G \neq \langle X \rangle$. Now $\langle X \rangle$ must be contained in some maximal subgroup of G, say M. But $g \in \varphi(G) \leq M$, so $G = \langle g, M \rangle = M$, which is another contradiction. □

Next we establish an important property the Frattini subgroup of a finite group.

(9.2.9) *If G is a finite group, then $\varphi(G)$ is nilpotent.*

Proof. The proof depends on a useful trick known as *the Frattini argument*. Let $F = \varphi(G)$ and let P be a Sylow p-subgroup of F. If $g \in G$, then $gPg^{-1} \leq F$ since $F \triangleleft G$, and also $|gPg^{-1}| = |P|$. Therefore gPg^{-1} is a Sylow subgroup of F, and as such must be conjugate to P in F by Sylow's Theorem. Thus $gPg^{-1} = xPx^{-1}$ for some x in F. This implies that $x^{-1}gP(x^{-1}g)^{-1} = P$, i.e., $x^{-1}g \in N_G(P)$ and $g \in FN_G(P)$. Thus the conclusion of the Frattini argument is that $G = FN_G(P)$. Now the non-generator property comes into play, allowing us to omit the elements of F one at a time, until eventually we get $G = N_G(P)$, i.e., $P \triangleleft G$. In particular $P \triangleleft F$, so that all the Sylow subgroups of F are normal and by (9.2.7) F is nilpotent. □

The Frattini subgroup of a finite p-group. The Frattini subgroup plays an especially significant role in the theory of finite p-groups. Suppose that G is a finite p-group. If M is a maximal subgroup of G, then, since G is nilpotent, M is subnormal and hence normal in G. Furthermore G/M cannot have any proper non-trivial subgroups by maximality of M. Thus $|G/M| = p$. Now define the p-th *power* of the group G to be

$$G^p = \langle g^p \mid g \in G \rangle.$$

Then $G^p G' \leq M$ for all M and so $G^p G' \leq \varphi(G)$.

On the other hand, G/G^pG' is a finite abelian group in which every p-th power is the identity, i.e., an elementary abelian p-group. We saw in (8.2.15) that such groups are direct products of groups of order p. This fact enables us to construct maximal subgroups of G/G^pG' by omitting all but one factor from the direct product. The resulting maximal subgroups of G/G^pG' clearly intersect in the identity subgroup, which implies that $\varphi(G) \leq G^pG'$. We have therefore proved:

(9.2.10) *If G is a finite p-group, then $\varphi(G) = G^pG'$.*

Next suppose that $V = G/G^pG'$ has order p^d; thus d is the dimension of V as a vector space over the field $\mathbb{Z}_p$. Consider an arbitrary set X of generators for G. Now the subset $\{xG^pG' \mid x \in X\}$ generates V as a vector space. By Exercise (8.2.10) there is a subset Y of X such that $\{yG^pG' \mid y \in Y\}$ is a basis of V. Of course $|Y| = d$. We now claim that Y generates G. Certainly we have that $G = \langle Y, G^pG' \rangle = \langle Y, \varphi(G) \rangle$. The non-generator property now shows that $G = \langle Y \rangle$.

Summing up these conclusions, we have the following basic result on finite p-groups.

(9.2.11) *Let G be a finite p-group and assume that $G/\varphi(G)$ has order p^d. Then every set of generators of G has a d-element subset that generates G. Thus G can be generated by d and no fewer elements.*

Example (9.2.2) A group G can be constructed as a semidirect product of a cyclic group $\langle a \rangle$ of order 2^n with a Klein 4-group $H = \langle x, y \rangle$ where $n \geq 3$, $xax^{-1} = a^{-1}$ and $yay^{-1} = a^{1+2^{n-1}}$. (For the semidirect product construction see 4.3.) Thus $|G| = 2^{n+2}$. Observe that $G' = \langle a^2 \rangle$, so $\varphi(G) = G^2G' = \langle a^2 \rangle$. Hence $G/\varphi(G)$ is elementary abelian of order $8 = 2^3$. Therefore G can be generated by 3 and no fewer elements: of course $G = \langle a, x, y \rangle$.

Exercises (9.2)

1. Let $M \triangleleft G$ and $N \triangleleft G$ where G is any group. If M and N are solvable, then MN is solvable.

2. Let $M \triangleleft G$ and $N \triangleleft G$ for any group G. If G/M and G/N are solvable, then $G/M \cap N$ is solvable.

3. A solvable group with a composition series is finite.

4. Let G be a finite group with two non-trivial elements a and b such that the orders $|a|, |b|, |ab|$ are relatively prime in pairs. Then G cannot be solvable. [Hint: put $H = \langle a, b \rangle$ and consider H/H'.]

5. Prove that if p, q, r are primes, then every group of order pqr is solvable. [Hint: assume that G is a simple group of order pqr where $p < q < r$ and show that $n_r = pq$, $n_q \geq r$ and $n_p \geq q$. Now count elements to obtain a contradiction].

6. Prove that if p and q are primes, then every group of order p^2q^2 is solvable. [Hint: follow the method of proof for groups of order p^mq in (9.2.4). Deal first with the case where each pair of Sylow p-subgroups intersects in 1. Then take two Sylow subgroups P_1 and P_2 such that $I = P_1 \cap P_2$ has order p and note that $I \triangleleft J = \langle P_1, P_2 \rangle$].

7. Establish the commutator identities $[x, y^{-1}] = y^{-1}([x, y]^{-1})y$ and $[x, yz] = [x, y](y[x, z]y^{-1})$.

8. Let G be a group and let $z \in Z_2(G)$. Show that the assignment $x \mapsto [z, x]$ is a homomorphism from G to $Z(G)$ whose kernel contains G'.

9. Let G be a group such that $Z_1(G) \neq Z_2(G)$. Use Exercise (9.2.8) to show that $G \neq G'$.

10. Find the upper central series of the group $G = \mathrm{Dih}(2^m)$ where $m \geq 2$. Hence compute the nilpotent class of G.

11. Let $n > 1$ and let $G = U_n(p)$, the group of upper unitriangular matrices, and define G_i to be the subgroup of all elements of G in which the first i superdiagonals consist of 0's where $0 \leq i < n$. Show that the G_i are the terms of a central series of G. What can you deduce about the nilpotent class of G?

12. Let G be a nilpotent group and let $1 \neq N \triangleleft G$. Prove that $N \cap Z(G) \neq 1$.

13. Let A be a largest abelian normal subgroup of G. Show that $C_G(A) = A$. [Hint: assume it is false and apply Exercise (9.2.12) to $C_G(A)/A \triangleleft G/A$].

14. If every abelian normal subgroup of a nilpotent group is finite, then the group is finite.

15. Find the Frattini subgroups of the groups A_n, S_n, $\mathrm{Dih}(2n)$ where n is odd.

16. Use (9.2.4) to show that a non-solvable group of order ≤ 100 must have order 60. (Note that the orders which need discussion are 72, 84 and 90).

17. Prove that A_5 is the only non-solvable group with order ≤ 100. [Hint: it is enough to show that a simple group of order 60 must have a subgroup of index 5. Consider the number of Sylow 2-subgroups.]

9.3 Theorems on finite solvable groups

The final section of this chapter will take us deeper into the theory of finite solvable groups and several famous theorems will be proved. Among these is the classification of finite abelian groups, one of the triumphs of 19th Century algebra, which states that any finite abelian group can be expressed as a direct product of cyclic groups. It is with this that we begin.

(9.3.1) *A finite abelian group is a direct product of cyclic groups with prime power orders.*

9.3 Theorems on finite solvable groups

The hardest part of the proof is accomplished in the following lemma, which is essentially a tool for splitting off cyclic direct factors.

(9.3.2) *Let A be a finite abelian p-group and suppose that a is an element of A with maximum order. Then $\langle a \rangle$ is a direct factor of A, i.e., $A = A_1 \times \langle a \rangle$ for some subgroup A_1.*

Proof. The first step is to choose a subgroup M which is maximal subject to the requirement $M \cap \langle a \rangle = 1$; thus $\langle M, a \rangle = M \times \langle a \rangle = A_0$ say. If it happens that A_0 equals A, then we are done. Assume therefore that $A \neq A_0$ and look for a contradiction. It will expedite matters if we choose an element x from $A - A_0$ with smallest order.

Now x^p has smaller order than x, so by choice of x we must have $x^p \in A_0$ and $x^p = ya^l$ where $y \in M$ and l is an integer. Recall that a has largest order of all elements of A – call this order p^n. Thus $1 = x^{p^n} = y^{p^{n-1}} a^{lp^{n-1}}$ and hence

$$a^{lp^{n-1}} = y^{-p^{n-1}} \in M \cap \langle a \rangle = 1.$$

It follows that p^n, the order of a, divides lp^{n-1}, i.e., p divides l. Write $l = pj$ with j an integer, so that $x^p = ya^{pj}$ and $y = (xa^{-j})^p \in M$. On the other hand, $xa^{-j} \notin M$ since $x \notin M_0 = M\langle a \rangle$. Hence $\langle xa^{-j} \rangle M$ properly contains M.

By maximality of M we have $(\langle xa^{-j} \rangle M) \cap \langle a \rangle \neq 1$, and so there exist integers k, m and an element y' of M such that $(xa^{-j})^m y' = a^k \neq 1$. This implies that $x^m \in M_0$. Now suppose that p divides m. Then, because $(xa^{-j})^p = y \in M$, it follows that $(xa^{-j})^m \in M$, and thus $a^k \in M \cap \langle a \rangle = 1$. But this is incorrect, so p cannot divide m. Hence $1 = \gcd\{p, m\} = pr + ms$ for some integers r and s. However $x^p \in A_0$ and $x^m \in A_0$, so that we reach the contradiction $x = x^{pr+ms} = (x^p)^r (x^m)^s \in A_0$. □

Proof of (9.3.1). This is now straightforward. Suppose that A is a finite abelian group. By (9.2.7) A is the direct product of its primary components; thus we can assume that A is a p-group. Let us argue by induction on $|A| > 1$. Choose an element of maximum order in A, say a. Then $A = A_1 \times \langle a \rangle$ for some subgroup A_1 by (9.3.1). Since $|A_1| < |A|$, the induction hypothesis tells us that A_1 is a direct product of cyclic groups. Hence the same is true of A. □

It is natural to ask whether the numbers and types of cyclic factors appearing in the direct decomposition of (9.3.1) are unique. The next result answers this question in the affirmative.

(9.3.3) *Let A and B be finite abelian groups and suppose that $m(p^i)$ and $n(p^i)$ are the numbers of cyclic direct factors of order p^i appearing in some cyclic direct decompositions of A and B respectively. Then A and B are isomorphic if and only if $m(p^i) = n(p^i)$ for all p and i.*

Proof. Of course, if $m(p^i) = n(p^i)$ for all p and i, it is clearly true that the groups are isomorphic. Conversely suppose that there is an isomorphism $\theta : A \to B$. Since θ must map the p-primary component of A to that of B, we may assume that A and B are p-groups.

Consider direct decompositions of A and B with m_i and n_i cyclic factors of order p^i respectively. Then, when we form A^p, all m_1 factors of order p are eliminated. Put $A[p] = \{a \in A \mid a^p = 1\}$ and note that this is a vector space over $\mathbb{Z}_p$. Then, counting direct factors, we have $m_1 = \dim(A[p]) - \dim(A^p[p])$. Similarly, on examining A^{p^2}, we find that $m_2 = \dim(A^p[p]) - \dim(A^{p^2}[p])$. And in general

$$m_i = \dim(A^{p^{i-1}}[p]) - \dim(A^{p^i}[p]).$$

Now θ certainly maps $A^{p^j}[p]$ to $B^{p^j}[p]$, so that

$$m_i = \dim(A^{p^{i-1}}[p]) - \dim(A^{p^i}[p]) = \dim(B^{p^{i-1}}[p]) - \dim(B^{p^i}[p]) = n_i. \qquad \square$$

The number of abelian groups of given order. The information we possess about the structure of finite abelian groups is sufficiently strong for us to make an exact count of the number of abelian groups of given order. First of all suppose that A is an abelian group of order p^l where p is a prime. Then by (9.3.1) A is with isomorphic with a direct product of l_1 copies of $\mathbb{Z}_p$, l_2 of $\mathbb{Z}_{p^2}$, etc., where $l = l_1 + 2l_2 + \cdots$ and $0 \le l_i \le l$. Moreover every such partition of the integer l yields an abelian group of order p^l. It follows via (9.3.3) that the number of isomorphism classes of abelian groups of order p^l equals $\lambda(l)$, the number of partitions of the integer l. Thus our discussion yields the following result.

(9.3.4) *Let n be a positive integer and write $n = p_1^{l_1} p_2^{l_2} \ldots p_k^{l_k}$ where the p_i are distinct primes and $l_i > 0$. Then the number of isomorphism types of abelian group of order n is $\lambda(l_1)\lambda(l_2)\ldots\lambda(l_k)$ where $\lambda(i)$ is the number of partitions of i.*

Example (9.3.1) Describe all abelian groups of order 600.

First of all write $600 = 2^3 \cdot 3 \cdot 5^2$. Then enumerate the partitions of the three exponents. The partitions of 3 are 3, $1+2$, $1+1+1$, those of 2 are 2, $1+1$, while there is just one partition of 1. So we expect there to be $3 \times 1 \times 2 = 6$ isomorphism types of group. They are

$$\mathbb{Z}_8 \oplus \mathbb{Z}_3 \oplus \mathbb{Z}_{5^2}, \quad \mathbb{Z}_8 \oplus \mathbb{Z}_3 \oplus \mathbb{Z}_5 \oplus \mathbb{Z}_5, \quad \mathbb{Z}_2 \oplus \mathbb{Z}_4 \oplus \mathbb{Z}_3 \oplus \mathbb{Z}_{5^2},$$

and

$$\mathbb{Z}_2 \oplus \mathbb{Z}_4 \oplus \mathbb{Z}_3 \oplus \mathbb{Z}_5 \oplus \mathbb{Z}_5, \quad \mathbb{Z}_2 \oplus \mathbb{Z}_2 \oplus \mathbb{Z}_2 \oplus \mathbb{Z}_3 \oplus \mathbb{Z}_{5^2}, \quad \mathbb{Z}_2 \oplus \mathbb{Z}_2 \oplus \mathbb{Z}_2 \oplus \mathbb{Z}_3 \oplus \mathbb{Z}_5 \oplus \mathbb{Z}_5.$$

For example the choice of partitions $1+2$, 1, $1+1$ correspond to the group $\mathbb{Z}_2 \oplus \mathbb{Z}_4 \oplus \mathbb{Z}_3 \oplus \mathbb{Z}_5 \oplus \mathbb{Z}_5$. Notice that first group on the list is the cyclic group of order 600.

The foregoing theory provides a very satisfactory description of finite abelian groups in terms of simple invariants, the numbers of cyclic direct factors of each prime power order. Such perfect classifications are unfortunately a rarity in algebra.

9.3 Theorems on finite solvable groups

Schur's splitting and conjugacy theorem. Suppose that N is a normal subgroup of a group G. A subgroup X such that $G = NX$ and $N \cap X = 1$ is called a *complement* of N in G. In this case G is said to *split over* N, and G is the semidirect product of N and X. A *splitting theorem* asserts that a group splits over some normal subgroup and such a theorem can be thought of as resolving a group into a product of potentially simpler groups. The most famous splitting theorem in group theory is undoubtedly:

(9.3.5) (Schur[5]) *Let A be an abelian normal subgroup of a finite group G such that $|A|$ and $|G:A|$ are relatively prime. Then G splits over A and all complements of A are conjugate in G.*

Proof. This falls into two parts.

(i) *Existence of a complement.* To begin the proof we choose an arbitrary transversal to A in G, say $\{t_x \mid x \in Q = G/N\}$ where $x = At_x$. Most likely this transversal will only be a subset of G. The idea behind the proof is to try to transform the transversal into one which is actually a subgroup. Let $x, y \in Q$: then $x = At_x$ and $y = At_y$, and in addition $At_{xy} = xy = At_x At_y = At_x t_y$. Hence it possible to write

$$t_x t_y = a(x, y) t_{xy}$$

for some $a(x, y) \in A$.

The associative law $(t_x t_y) t_z = t_x (t_y t_z)$ imposes a condition on the elements $a(x, y)$. For, applying the above equation twice, we obtain

$$(t_x t_y) t_z = a(x, y) a(xy, z) t_{xyz},$$

and similarly

$$t_x(t_y t_z) = t_x a(y, z) t_{yz} = (t_x a(y, z) t_x^{-1}) t_x t_{yz} = (t_x a(y, z) t_x^{-1}) a(x, yz) t_{xyz}.$$

Now conjugation by t_x induces an automorphism in A which depends only on x: indeed, if $a, b \in A$, then $(bt_x) a (bt_x)^{-1} = t_x a t_x^{-1}$ since A is abelian. Denote this automorphism by the assignment $a \mapsto {}^x a$. Then on equating $(t_x t_y) t_z$ and $t_x(t_y t_z)$ and cancelling t_{xyz}, we arrive at the equation

$$a(x, y) a(xy, z) = {}^x a(y, z) a(x, yz) \quad (*)$$

which is valid for all $x, y, z, \in Q$. Such a function $a : Q \times Q \to A'$ is called a *factor set* or *2-cocycle*.

Next we form the product of all the elements $a(x, y)$ for $y \in Q$, with x fixed, and call this element b_x. Now take the product of the equations $(*)$ above for all z in Q with x and y fixed. There results the equation

$$a(x, y)^m b_{xy} = {}^x b_y b_x, \quad (**)$$

---
[5] Issai Schur (1875–1941)

where $m = |Q| = |G : A|$. Note here that the product of all the ${}^xa(y, z)$ is ${}^xb_y$ and the product of all the $a(x, yz)$ is b_x.

Since m is relatively prime to $|A|$, every element of A is an m-th power and the mapping in which $a \mapsto a^m$ is an automorphism of A. Thus we can write b_x as an m-th power, say $b_x = c_x^{-m}$ where $c_x \in A$. Substituting for b_x in equation (**), we get $(a(x, y)c_{xy}^{-1})^m = (({}^xc_yc_x)^{-1})^m$, from which it follows that

$$c_{xy} = c_x\,{}^xc_y a(x, y).$$

We are now ready to form the new transversal; put $s_x = c_x t_x$. Then

$$s_x s_y = c_x t_x c_y t_y = c_x ({}^xc_y) t_x t_y = c_x({}^xc_y) a(x, y) t_{xy} = c_{xy} t_{xy} = s_{xy}.$$

This proves that the transversal $H = \{s_x | x \in Q\}$ is indeed a subgroup. Since $G = AH$ and $A \cap H = 1$, we have found a complement H for A in G.

(ii) *Conjugacy of complements.* Let H and H^* be two complements of A in G. Then for each x in Q we can write $x = As_x = As_x^*$ where s_x and s_x^* belong to H and H^* respectively, and $H = \{s_x \mid x \in Q\}$ and $H^* = \{s_x^* \mid x \in Q\}$. Here s_x and s_x^* are related by an equation of the form

$$s_x^* = d_x s_x$$

where $d_x \in A$. Since $As_{xy} = xy = As_x As_y = As_x s_y$, we have $s_x s_y = s_{xy}$, and also $s_x^* s_y^* = s_{xy}^*$. In the last equation replace s_x^* by $d_x s_x$ and s_y^* by $d_y s_y$ to get

$$d_{xy} = d_x({}^xd_y)$$

for all $x, y \in Q$. Such a function $d : Q \to A$ is called a *derivation* or *1-cocycle*.

Let d denote the product of all the d_x's for $x \in Q$. Now take the product of all the equations $d_{xy} = d_x\,{}^xd_y$ for $y \in Q$ with x fixed. This leads to $d = d_x^m\,{}^xd$. Writing $d = e^m$ with $e \in A$, we obtain $e^m = (d_x\,{}^xe)^m$ and hence $e = d_x({}^xe)$. Thus

$$d_x = e({}^xe)^{-1}.$$

Since ${}^xe = s_x e s_x^{-1}$, we have

$$s_x^* = d_x s_x = e({}^xe)^{-1} s_x = e(s_x e^{-1} s_x^{-1}) s_x = e s_x e^{-1}.$$

Therefore $H^* = eHe^{-1}$ and all complements of A are conjugate. □

Hall's theorem on finite solvable groups. To illustrate the usefulness of Schur's splitting theorem we make a final foray into finite solvable group theory by proving a celebrated result.

(9.3.6) (P. Hall[6]) *Let G be a finite solvable group and write its order in the form mn where m and n are relatively prime integers. Then G has a subgroup of order m and all subgroups of this order are conjugate.*

Proof. (i) *Existence*. We argue by induction on $|G| > 1$. By (9.2.3) G has a non-trivial abelian normal subgroup A. Since A is the direct product of its primary components, we can assume that A is a p-group, with $|A| = p^k$, say. There are two cases to consider.

Suppose first that p does not divide m. Then $p^k \mid n$ because m and n are relatively prime. Since $|G/A| = m \cdot (n/p^k)$, the induction hypothesis may be applied to the group G/A to show that there is a subgroup of order m, say K/A. Now $|A|$ is relatively prime to $|K : A|$, so (9.3.5) may be applied to K; thus K splits over A and hence K has a subgroup of order m.

Now assume that $p \mid m$; then $p^k \mid m$ since p cannot divide n. We have $|G/A| = (m/p^k) \cdot n$, so by induction once again G/A has a subgroup of order m/p^k, say K/A. Then $|K| = |A| \cdot |K/A| = p^k(m/p^k) = m$, as required.

(ii) *Conjugacy*. Let H and H^* be two subgroups of order m, and choose A as in (i). If p does not divide m, then $A \cap H = 1 = A \cap H^*$, and AH/A and AH^*/A are subgroups of G/A with order m. By induction on $|G|$ these subgroups are conjugate and thus $AH = A(gH^*g^{-1})$ for some $g \in G$. By replacing H^* by gH^*g^{-1}, we can assume that $AH = AH^*$. But now H and H^* are two complements of A in HA, and Schur's theorem guarantees that they are conjugate.

Finally, suppose that p divides m. Then p does not divide $n = |G : H| = |G : H^*|$. Since $|AH : H|$ is a power of p and also divides n, we have $|AH : H| = 1$ and thus $A \leq H$. Similarly $A \leq H^*$. By induction H/A and H^*/A are conjugate in G/A, as must H and H^* be in G.

Hall π-subgroups. Let π denote a set of primes and let π' be the complementary set of primes. A finite group is called a *π-group* if its order is a product of primes in π. Now let G be a finite solvable group and write $|G| = mn$ where m is the product of the prime powers dividing the order of G for primes which belong to the set π and n is a product of the remaining prime powers dividing $|G|$. Then Hall's theorem tells us that G has a subgroup of order m and index n. Thus H is a π-group and $|G : H|$ is a product of primes in π': such a subgroup is called a *Hall π-subgroup* of G. Thus Hall's theorem asserts that Hall π-subgroups exist in a finite soluble group for any set of primes π, and any two of these are conjugate.

Hall's theorem can be regarded as an extension of Sylow's Theorem since if $\pi = \{p\}$, a Hall π-subgroup is simply a Sylow p-subgroup. While Sylow's Theorem is valid for any finite group, Hall subgroups need not exist in insolvable groups. For example A_5 has order $60 = 3 \cdot 20$, but it has no subgroups of order 20, as the reader should verify. Hall's theorem is the starting point for a rich theory of finite solvable groups which has been developed over the last seventy years.

---

[6]Philip Hall (1904–1982).

Exercises (9.3)

1. Describe all abelian groups of order 648 and 2250.

2. How many abelian groups are there of order 64?

3. Let A be a finite abelian group. Prove that A can be written in the form

$$A = \langle a_1 \rangle \times \langle a_2 \rangle \times \cdots \times \langle a_k \rangle$$

where $|a_i|$ divides $|a_{i+1}|$.

4. Give an example of a finite group G with an abelian normal subgroup A such that G does not split over A.

5. If G is a finite solvable group, a smallest non-trivial normal subgroup of G is an elementary abelian p-group for some p dividing $|G|$.

6. If M is a maximal subgroup of a finite solvable group G, then $|G : M|$ is a prime power. [Hint: use induction on $|G|$ to reduce to the case where M contains no non-trivial normal subgroups of G. Then choose a smallest non-trivial normal subgroup A of G. Apply Exercise (9.3.5).]

7. For which sets of primes π does the group A_5 have Hall π-subgroups?

8. Let G be a finite solvable group and let p be a prime dividing the order of G. Prove that G has a maximal subgroup with index a power of p. [Hint: apply Hall's theorem].

9. Let G be a finite solvable group and let p be a prime dividing the order of G. Prove that p divides $|G/\varphi(G)|$.

10. Let G be a finite group and let H be a Hall π-subgroup of a solvable normal subgroup L. Use Hall's Theorem to prove that $G = LN_G(H)$.

Chapter 10
Introduction to the theory of fields

The theory of fields is one of the most attractive parts of algebra. It contains many powerful theorems on the structure of fields, for example, the Fundamental Theorem of Galois Theory, which establishes a correspondence between subfields of a field and subgroups of the Galois group. In addition field theory can be applied to a wide variety of problems, some of which date from classical antiquity. Among the applications to be described here and in the following chapter are: ruler and compass constructions, solution of equations by radicals, mutually orthogonal latin squares and Steiner systems. In short field theory is algebra at its best – deep theorems with convincing applications to problems which might otherwise be intractible.

10.1 Field extensions

Recall from 7.4 that a *subfield* of a field F is a subset K containing 0 and 1 which is closed under addition, forming negatives, multiplication and inversion. The following is an immediate consequence of the definition of a subfield.

(10.1.1) *The intersection of any set of subfields of a field is a subfield.*

Now suppose that X is a non-empty subset of a field F. By (10.1.1) the intersection of all the subfields of F that contain X is a subfield, evidently the smallest subfield containing X. We call this the *subfield of F generated by X*. It is easy to describe the elements of the subfield generated by a given subset.

(10.1.2) *If X is a subset of a field F, then the subfield generated by X consists of all elements $f(x_1, \ldots, x_m) g(y_1, \ldots, y_n)^{-1}$ where $f \in \mathbb{Z}[t_1, \ldots, t_m]$, $g \in \mathbb{Z}[t_1, \ldots, t_n]$, $x_i, y_j \in X$ and $g(y_1, \ldots, y_n) \neq 0$.*

To prove this first observe that the set S of elements specified is a subfield of F containing X. Then note that any subfield of F which contains X must also contain all the elements of S.

Prime subfields. In a field F one can form the intersection of *all* the subfields. This is the unique smallest subfield of F, and it is called the *prime subfield* of F. A field which equals its prime subfield is called a *prime field*. It is easy to identify the prime fields.

(10.1.3) *A prime field of characteristic 0 is isomorphic with $\mathbb{Q}$ and one of prime characteristic p is isomorphic with $\mathbb{Z}_p$. Conversely, $\mathbb{Q}$ and $\mathbb{Z}_p$ are prime fields.*

Proof. Assume that F is a prime field and put $I = \langle 1_F \rangle = \{n 1_F | n \in \mathbb{Z}\}$. Suppose first that F has characteristic 0, so I is infinite cyclic. We define a surjective mapping $\alpha : \mathbb{Q} \to F$ by the rule $\alpha(\frac{m}{n}) = (m 1_F)(n 1_F)^{-1}$ where of course $n \neq 0$. It is easily checked that α is a ring homomorphism, and its kernel is therefore an ideal of $\mathbb{Q}$. Now 0 and $\mathbb{Q}$ are the only ideals of $\mathbb{Q}$ and $\alpha(1) = 1_F \neq 0_F$, so $\text{Ker}(\alpha) \neq \mathbb{Q}$. It follows that $\text{Ker}(\alpha) = 0$ and $\mathbb{Q} \simeq \text{Im}(\alpha)$. Since F is a prime field and $\text{Im}(\alpha)$ is a subfield, $\text{Im}(\alpha) = F$ and α is an isomorphism. Thus $F \simeq \mathbb{Q}$.

Now suppose that F has prime characteristic p, so that $|I| = p$. In this situation we define $\alpha : \mathbb{Z} \to F$ by $\alpha(n) = n 1_F$. So $\alpha(n) = 0_F$ if and only if $n 1_F = 0$, i.e., p divides n. Thus $\text{Ker}(\alpha) = p\mathbb{Z}$ and $\text{Im}(\alpha) \simeq \mathbb{Z}/p\mathbb{Z} = \mathbb{Z}_p$. Hence $\mathbb{Z}_p$ is isomorphic with a subfield of F, and since F is prime, we have $\mathbb{Z}_p \simeq F$. We leave the reader to check that $\mathbb{Q}$ and $\mathbb{Z}_p$ are prime fields. □

Field extensions. Consider two fields F and E and assume that there is an injective ring homomorphism $\alpha : F \to E$. Then F is isomorphic with $\text{Im}(\alpha)$, which is a subfield of E: under these circumstances we shall say that E is an *extension of F*. Often we will prefer to assume that F is actually a subfield of E. This is usually a harmless assumption since F can be replaced by the isomorphic field $\text{Im}(\alpha)$. Notice that by (10.1.3) every field is an extension of either $\mathbb{Z}_p$ or $\mathbb{Q}$.

If E is an extension of F, then E can be regarded as a vector space over F by using the field operations of E. This simple idea is critical since it allows us to define the *degree of E over F* as

$$(E : F) = \dim_F(E),$$

assuming that this dimension is finite. Then E is called a *finite extension* of F.

Simple extensions. Let F be a subfield and X a non-empty subset of a field E. The subfield of E generated by $F \cup X$ is denoted by

$$F(X).$$

It follows readily from (10.1.2) that $F(X)$ consists of all elements of the form

$$f(x_1, \ldots, x_m) g(y_1, \ldots, y_n)^{-1}$$

where $f \in F[t_1, \ldots, t_m]$, $g \in F[t_1, \ldots, t_n]$, $x_i, y_j \in X$ and $g(y_1, \ldots, y_n) \neq 0$. If $X = \{x_1, x_2, \ldots, x_l\}$, we write

$$F(x_1, x_2, \ldots, x_l)$$

instead of $F(\{x_1, x_2, \ldots, x_l\})$. The most interesting case for us is when $X = \{x\}$ and a typical element of $F(x)$ has the form $f(x)g(x)^{-1}$ where $f, g \in F[t]$ and $g(x) \neq 0$.

10.1 Field extensions

If $E = F(x)$ for some $x \in E$, then we call E a *simple extension* of F. We proceed at once to determine the structure of simple extensions.

(10.1.4) *Let $E = F(x)$ be a simple extension of a field F. Then one of the following must hold:*

(a) $f(x) \neq 0$ for all $0 \neq f \in F[t]$ and $E \simeq F\{t\}$, *the field of rational functions in t over F;*

(b) $f(x) = 0$ *for some monic irreducible polynomial* $f \in F[t]$ *and* $E \simeq F[t]/(f)$, *so that* $(E:F) = \deg(f)$.

Proof. We may assume that $F \subseteq E$. Define a mapping $\theta : F[t] \to E$ by evaluation at x, i.e., $\theta(f) = f(x)$. This is a ring homomorphism whose kernel is an ideal of $F[t]$, say I.

Suppose first that $I = 0$, i.e., $f(x) = 0$ implies that $f = 0$. Then θ can be extended to a function $\alpha : F\{t\} \to E$ by the rule $\alpha\left(\frac{f}{g}\right) = f(x)g(x)^{-1}$; this function is also a ring homomorphism. Notice that $\alpha\left(\frac{f}{g}\right) = 0$ implies that $f(x) = 0$ and hence $f = 0$. Therefore $\operatorname{Ker}(\alpha) = 0$ and $F\{t\}$ is isomorphic with $\operatorname{Im}(\alpha)$, which is a subfield of E. Now $\operatorname{Im}(\alpha)$ contains F and x since $\alpha(a) = a$ if $a \in F$ and $\alpha(t) = x$. However E is a smallest field containing F and x, so $E = \operatorname{Im}(\alpha) \simeq F\{t\}$.

Now suppose that $I \neq 0$. Since $F[t]/I$ is isomorphic with a subring of E, it is a domain and I is a prime ideal. By (7.2.6) $I = (f)$ where f is a monic irreducible polynomial in $F[t]$. Thus $F[t]/I$ is a field, which is isomorphic with $\operatorname{Im}(\theta)$, a subfield of E containing F and x for reasons given above. Therefore $F[t]/I \simeq \operatorname{Im}(\theta) = E$. □

Algebraic elements. Consider an arbitrary field extension K of F and let $x \in E$. There are two possible forms for $F(x)$, as indicated in (10.1.4). If $f(x) \neq 0$ whenever $0 \neq f \in F[t]$, then $F(x) \simeq F\{t\}$ and x is said to be *transcendent over F*.

The other possibility is that x is a root of some monic irreducible polynomial f in $F[t]$. In this case $F(x) \simeq F[t]/(f)$ and x is said to be *algebraic over F*. The polynomial f is the unique monic irreducible polynomial over F which has x as a root: for if g is another one, then $g \in (f)$ and $f | g$, so that $f = g$ by irreducibility and monicity. We call f the *irreducible polynomial of f over F*,

$$\operatorname{Irr}_F(x) :$$

thus $F(x) \simeq F[t]/(\operatorname{Irr}_F(x))$.

Now assume that $f = \operatorname{Irr}_F(x)$ has degree n. For any g in $F[t]$ we can write $g = fq + r$ where $q, r \in F[t]$ and $\deg(r) < n$, by the Division Algorithm for $F[t]$, (see (7.1.3)). Then $g + (f) = r + (f)$, which shows that $F(x)$ is generated as an F-vector space by $1, x, x^2, \ldots, x^{n-1}$. But these elements are linearly independent over F. For if $a_0 + a_1 x + \cdots + a_{n-1} x^{n-1} = 0$ with $a_i \in F$, then $g(x) = 0$ where $g = a_0 + a_1 t + \cdots + a_{n-1} t^{n-1}$, and hence $f | g$. But $\deg(g) \leq n - 1$, which can only mean that $g = 0$ and all the a_i are zero. It follows that $1, x, x^2, \ldots, x^{n-1}$ form an F-basis of the vector space $F(x)$ and hence $(F(x) : F) = n = \deg(f)$.

We sum up these conclusions in:

(10.1.5) *Let $E = F(x)$ be a simple field extension.*

(a) *If x is transcendent over F, then $E \simeq F\{t\}$.*

(b) *If x is algebraic over F, then $E \simeq F[t]/(\mathrm{Irr}_F(x))$ and $(E : F) = \deg(\mathrm{Irr}_F(x))$.*

Example (10.1.1) Show that $\sqrt{3} - \sqrt{2}$ is algebraic over $\mathbb{Q}$. Find its irreducible polynomial and hence the degree of $\mathbb{Q}(\sqrt{3} - \sqrt{2})$ over $\mathbb{Q}$.

Put $a = \sqrt{3} - \sqrt{2}$. The first move is to find a rational polynomial with a as a root. Now $a^2 = 5 - 2\sqrt{6}$, so $(a^2 - 5)^2 = 24$ and $a^4 - 10a^2 + 1 = 0$. Hence a is a root of $f = t^4 - 10t^2 + 1$ and thus a is algebraic over $\mathbb{Q}$. If we can show that f is irreducible over $\mathbb{Q}$, it will follow that $\mathrm{Irr}_\mathbb{Q}(a) = f$ and $(\mathbb{Q}(a) : \mathbb{Q}) = 4$. By Gauss's Lemma (7.3.7) it is enough to show that f is $\mathbb{Z}$-irreducible.

Now clearly f has no integer roots since ± 1 are the only candidates and neither of these is a root. Thus, if f is reducible, there must be a decomposition of the form

$$f = (t^2 + at + b)(t^2 + a_1 t + b_1)$$

where a, b, a_1, b_1 are integers. Equating coefficients of $1, t^3, t^2$ on both sides, we arrive at the equations

$$bb_1 = 1, \quad a + a_1 = 0, \quad aa_1 + b + b_1 = -10.$$

Hence $b = b_1 = \pm 1$ and $a_1 = -a$, so that $-a^2 \pm 2 = -10$. Since this equation has no integer solutions, f is irreducible.

Algebraic extensions. Let E be an extension of a field F. If every element of E is algebraic over F, then E is called an *algebraic extension* of F. Extensions of finite degree are an important source of algebraic extensions.

(10.1.6) *A field extension of finite degree is algebraic.*

Proof. Let E be a finite extension of F and let $x \in E$. By hypothesis E has finite dimension, say n, as a vector space over F; consequently the set $\{1, x, x^2, \ldots, x^n\}$ is linearly dependent and there are elements $a_0, a_1, \ldots, a_n$ of F, not all zero, such that $a_0 + a_1 x + a_2 x^2 + \cdots + a_n x^n = 0$. So x is a root of the non-zero polynomial $a_0 + a_1 t + \cdots + a_n t^n$ and is therefore algebraic over F. □

The next result is very useful in calculations with degrees.

(10.1.7) *Let $F \subseteq K \subseteq E$ be successive field extensions. If K is finite over F and E is finite over K, then E is finite over F and $(E : F) = (E : K) \cdot (K : F)$.*

10.1 Field extensions   189

Proof. Let $\{x_1, \ldots, x_m\}$ be an F-basis of K and $\{y_1, \ldots, y_n\}$ a K-basis of E. Then each $e \in E$ can be written $e = \sum_{i=1}^{n} k_i y_i$ where $k_i \in K$. Also each k_i can be written $k_i = \sum_{j=1}^{m} f_{ij} x_j$ with $f_{ij} \in F$. Therefore $e = \sum_{i=1}^{n} \sum_{j=1}^{m} f_{ij} x_j y_i$ and it follows that the elements $x_j y_i$ generate the F-vector space E.

Now assume there is a linear relation $\sum_{i=1}^{n} \sum_{j=1}^{m} f_{ij} x_j y_i = 0$ where $f_{ij} \in F$. Then $\sum_{i=1}^{n} \left(\sum_{j=1}^{m} f_{ij} x_j \right) y_i = 0$, so that $\sum_{j=1}^{m} f_{ij} x_j = 0$ for all i since the y_i are K-linearly independent. Finally, $f_{ij} = 0$ for all i and j by linear independence of the x_j over F. Consequently the elements $x_j y_i$ form an F-basis of E and $(E : F) = nm = (E : K) \cdot (K : F)$. □

Corollary (10.1.8) *Let $F \subseteq K \subseteq E$ be successive field extensions with E algebraic over K and K algebraic over F. Then E is algebraic over F.*

Proof. Let $x \in E$. Then x is algebraic over K; let its irreducible polynomial be $f = a_0 + a_1 t + \cdots + t^n$ where $a_i \in K$. Then a_i is algebraic over F and hence over $K_i = F(a_0, a_1, \ldots a_{i-1})$. Thus $(K_i : K_{i-1})$ is finite for $i = 0, 1, \ldots, n$, where $K_0 = F$, and so by (10.1.7) $(K_n : F)$ is finite. Now x is algebraic over K_n, so that $(K_n(x) : K_n)$ is finite and therefore $(K_n(x) : F)$ is finite. It follows via (10.1.6) that x is algebraic over F. □

Algebraic and transcendental numbers. Next let us consider the complex field $\mathbb{C}$ as an extension of the rational field $\mathbb{Q}$. If $x \in \mathbb{C}$ is algebraic over $\mathbb{Q}$, call x an *algebraic number*: otherwise x is a *transcendental number*. Thus the algebraic numbers are those complex numbers which are roots of non-zero rational polynomials in t.

(10.1.9) *The algebraic numbers form a subfield of $\mathbb{C}$.*

Proof. Let a and b be algebraic numbers. It is sufficient to show that $a \pm b$, ab and ab^{-1} (if $b \neq 0$) are algebraic numbers. To see this note that $(\mathbb{Q}(a) : \mathbb{Q})$ is finite by (10.1.5). Also $\mathbb{Q}(a, b) = (\mathbb{Q}(a))(b)$ is finite over $\mathbb{Q}(a)$ for the same reason. Therefore $(\mathbb{Q}(a, b) : \mathbb{Q})$ is finite by (10.1.7), and hence $\mathbb{Q}(a, b)$ is algebraic over $\mathbb{Q}$ by (10.1.6). □

The next result shows that not every complex number is an algebraic number.

(10.1.10) *There are countably many algebraic numbers, but uncountably many complex numbers.*

Proof. Of course $\mathbb{C}$ is uncountable by (1.4.7). To see that there are countably many algebraic numbers, note that $\mathbb{Q}[t]$ is countable since it is a countable union of countable sets – cf. Exercise (1.4.5). Also each non-zero polynomial in $\mathbb{Q}[t]$ has finitely many roots. It follows that there are only countably many such roots, and these are precisely the algebraic numbers. □

We have just demonstrated the existence of transcendental numbers, but without giving any examples. Indeed it is a good deal harder to find specific examples. The best known transcendental numbers are the numbers π and e. The fact that π is transcendental underlies the impossibility of "squaring the circle" – for this see 10.2. A good reference for the transcendence of π, e and other numbers is [8].

A subfield of $\mathbb{C}$ which is a finite extension of $\mathbb{Q}$ is called an *algebraic number field*: the elements of algebraic number fields constitute all the algebraic numbers. The theory of algebraic number fields is well-developed and is one of the most active areas of algebra.

Exercises (10.1)

1. Give examples of infinite extensions of $\mathbb{Q}$ and of $\mathbb{Z}_p$.

2. Let $a = 2^{\frac{1}{p}}$ where p is a prime. Prove that $(\mathbb{Q}(a) : \mathbb{Q}) = p$ and that $\mathbb{Q}(a)$ has only two subfields.

3. Let n be an arbitrary positive integer. Construct an algebraic number field of degree n over $\mathbb{Q}$.

4. Let a be a root of $t^6 - 4t + 2 \in \mathbb{Q}[t]$. Prove that $(\mathbb{Q}(a) : \mathbb{Q}) = 6$.

5. Let p and q be distinct primes and let $F = \mathbb{Q}(\sqrt{p}, \sqrt{q})$. Prove the following statements. (a) $(F : \mathbb{Q}) = 4$; (b) $F = \mathbb{Q}(\sqrt{p} + \sqrt{q})$; (c) the irreducible polynomial of $\sqrt{p} + \sqrt{q}$ over $\mathbb{Q}$ is $t^4 - 2(p+q)t^2 + (p-q)^2$.

6. Let K be a finite extension of a field F and F_1 a subfield such that $F \subseteq F_1 \subseteq K$. Prove that F_1 is finite over F and K is finite over F_1.

7. Prove that every non-constant element of $\mathbb{Q}\{t\}$ is transcendent over $\mathbb{Q}$.

8. Let $a = 3^{\frac{1}{2}} - 2^{\frac{1}{3}}$. Show that $(\mathbb{Q}(a) : \mathbb{Q}) = 6$ and find $\mathrm{Irr}_{\mathbb{Q}}(a)$.

9. Let p be a prime and let $a = e^{2\pi i/p}$, a primitive p-th root of unity. Prove that $\mathrm{Irr}_{\mathbb{Q}}(a) = 1 + t + t^2 + \cdots + t^{p-1}$ and $(\mathbb{Q}(a) : \mathbb{Q}) = p - 1$.

10.2 Constructions with ruler and compass

One of the most striking applications of field theory is to settle certain famous geometrical problems dating back to classical Greece. Each problem asks whether it is possible to construct a geometrical object *using ruler and compass only*. Here one has to keep in mind that to the ancient Greeks only mathematical objects constructed by such means had any reality since Greek mathematics was based on geometry. We shall describe four constructional problems and then translate them to field theory.

10.2 Constructions with ruler and compass

(a) Duplication of the cube. A cube of side one unit is given. The problem is to construct a cube with double the volume using ruler and compass. This problem is said to have arisen when the priests in a certain temple were commanded to double the volume of the altar, which had the shape of a cube.

(b) Squaring the circle. Is it possible to construct, using ruler and compass, a square whose area equals that of a circle with radius one unit? This is perhaps the most notorious of the ruler and compass problems. It is of course really a question about the number π.

(c) Trisection of an angle. Another notorious problem asks if it is always possible to trisect a given angle using ruler and compass.

(d) Construction of a regular n-gon. Here the problem is to construct by ruler and compass a regular n-sided plane polygon with side equal to one unit where $n \geq 3$.

These problems defied the efforts of mathematicians for more than 2000 years despite many ingenious attempts to solve them. It was only with the rise of abstract algebra in the 18th and 19th centuries that it was realized that all four problems have negative answers.

Constructibility. Our first move must be to analyze precisely what is meant by a ruler and compass construction. Let S be a set of points in the plane containing the points $O(0, 0)$ and $I(1, 0)$; note that O and I are one unit apart. A point P in the plane is said to be *constructible from S by ruler and compass* if there is a finite sequence of points $P_0, P_1, \ldots, P_n = P$ with P_0 in S where P_{i+1} is obtained from $P_0, \ldots, P_i$ by applying an operation of the following types:

(a) draw a straight line joining two of $P_0, P_1, \ldots, P_i$;

(b) draw a circle with center one of $P_0, P_1, \ldots, P_i$ and radius equal to the distance between two of these points.

Then P_{i+1} is to be a point of intersection of two lines, of a line and a circle or of two circles, where the lines and circles are as described as in (a) and (b).

Finally, a real number r is said to be *constructible from S* if the point $(r, 0)$ is constructible from S. The reader should realize that these definitions are designed to express precisely our intuitive idea of a construction with ruler and compass. Each of the four problems asks whether a certain real number is constructible from some given set of points. For example, in the problem of duplicating a cube of side 1, take S to be the set $\{O, I\}$: the question is whether $2^{\frac{1}{3}}$ is constructible from S.

We begin by showing that the real numbers which are constructible from a given set of points form a field, which explains why field theory is relevant to constructional problems.

10 Introduction to the theory of fields

(10.2.1) *Let S be a set of points in the plane containing $O(0, 0)$ and $I(1, 0)$ and let S^* be the set of all real numbers constructible from S. Then S^* is a subfield of $\mathbb{R}$. Also $\sqrt{a} \in S^*$ whenever $a \in S^*$ and $a > 0$.*

Proof. This is entirely elementary plane geometry. Let $a, b \in S^*$; we have first to prove that $a \pm b$, ab and a^{-1} (if $a \neq 0$) belong to S^*. Keep in mind here that by hypothesis a and b can be constructed.

To construct $a \pm b$, where say $a \geq b$, draw the circle with center $A(a, 0)$ and radius b. This intersects the x-axis at the points $(a + b, 0)$ and $(a - b, 0)$. Hence $a + b$ and $a - b$ are constructible from S and belong to S^*.

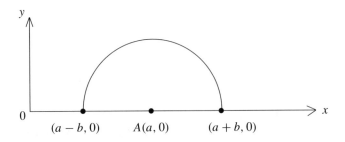

It is a little harder to construct ab. Assume that $a \leq 1 \leq b$: in other cases the procedure is similar. Let A and B be the points $(a, 0)$ and $(b, 0)$. Mark a point $B'(0, b)$ on the y-axis; thus $|OB'| = |OB|$. Draw the line IB' and then draw AC' parallel to

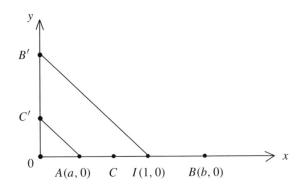

IB' with C' on the y-axis: elementary geometry tells us how to do this. Mark C on the x-axis so that $|OC| = |OC'|$. By similar triangles $|OC'|/|OB'| = |OA|/|OI|$; therefore we have $|OC| = |OC'| = |OA|\cdot|OB'| = ab$. Hence $(ab, 0)$ is constructible and $ab \in S^*$.

Next we show how to construct a^{-1} where, say, $a > 1$. Mark the point $I'(0, 1)$ on the y-axis. Draw the line IC' parallel to AI' with C' on the y-axis. Mark C on the x-axis so that $|OC| = |OC'|$. Then $|OC'|/|OI'| = |OI|/|OA|$, so $|OC'| = a^{-1}$ and $|OC| = a^{-1}$. Thus $(a^{-1}, 0)$ is constructible and $a^{-1} \in S^*$.

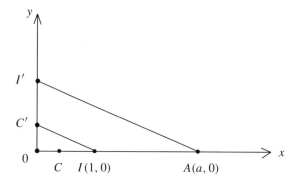

Finally, let $a \in S^*$ where $a > 0$. We must show how to construct the point $(\sqrt{a}, 0)$: it will then follow that $\sqrt{a} \in S^*$. Mark the point $A_1(a+1, 0)$.

Let C be the mid-point of the line segment OA_1; then C is the point $(\frac{a+1}{2}, 0)$ and it is clear how to construct this. Now draw the circle with center C and radius $|OC|$.

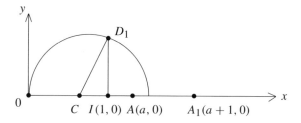

Then draw the perpendicular to the x-axis through the point I and let it meet the upper semicircle at D_1. Mark D on the x-axis so that $|OD| = |ID_1|$. Then

$$|OD|^2 = |ID_1|^2 = |D_1C|^2 - |IC|^2 = \left(\frac{a+1}{2}\right)^2 - \left(\frac{a-1}{2}\right)^2 = a.$$

Hence $|OD| = \sqrt{a}$ and $(\sqrt{a}, 0)$ is constructible. □

It is now possible to discuss the field theoretic implications of constructibility.

(10.2.2) *Let S be a set of points in the plane containing $O(0, 0)$ and $I(1, 0)$, and denote by F the subfield of $\mathbb{R}$ generated by the coordinates of the points of S. Let a be any real number. If a is constructible from S, then $(F(a) : F)$ is a power of 2.*

Proof. Let P be the point $(a, 0)$. Since P is constructible from S, there is by definition a sequence of points $P_0, P_1, \ldots, P_n = P$ with $P_0 \in S$, where P_{i+1} is obtained from $P_0, P_1, \ldots, P_i$ by intersecting lines and circles as explained above. Let P_i be the point (a_i, b_i) and put $E_i = F(a_1, \ldots, a_i, b_1, \ldots, b_i)$ and $E_0 = F$. Then $F(a) \subseteq E_n = E$ since $(a, 0) = (a_n, b_n)$. If P_{i+1} is the point of intersection of two lines whose equations have coefficients over E_i, then a_{i+1} and b_{i+1} are in E_i, as we see by solving the equations, i.e., $E_i = E_{i+1}$. If P_{i+1} is a point of intersection of a line and a circle whose equations have coefficients in E_i, then a_{i+1} is a root of a quadratic equation over E_i. Hence $(E_i(a_{i+1}) : E_i) \leq 2$. Clearly we can solve for b_{i+1} in terms of a_{i+1}, so $b_{i+1} \in E_i(a_{i+1})$ and in fact $E_{i+1} = E_i(a_{i+1})$. Therefore $(E_{i+1} : E_i) \leq 2$. If P_{i+1} is a point of intersection of two circles over F_i, subtract the equations of the circles (in standard form) to realize P_{i+1} as a point of intersection of a line and a circle. Thus $(E_{i+1} : E_i) \leq 2$ in all cases. It follows that $(E : F) = \prod_{i=0}^{n-1}(E_{i+1} : E_i)$ is a power of 2, and so is $(F(a) : F)$ since $(E : F) = (E : F(a))(F(a) : F)$ by (10.1.7). $\square$

The first two ruler and compass problems can now be resolved.

(10.2.3) *It is impossible to duplicate a cube of side 1 or to square a circle of radius 1 by ruler and compass.*

Proof. Let S consist of the points $O(0,0)$ and $I(1,0)$. In the case of the cube, constructibility would imply that $(\mathbb{Q}(2^{\frac{1}{3}}) : \mathbb{Q})$ is a power of 2 by (10.2.2). But $(\mathbb{Q}(2^{\frac{1}{3}}) : \mathbb{Q}) = 3$ since $\text{Irr}_\mathbb{Q}(2^{\frac{1}{3}}) = t^3 - 2$, a contradiction.

If it were possible to square the circle, then $\sqrt{\pi}$ would be constructible from S. By (10.2.2) this would mean that $(\mathbb{Q}(\sqrt{\pi}) : \mathbb{Q})$ is a power of 2, as would $(\mathbb{Q}(\pi) : \mathbb{Q})$ be since $(\mathbb{Q}(\pi) : \mathbb{Q}(\sqrt{\pi})) \leq 2$. But in fact π is transcendental over $\mathbb{Q}$ by a famous result of Lindemann[1], so $(\mathbb{Q}(\pi) : \mathbb{Q})$ is actually infinite. Therefore it is impossible to square the circle. $\square$

With a little more effort we can determine which angles can be trisected.

(10.2.4) *An angle α can be trisected by ruler and compass if and only if the polynomial $4t^3 - 3t - \cos\alpha$ is reducible over the field $\mathbb{Q}(\cos\alpha)$.*

Proof. In this problem the angle α is given, and so we can find its cosine by constructing a right angled triangle with angle α and hypotenuse 1. Now let S consist of the points $O, I, (\cos\alpha, 0)$. Let $F = \mathbb{Q}(\cos\alpha)$ and put $\theta = \frac{1}{3}\alpha$. The problem is to decide if θ, or equivalently $\cos\theta$, is constructible from S. If this is the case, then $(F(\cos\theta) : F)$ must be a power of 2.

At this point we need to recall the well-known trigonometric identity

$$\cos 3\theta = 4\cos^3\theta - 3\cos\theta.$$

Hence $4\cos^3\theta - 3\cos\theta - \cos\alpha = 0$, and $\cos\theta$ is a root of the polynomial $f = 4t^3 - 3t - \cos\alpha \in F[t]$. If θ is constructible, then $\text{Irr}_F(\cos\alpha)$ has degree a power of 2. Hence f must be reducible.

Conversely, suppose that f is reducible, so that $\cos\theta$ is a root of a linear or quadratic polynomial over F; thus $\cos\theta$ has the form $u + v\sqrt{w}$ where $u, v, w \in F$ and $w \geq 0$. Since $F \subseteq S^*$, it follows via (10.2.1) that $\sqrt{w} \in S^*$. Hence $\cos\theta \in S^*$ and $\cos\theta$ is constructible from S, as required. $\square$

---

[1] Carl Ferdinand von Lindemann (1852–1939).

10.3 Finite fields

Example (10.2.1) The angle $\frac{\pi}{4}$ is trisectible by ruler and compass.

Since $\cos \frac{\pi}{4} = \frac{1}{\sqrt{2}}$, the polynomial f equals $4t^3 - 3t - \frac{1}{\sqrt{2}}$, which has the root $-\frac{1}{\sqrt{2}}$ in $\mathbb{Q}(\cos(\pi/4)) = \mathbb{Q}(\sqrt{2})$ and so is reducible. Now apply (10.2.4).

Example (10.2.2) The angle $\frac{\pi}{3}$ is not trisectible by ruler and compass.

In this case $\cos \frac{\pi}{3} = \frac{1}{2}$ and $f = 4t^3 - 3t - \frac{1}{2}$. This polynomial is irreducible over $\mathbb{Q}(\frac{1}{2}) = \mathbb{Q}$ since it has no rational roots. Hence $\frac{\pi}{3}$ is not trisectible.

A complete discussion of the problem of constructing a regular n-gon requires Galois theory, so it will be deferred until 11.3.

Exercises (10.2)

1. Complete the proof that $ab \in S^*$ in (10.2.1) by dealing with the cases $1 \leq a \leq b$, and $a \leq b \leq 1$.

2. A cube of side a can be duplicated if and only if $2a$ is the cube of a rational number.

3. Suppose the problem is to double the surface area of a cube of side 1. Can this be done by ruler and compass?

4. Determine which of the following angles are trisectible:
   (i) $\frac{\pi}{2}$, (ii) $\frac{\pi}{6}$, (iii) $\frac{\pi}{12}$.

5. Let p be a prime and suppose that $a = e^{2\pi i/p}$ is constructible from O and I. Show that p must have the form $2^{2^c} + 1$ for some integer c, i.e., p is a *Fermat prime*. (The known Fermat primes occur for $1 \leq c \leq 4$).

10.3 Finite fields

It was shown in (8.2.16) that the order of a finite field is always a power of a prime. More precisely, if F is a field of characteristic p and $(F : \mathbb{Z}_p) = n$, then $|F| = p^n$. Our main purpose here is to show that there are fields with arbitrary prime power order and that fields of the same order are isomorphic.

We begin by identifying finite fields as the splitting fields of certain polynomials. Let F be a field of order $q = p^n$ where p is a prime, namely, the characteristic of F. The multiplicative group $U(F)$ has order $q - 1$ and Lagrange's Theorem shows that the order of every element of $U(F)$ divides $q - 1$. This means that $a^{q-1} = 1$ for every $a \neq 0$ in F, and so $a^q - a = 0$. Now the zero element also satisfies this last equation. Consequently every element of F is a root of the polynomial $t^q - t \in \mathbb{Z}_p[t]$. Since $t^q - t$ cannot have more than q roots, we conclude that the elements of F constitute all the roots of $t^q - t$, so that F is a splitting field of $t^q - t$.

The foregoing discussion suggests that we may be able to establish the existence of finite fields by using splitting fields. The main result is:

(10.3.1) *Let $q = p^n$ where p is a prime and $n > 0$. Then:*
  (i) *a splitting field of the polynomial $t^q - t \in \mathbb{Z}_p[t]$ has order q;*
  (ii) *if F is any field of order q, then F is a splitting field of $t^q - t$ over $\mathbb{Z}_p$.*

Proof. We have already proved (ii), so let us consider the situation of (i) and write F for a splitting field of $t^q - t$. Define $S = \{a \in F \mid a^q = a\}$, i.e., the set of roots of $t^q - t$ in F. First we show that S is a subfield of F. For this purpose let $a, b \in S$. Recall that p divides $\binom{p}{i}$ if $1 \leq i < p$ by (2.3.3); therefore the Binomial Theorem for the field F – see Exercise (6.1.6) – shows that $(a \pm b)^p = a^p \pm b^p$. Taking further powers of p, we conclude that

$$(a \pm b)^q = a^q \pm b^q = a \pm b,$$

which shows that $a \pm b \in S$. Also $(ab)^q = a^q b^q = ab$ and $(a^{-1})^q = (a^q)^{-1} = a^{-1}$ if $a \neq 0$, so it follows that $ab \in S$ and $a^{-1} \in S$. Thus S is a subfield of F.

Next all the roots of the polynomial $t^q - t$ are different. For

$$(t^q - t)' = qt^{q-1} - 1 = -1,$$

so that $t^q - t$ and $(t^q - t)'$ are relatively prime; by (7.4.7) $t^q - t$ has no repeated roots and it follows that $|S| = q$. Finally, since F is a splitting field of $t^q - t$, it is generated by $\mathbb{Z}_p$ and the roots of $t^q - t$. Therefore $F = S$ and $|F| = q$. □

Our next object is to show that two fields of the same finite order are isomorphic. Since every finite field has been identified as a splitting field, our strategy will be to prove that any two splitting fields of a polynomial are isomorphic, plainly a result of independent interest. In proving this we will employ a useful lemma which tells us how to extend an isomorphism between fields.

(10.3.2) *Let $E = F(x)$ and $E^* = F^*(x^*)$ be simple algebraic extensions of fields F and F^*. Further assume there is an isomorphism $\alpha: F \to F^*$ such that $\alpha(\mathrm{Irr}_F(x)) = \mathrm{Irr}_{F^*}(x^*)$. Then there exists an isomorphism $\theta: E \to E^*$ such that $\theta|_F = \alpha$ and $\theta(x) = x^*$.*

Notice that in the statement of the result α has been extended in the obvious way to a ring isomorphism $\alpha: F[t] \to F^*[t]$, by the rule

$$\alpha\left(\sum_{i=1}^{m} a_i t^i\right) = \sum_{i=1}^{m} \alpha(a_i) t^i$$

where $a_i \in F$.

Proof of (10.3.2). Put $f = \mathrm{Irr}_F(x)$ and $f^* = \mathrm{Irr}_{F^*}(x^*)$. Then by hypothesis $\alpha(f) = f^*$. This fact permits us to define a mapping

$$\theta_0: F[t]/(f) \to F^*[t]/(f^*)$$

by $\theta_0(g+(f)) = \alpha(g)+(f^*)$; this is a well defined isomorphism. Next recall that by (10.1.4) $F(x) \simeq F[t]/(f)$ and $F^*(x^*) \simeq F^*[t]/(f^*)$ via assignments

$$g(x) \mapsto g+(f) \quad \text{and} \quad g^*(x^*) \mapsto g^*+(f^*) \quad (g \in F[t], g^* \in F^*[t]).$$

Composition of these mappings with θ_0 yields an isomorphism $\theta : F(x) \to F^*(x^*)$ where $\theta(g(x)) = \alpha g(x^*)$, as indicated below.

$$F(x) \to F[t]/(f) \xrightarrow{\theta_0} F^*[t]/(f^*) \to F^*(x^*).$$

Further $\theta|_F = \alpha$ and $\theta(x) = x^*$, so the proof is complete. □

The uniqueness of splitting fields is a special case of the next result.

(10.3.3) *Let $\alpha : F \to F^*$ be an isomorphism of fields, and let $f \in F[t]$ and $f^* = \alpha(f) \in F^*[t]$. If E and E^* are splitting fields of f and f^* respectively, there is an isomorphism $\theta : E \to E^*$ such that $\theta|_F = \alpha$.*

Proof. Argue by induction on $n = \deg(f)$. If $n = 1$, then $E = F$, $E^* = F^*$ and $\theta = \alpha$. Assume that $n > 1$. Let a be a root of f in E and put $g = \text{Irr}_F(a)$. Choose any root a^* of $g^* = \alpha(g) \in F^*[t]$. Then $g^* = \text{Irr}_{F^*}(a^*)$. By (10.3.2) we may extend α to an isomorphism $\theta_1 : F(a) \to F^*(a^*)$ such that $\theta_1|_F = \alpha$ and $\theta_1(a) = a^*$.

Now regard E and E^* as splitting fields of the polynomials $f/(t-a)$ and $f^*/(t-a^*)$ over $F(a)$ and $F^*(a^*)$ respectively. By induction on n we may extend θ_1 to an isomorphism $\theta : E \to E^*$; furthermore $\theta|_F = \theta_1|_F = \alpha$. □

Corollary (10.3.4) *Let f be a non-constant polynomial over a field F. Then up to isomorphism f has a unique splitting field.*

This follows from (10.3.3) by taking $F = F^*$ and α to be the identity map. Since a finite field of order q is a splitting field of $t^q - t$, we deduce at once:

(10.3.5) (E. H. Moore[2]) *Two fields of the same order are isomorphic.*

It is customary to write

$$\text{GF}(q)$$

for the unique field of order q. Here "GF" stands for *Galois field*.

It is a very important fact about finite fields that their multiplicative groups are cyclic. Somewhat more generally we will prove:

(10.3.6) *If F is any field, every finite subgroup of $U(F)$, the multiplicative group of F, is cyclic. If F has finite order q, then $U(F)$ is a cyclic group of order $q-1$.*

---

[2]Eliakim Hastings Moore (1862–1932).

Proof. Let X be a finite subgroup of $U(F)$. Then X is a finite abelian group, so by (9.2.7) $X = P_1 \times P_2 \times \cdots \times P_k$ where P_i is a finite p_i-group and $p_1, p_2, \ldots, p_k$ are different primes. Choose an element x_i of P_i with maximum order, say $p_i^{\ell_i}$, and let $x = x_1 x_2 \ldots x_k$. Now $x^m = x_1^m x_2^m \ldots x_k^m$ and the x_i have relative prime orders. Thus $x^m = 1$ if and only if $x_i^m = 1$, i.e., $p_i^{\ell_i}$ divides m for all i. It follows that x has order $d = p_1^{\ell_1} p_2^{\ell_2} \ldots p_k^{\ell_k}$ and $|X| \geq |x| = d$.

Now let y be any element of X and write $y = y_1 y_2 \ldots y_k$ with $y_i \in P_i$. Then $y_i^{p_i^{\ell_i}} = 1$ since $p_i^{\ell_i}$ is the largest order of an element of P_i. Therefore $y_i^d = 1$ for all i and $y^d = 1$. It follows that every element of X is a root of the polynomial $t^d - 1$, and hence $|X| \leq d$. Finally $|X| = d = |x|$ and so $X = \langle x \rangle$. □

The last result provides another way to represent the elements of a field F of order q. If $U(F) = \langle a \rangle$, then $F = \{0, 1, a, a^2, \ldots, a^{q-2}\}$ where $a^{q-1} = 1$. This representation is useful for computational purposes.

Corollary (10.3.7) *Every finite field F is a simple extension of its prime subfield.*

For if $U(F) = \langle a \rangle$, then obviously $F = \mathbb{Z}_p(a)$ where p is the characteristic of F.

Example (10.3.1) Let $F = GF(27)$ be the Galois field of order 27. Exhibit F as a simple extension of $GF(3)$ and find a generator of $U(F)$.

The field F may be realized as the splitting field of the polynomial $t^{27} - t$, but it is simpler to choose an irreducible polynomial of degree 3 over $GF(3)$, for example $f = t^3 - t + 1$. Since this polynomial is irreducible over $GF(3)$, $F = (GF(3)[t])/(f)$ is a field of order 3^3, which must be $GF(27)$. Put $x = t + (f)$. Then, because f has degree 3, each element of F has the unique form $a_0 + a_1 t + a_2 t^2 + (f)$ with $a_i \in GF(3)$, i.e., $a_0 + a_1 x + a_2 x^2$. Thus $F = GF(3)(x)$, and $\mathrm{Irr}_{GF(3)}(x) = f = t^3 - t + 1$.

Next we argue that $U(F) = \langle x \rangle$. Since $|U(F)| = 26$, it is enough to prove that $|x| = 26$. Certainly $|x|$ divides 26, so it suffices to show that $x^2 \neq 1$ and $x^{13} \neq 1$. The first statement is true because $f \nmid t^2 - 1$. To prove that $x^{13} \neq 1$, use the relation $x^3 = x - 1$ to compute $x^{12} = (x-1)^4 = x^2 + 2$ and $x^{13} = -1 \neq 1$.

Exercises (10.3)

1. Let F be a field of order p^m and K a subfield of F where p is a prime. Prove that $|K| = p^d$ where d divides m.

2. If F is a field of order p^m and d is a positive divisor of m, show that F has exactly one subfield of order p^d. [Hint: $t^{p^d} - t$ divides $t^{p^m} - t$].

3. Find an element of order 7 in the multiplicative group of $GF(8)$.

4. Find elements of order 3, 5 and 15 in the multiplicative group of $GF(16)$.

5. Prove that $t^{p^n} - t \in \mathrm{GF}(p)[t]$ is the product of all the distinct monic irreducible polynomials with degree dividing n.

6. Let $\psi(n)$ denote the number of monic irreducible polynomials of degree n in $\mathrm{GF}(p)[t]$ where p is a fixed prime.
   (i) Prove that $p^n = \sum_{d|n} \psi(d)$ where the sum is over all positive divisors d of n.
   (ii) Deduce that $\psi(n) = \frac{1}{n} \sum_{d|n} \mu(d) p^{n/d}$ where μ is the Möbius function defined as follows: $\mu(1) = 1$, $\mu(n)$ equals $(-1)^r$ where r is the number of distinct prime divisors of n if n is square-free and $\mu(n) = 0$ otherwise. [You will need the Möbius[3] Inversion Formula: if $f(n) = \sum_{d|n} g(d)$, then $g(n) = \sum_{d|n} \mu(d) f(n/d)$. For an account of the Möbius function see 11.3 below].

7. Find all monic irreducible polynomials over $\mathrm{GF}(2)$ with degrees 2, 3, 4 and 5, using Exercise (10.3.6) to check your answer.

10.4 Applications to latin squares and Steiner triple systems

In this section we describe two applications of finite fields to combinatorics which demonstrate the efficacy of algebraic methods in solving difficult combinatorial problems.

Latin squares. A *latin square of order n* is an $n \times n$ matrix with entries in a set of n symbols such that each symbol occurs exactly once in each row and exactly once in each column. Examples of latin squares are easily found.

Example (10.4.1) (i) The matrices $\begin{bmatrix} a & b \\ b & a \end{bmatrix}$ and $\begin{bmatrix} a & b & c \\ b & c & a \\ c & a & b \end{bmatrix}$ are latin squares of orders 2 and 3.

(ii) Let $G = \{g_1, g_2, \ldots, g_n\}$ be a group of order n. Then the multiplication table of G – see 3.3 – is a latin square of order n. For if the first row is $g_1, g_2, \ldots, g_n$, the entries of the ith row are $g_i g_1, g_i g_2, \ldots, g_i g_n$, and these are clearly all different. The same argument applies to the columns.

On the other hand, not every latin square determines the multiplication table of a group since the associative law might not hold. In fact a latin square determines a more general algebraic structure called a *quasigroup* – for this concept see Exercises (10.4.4) and (10.4.5) below.

While latin squares often occur in puzzles, they have a more serious use in the design of statistical experiments. An example will illustrate how this arises.

---
[3] August Ferdinand Möbius (1790–1868).

10 Introduction to the theory of fields

Example (10.4.2) Five types of washing powder P_1, P_2, P_3, P_4, P_5 are to be tested in five machines A, B, C, D, E over five days D_1, D_2, D_3, D_4, D_5. Each washing powder is to be used once each day and tested once on each machine. How can this be done?

Here the object is to allow for differences in the machines and in the water supply on different days, while keeping the number of tests to a minimum. A schedule of tests can be given in the form of a latin square of order 5 whose rows correspond to the washing powders and whose columns correspond to the days; the symbols are the machines. For example, we could use the latin square

$$\begin{bmatrix} A & B & C & D & E \\ B & C & D & E & A \\ C & D & E & A & B \\ D & E & A & B & C \\ E & A & B & C & D \end{bmatrix}.$$

This would mean, for example, that washing powder P_3 is to be used on day D_4 in machine A. There are of course many other possible schedules.

The number of latin squares. Let $L(n)$ denote the number of latin squares of order n which can be formed from a given set of n symbols. It is clear that the number $L(n)$ must increase rapidly with n. A rough upper bound for $L(n)$ can be found using our count of derangements in (3.1.11).

(10.4.1) *The number $L(n)$ of latin squares of order n that can be formed from n given symbols satisfies the inequality*

$$L(n) \leq (n!)^n \left(1 - \frac{1}{1!} + \frac{1}{2!} - \cdots + \frac{(-1)^n}{n!}\right)^{n-1},$$

and so $L(n) = O((n!)^n/e^{n-1})$.

Proof. Taking the symbols to be $1, 2, \ldots, n$, we note that each row of a latin square of order n corresponds to a permutation of $\{1, 2, \ldots, n\}$, i.e., to an element of the symmetric group S_n. Thus there are $n!$ choices for the first row. Now rows 2 through n must be derangements of row 1 since no column can have a repeated element. Recall from (3.1.11) that the number of derangements of n symbols is

$$d_n = n!\left(1 - \frac{1}{1!} + \frac{1}{2!} + \cdots + \frac{(-1)^n}{n!}\right).$$

Hence rows 2 through n of the latin square can be chosen in at most $(d_n)^{n-1}$ ways. Therefore we have

$$L(n) \leq (n!)(d_n)^{n-1} = (n!)^n \left(1 - \frac{1}{1!} + \frac{1}{2!} - \cdots + \frac{(-1)^n}{n!}\right)^{n-1}. \qquad \square$$

10.4 Applications to latin squares and Steiner triple systems

It can be shown that a lower bound for $L(n)$ is $\prod_{i=1}^{n} i!$ and also that $\log L(n) = n^2 \log n + O(n^2)$ – for details see [2].

Mutually orthogonal latin squares. Suppose that $A = [a_{ij}]$ and $B = [b_{ij}]$ are two latin squares of order n. Then A and B are called *mutually orthogonal latin squares* (or MOLS) if the n^2 ordered pairs (a_{ij}, b_{ij}) are all distinct.

Example (10.4.3) The latin squares

$$\begin{bmatrix} a & b & c \\ b & c & a \\ c & a & b \end{bmatrix} \quad \text{and} \quad \begin{bmatrix} \alpha & \beta & \gamma \\ \gamma & \alpha & \beta \\ \beta & \gamma & \alpha \end{bmatrix}$$

are mutually orthogonal, as can be seen by listing the nine pairs of entries. On the other hand, there are no pairs of MOLS of order 2 since these would have to be of the form

$$\begin{bmatrix} a & b \\ b & a \end{bmatrix}, \quad \begin{bmatrix} a' & b' \\ b' & a' \end{bmatrix}$$

and the pair (a, a') is repeated.

Why are mutually orthogonal latin squares significant? In fact they too have a statistical use, as can be seen from an elaboration of the washing powder example.

Example (10.4.4) Suppose that in Example (10.4.2) there are also five washing machine operators $\alpha, \beta, \gamma, \delta, \epsilon$. Each operator is to test each powder once and to carry out one test per day. In addition, for reasons of economy, we do not want to repeat the same combination of machine and operator for any powder and day.

What is asked for here is a pair of MOLS of order 5. A latin square with the schedule of machines was given in Example (10.4.2). By a little experimentation another latin square can be found such that the pair are mutually orthogonal; this is

$$\begin{bmatrix} \alpha & \beta & \gamma & \delta & \epsilon \\ \gamma & \delta & \epsilon & \alpha & \beta \\ \epsilon & \alpha & \beta & \gamma & \delta \\ \beta & \gamma & \delta & \epsilon & \alpha \\ \delta & \epsilon & \alpha & \beta & \gamma \end{bmatrix}.$$

A direct enumeration of the 25 pairs of entries from the two latin squares shows that all are different. The two latin squares tell us the schedule of operations: thus, for example, powder P_3 is to be used on day D_4 by operator γ in machine A.

We are interested in determining the maximum number of MOLS of order n, say

$$f(n).$$

In the first place there is an easy upper bound for $f(n)$.

(10.4.2) *If $n \geq 1$, then $f(n) \leq n - 1$.*

Proof. Suppose that $A_1, A_2, \ldots, A_r$ are MOLS of order n, and let the $(1, 1)$ entry of A_i be a_i. Consider row 2 of A_1. It has an a_1 in the $(2, i_1)$ position for some $i_1 \neq 1$ since there is already an a_1 in the first column. So there are $n - 1$ possibilities for i_1. Next in A_2 there is an a_2 in row 2, say as the $(2, i_2)$ entry where $i_2 \neq 1$; also $i_2 \neq i_1$ since the pair (a_1, a_2) has already occurred and cannot be repeated. Hence there are $n - 2$ possibilities for i_2. Continuing this line of argument until A_r is reached, we conclude that a_r is the $(2, i_r)$ entry of A_r where there are $n - r$ possibilities for i_r. Therefore $n - r > 0$ and $r \leq n - 1$, as required. □

The question to be addressed is whether $f(n) > 1$ for all $n > 2$; note that $f(2) = 1$ since, as already observed, there cannot exist two MOLS of order 2.

The intervention of field theory. The mere existence of finite fields of every prime power order is enough to make a decisive advance in the construction of MOLS of prime power order.

(10.4.3) *Let p be a prime and m a positive integer. Then $f(p^m) = p^m - 1$.*

Proof. Let F be a field of order p^m, which exists by (10.3.1). For each $a \neq 0$ in F define a $p^m \times p^m$ matrix $A(a)$ over F, with rows and columns labelled by the elements of F written in some fixed order; the (u, v) entry of $A(a)$ is to be computed using the formula

$$(A(a))_{u,v} = ua + v$$

where $u, v \in F$. In the first place each $A(a)$ is a latin square of order p^m. For $ua + v = u'a + v$ implies that $ua = u'a$ and $u = u'$ since $0 \neq a \in F$. Also $ua + v = ua + v'$ implies that $v = v'$.

Next we must show that $A(a)$'s are mutually orthogonal. Suppose that $A(a_1)$ and $A(a_2)$ are not orthogonal where $a_1 \neq a_2$: then

$$(ua_1 + v, ua_2 + v) = (u'a_1 + v', u'a_2 + v')$$

for some $u, v, u', v' \in F$. Then of course $ua_1 + v = u'a_1 + v'$ and $ua_2 + v = u'a_2 + v'$. Subtracting the second of these equations from the first results in $u(a_1 - a_2) = u'(a_1 - a_2)$. Since $a_1 - a_2 \neq 0$ and F is a field, it follows that $u = u'$ and hence $v = v'$. Thus we have constructed $p^m - 1$ MOLS of order p^m, which is the maximum number by (10.4.2). This completes the proof. □

Example (10.4.5) Construct three MOLS of order 4.

Here we start with a field F of order 4, constructed from the irreducible polynomial $t^2 + t + 1 \in \mathbb{Z}_2[t]$. If a is a root of this polynomial, then $F = \{0, 1, a, 1 + a\}$ where

10.4 Applications to latin squares and Steiner triple systems

$a^2 = a + 1$. Now construct the three MOLS $A(1)$, $A(a)$, $A(1+a)$, using the formula indicated in the proof of (10.4.4):

$$A(1) = \begin{bmatrix} 0 & 1 & a & 1+a \\ 1 & 0 & 1+a & a \\ a & 1+a & 0 & 1 \\ 1+a & a & 1 & 0 \end{bmatrix}, \quad A(a) = \begin{bmatrix} 0 & 1 & a & 1+a \\ a & 1+a & 0 & 1 \\ 1+a & a & 1 & 0 \\ 1 & 0 & 1+a & a \end{bmatrix},$$

$$A(1+a) = \begin{bmatrix} 0 & 1 & a & 1+a \\ 1+a & a & 1 & 0 \\ 1 & 0 & 1+a & a \\ a & 1+a & 0 & 1 \end{bmatrix}.$$

To construct MOLS whose order is not a prime power one can use a direct product construction. Let A and B be latin squares of orders m and n respectively. The *direct product* $A \times B$ is an $mn \times mn$ matrix whose entries are pairs of elements $(a_{ij}, b_{i'j'})$. It is best to visualize the matrix in block form:

$$\begin{bmatrix} (a_{11}, B) & (a_{12}, B) & \cdots & (a_{1m}, B) \\ (a_{21}, B) & (a_{22}, B) & \cdots & (a_{2m}, B) \\ \vdots & \vdots & \vdots & \vdots \\ (a_{m1}, B) & (a_{m2}, B) & \cdots & (a_{mm}, B) \end{bmatrix}$$

where (a_{ij}, B) means that a_{ij} is paired with each entry of B in the natural matrix order. It is easy to see that $A \times B$ is a latin square of order mn.

Example (10.4.6) Let $A = \begin{bmatrix} a & b \\ c & d \end{bmatrix}$ and $B = \begin{bmatrix} \alpha & \beta & \gamma \\ \beta & \gamma & \delta \\ \gamma & \delta & \alpha \end{bmatrix}$. Then

$$A \times B = \begin{bmatrix} (a,\alpha) & (a,\beta) & (a,\gamma) & (b,\alpha) & (b,\beta) & (b,\gamma) \\ (a,\beta) & (a,\gamma) & (a,\delta) & (b,\beta) & (b,\gamma) & (b,\delta) \\ (a,\gamma) & (a,\delta) & (a,\alpha) & (b,\gamma) & (b,\delta) & (b,\alpha) \\ (c,\alpha) & (c,\beta) & (c,\gamma) & (d,\alpha) & (d,\beta) & (d,\gamma) \\ (c,\beta) & (c,\gamma) & (c,\delta) & (d,\beta) & (d,\gamma) & (d,\delta) \\ (c,\gamma) & (c,\delta) & (c,\alpha) & (d,\gamma) & (d,\delta) & (d,\alpha) \end{bmatrix},$$

a latin square of order 6.

Now suppose we have MOLS $A_1, A_2, \ldots, A_r$ of order m and $B_1, B_2, \ldots, B_s$ of order n where $r \leq s$; then the latin squares $A_1 \times B_1, A_2 \times B_2, \ldots, A_r \times B_r$ have order mn and they are mutually orthogonal, as a check of the entry pairs shows. On the basis of this observation we can state:

(10.4.4) *If $n = n_1 n_2$, then $f(n) \geq \min\{f(n_1), f(n_2)\}$.*

This result can be used to give further information about the integer $f(n)$. Let $n = p_1^{e_1} \ldots p_k^{e_k}$ be the primary decomposition of n. Then

$$f(n) \geq \min\{p_i^{e_i} - 1 \mid i = 1, \ldots, k\}$$

by (10.4.3) and (10.4.4). Therefore $f(n) > 1$ provided that $p_i^{e_i} \neq 2$ for all i. This means that either n is odd or it is divisible by 4, i.e., $n \not\equiv 2 \pmod{4}$. Hence we have:

(10.4.5) *If $n \not\equiv 2 \pmod{4}$, there exist at least two mutually orthogonal latin squares of order n.*

In 1782 Euler conjectured that the converse is true, i.e. if $n \equiv 2 \pmod 4$, there cannot be a pair of $n \times n$ MOLS. Indeed in 1900 Tarry[4] was able to confirm that there does not exist a pair of 6×6 MOLS; thus $f(6) = 1$. However in the end it turned out that Euler was wrong; for in a remarkable piece of work Bose, Shrikhande and Parker were able to prove that there is a pair of $n \times n$ MOLS for all even integers $n \neq 2, 6$.

The case $n = 6$ is Euler's celebrated *Problem of the Thirty Six Officers*. Euler asked if it is possible for thirty six officers of six ranks and six regiments to march in six rows of six, so that each row and each column contain exactly one officer of each rank and exactly one of each regiment, *no combination of rank and regiment being repeated*. Of course Euler was really asking if there are two mutually orthogonal latin squares of order 6, the symbols of the first latin square being the ranks and those of the second the regiments of the officers. By Tarry's result the answer is no!

Steiner triple systems. Another striking use of finite fields is to construct combinatorial objects known as Steiner[5] triple systems. We begin with a brief explanation of these. A *Steiner triple system of order n* is a pair $(X, \mathcal{T})$ where X is a set with n elements (called *points*) and $\mathcal{T}$ is a set of 3-element subsets of X (called *triples*) such that every pair of points occurs in exactly one triple. Steiner triple systems belong to a general class of combinatorial objects called designs which are widely used in statistics.

Example (10.4.7) A Steiner system of order 7.

Consider the diagram consisting of a triangle with the three medians drawn. Let X be the set of 7 points consisting of the vertices, the midpoints of the sides and the centroid, labelled $1, 2, \ldots, 7$ for convenience. Let the triples be the sets of 3 points lying on each line and on the circle 456. Thus $\mathcal{T} = \{175, 276, 374, 142, 253, 361, 456\}$ where we have written abc for $\{a, b, c\}$. It is clear from the diagram that each pair of points occurs in a unique triple.

The question we shall consider is this: for which positive integers n do there exist Steiner triple systems of order n? It is quite easy to derive some necessary conditions on n; these will follow from the next result.

---

[4] Gaston Tarry (1843–1913).
[5] Jakob Steiner (1796–1863).

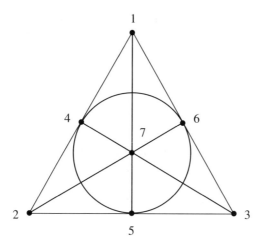

(10.4.6) *Suppose that $(X, \mathcal{T})$ is a Steiner triple system of order n. Then*

(i) *each point belongs to exactly $\frac{n-1}{2}$ triples;*

(ii) *the number of triples is $\frac{n(n-1)}{6}$.*

Proof. (i) Let $x, y \in X$ with x fixed. The idea behind the proof is to count in two different ways the pairs (y, T) such that $x, y \in T$, $y \neq x$, $T \in \mathcal{T}$. There are $n - 1$ choices for y; then, once y has been chosen, there is a unique $T \in \mathcal{T}$ containing x and y. So the number of pairs is $n - 1$. On the other hand, let r denote the number of triples in $\mathcal{T}$ to which x belongs. Once a T in $\mathcal{T}$ containing x has been chosen, there are two choices for y in T. Thus the number of pairs is $2r$. Therefore $2r = n - 1$ and $r = \frac{n-1}{2}$.

(ii) In a similar vein we count in two different ways the pairs (x, T) where $x \in T \in \mathcal{T}$. If t is the total number of triples, the number of pairs is $3t$ since there are three choices for x in T. On the other hand, we may also choose x in n ways and a triple T containing x in $\frac{n-1}{2}$ ways by (i). Therefore $3t = \frac{n(n-1)}{2}$ and $t = \frac{1}{6}n(n - 1)$. □

From this result we deduce a necessary condition on n for a Steiner triple system of order n to exist.

Corollary (10.4.7) *If there exists a Steiner triple system of order n, then $n \equiv 1$ or $3 \pmod 6$.*

Proof. In the first place $\frac{n-1}{2}$ must be an integer, so n is odd. Then we can write $n = 6k + \ell$ where $\ell = 1, 3$ or 5. If $\ell = 5$, then $\frac{1}{6}n(n - 1) = \frac{1}{3}(6k + 5)(3k + 2)$, which is not an integer. Hence $l = 1$ or 3 and $n \equiv 1$ or $3 \pmod 6$. □

206   10 Introduction to the theory of fields

A fundamental theorem on Steiner triple systems asserts that the converse of (10.4.7) is true. *If $n \equiv 1$ or $3 \pmod 6$, then there is a Steiner triple system of order n.* We shall prove a special case of this to illustrate how finite field theory can be applied.

(10.4.8) *If q is a prime power such that $q \equiv 1 \pmod 6$, then there is a Steiner triple system of order q.*

Proof. Let F be a finite field of order q. Recall from (10.3.6) that $U(F)$ is a cyclic group of order $q - 1$. Since $6 \mid q - 1$ by hypothesis, it follows from (4.1.6) that $U(F)$ contains an element z of order 6. Thus $|U(F) : \langle z \rangle| = \frac{q-1}{6}$. Choose a transversal to $\langle z \rangle$ in $U(F)$, say $\{t_1, t_2, \ldots, t_{\frac{q-1}{6}}\}$. Now define subsets

$$T_i = \{0, t_i, t_i z\}$$

for $i = 1, 2, \ldots, \frac{q-1}{6}$.

The points of our Steiner triple system are the elements of the field F, while the set of triples is designated as

$$\mathcal{T} = \left\{ a + T_i \mid a \in F, \ i = 1, 2, \ldots, \frac{q-1}{6} \right\}.$$

Here $a + T_i$ denotes the set $\{a + x \mid x \in T_i\}$. We claim that $(X, \mathcal{T})$ is a Steiner triple system. First we make an observation. Define D_i to be the set of all differences of two elements in T_i; thus $D_i = \{0, \pm t_i, \pm t_i z, \pm t_i(1 - z)\}$. Now z has order 6 and $0 = z^6 - 1 = (z^3 - 1)(z + 1)(z^2 - z + 1)$, so that $z^2 - z + 1 = 0$ and $z^2 = z - 1$. Hence $z^3 = -1$, $z^4 = -z$, $z^5 = 1 - z$. From these equations it follows that D_i is just the coset $t_i \langle z \rangle = \{t_i z^k \mid 0 \le k \le 5\}$ with 0 adjoined.

To show that we have a Steiner triple system, we need to prove that any two distinct elements x and y of F belong to a unique triple $a + T_i$. Let $f = x - y \in U(F)$. Then f belongs to a unique coset $t_i \langle z \rangle$, and by the observation above $f \in D_i$, so that f is expressible as the difference between two elements in the set T_i, say $f = u_i - v_i$. Writing $a = y - v_i$, we have $x = f + y = (y - v_i) + u_i \in a + T_i$ and $y = (y - v_i) + v_i \in a + T_i$.

Now suppose that x and y belong to another triple $b + T_j$, with $x = b + d_j$ and $y = b + e_j$ where $d_j, e_j \in T_j$. Then $0 \ne f = x - y = d_j - e_j$ and hence $f \in D_j = t_j \langle z \rangle$, which means that $j = i$. Also there is clearly only one way to write f as the difference between two elements of T_i. Therefore $d_i = u_i$ and $e_i = v_i$, from which it follows that $a = b$. The proof is now complete. □

The construction just described produces Steiner triple systems of order 7, 13, 19, 25 etc.: trivially there are Steiner triple systems of orders 1 and 3. But there are no Steiner systems of orders 2, 4, 5 or 6 by (10.4.7). The construction of a Steiner triple system of order 9 is indicated in Exercise (10.4.6) below.

Exercises (10.4)

1. Show $L(1) = 1$, $L(2) = 1$, $L(3) = 12$.

2. Construct explicitly (a) four 5×5 MOLS; (b) eight 9×9 MOLS.

3. Show that there are at least 48 MOLS of order 6125.

4. A *quasigroup* is a set Q together with a binary operation $(x, y) \mapsto xy$ such that, given $x, y \in Q$, there is a unique $u \in Q$ such that $ux = y$ and a unique $v \in Q$ such that $xv = y$. Prove that the multiplication table of a finite quasigroup is a latin square.

5. Prove that every latin square determines a finite quasigroup.

6. Construct a Steiner triple system of order 9 using the following geometric procedure. Start with a 3×3 array of 9 points. Draw all horizontals, verticals and diagonals in the figure. Then draw four curves connecting exterior points.

7. (*Kirkman's[6] schoolgirl problem*). Show that it is possible for nine schoolgirls to walk in three groups of three for four successive days so that each pair of girls walks together on exactly one day. [Hint: use Exercise (10.4.6)].

8. Let n be a positive integer such that $n \equiv 3 \pmod{6}$. Assuming the existence of Steiner triple systems of order n, generalize the preceding problem by showing that it is possible for n schoolgirls to walk in $\frac{n}{3}$ groups of three on $\frac{n-1}{2}$ days without two girls walking together on more than one day.

9. Use the method of (10.4.8) to construct a Steiner triple system of order 13.

10. Construct a Steiner triple system of order 25 by starting with the field

$$\mathbb{Z}_5[t]/(t^2 - t + 1).$$

[Note that a root of $t^2 - t + 1$ has order 6].

---

[6]Thomas Penyngton Kirkman (1806–1895).

Chapter 11
Galois theory

In this chapter the Galois group of a field extension is introduced. This establishes the critical link between field theory and group theory in which subfields correspond to subgroups of the Galois group. A major application is to the classical problem of solving polynomial equations by radicals, which is an excellent illustration of the rich rewards that can be reaped when connections are made between different mathematical theories.

11.1 Normal and separable extensions

We begin by introducing two special types of field extension, leading up to the concept of a Galois extension. Let E be a field extension of a subfield F. Then E is said to be *normal* over F if it is algebraic over F and if every irreducible polynomial in $F[t]$ which has a root in E has all its roots in E; thus the polynomial is a product of linear factors over E.

Example (11.1.1) Consider the field $E = \mathbb{Q}(a)$ where $a = 2^{1/3}$. Then E is algebraic over $\mathbb{Q}$ since $(E : \mathbb{Q})$ is finite, but it is not normal over $\mathbb{Q}$. This is because $t^3 - 2$ has one root a in E but not the other two roots $a\omega$, $a\omega^2$ where $\omega = e^{2\pi i/3}$.

Example (11.1.2) Let E be an extension of a field F with $(E : F) = 2$. Then E is normal over F.

In the first place E is algebraic over F. Suppose that $x \in E$ is a root of some monic irreducible polynomial $f \in F[t]$. Then $f = \mathrm{Irr}_F(x)$ and

$$\deg(f) = (F(x) : F) \leq (E : F) = 2,$$

which means that $\deg(f) = 1$ or 2. In the first case x is the only root of f. Suppose that $\deg(f) = 2$ with say $f = t^2 + at + b$ and $a, b \in F$; if x' is another root of f, then $xx' = b \in F$, so that $x' \in E$. Therefore E is normal over F.

That there is a close connection between normal extensions and splitting fields of polynomials is demonstrated by the following fundamental result.

(11.1.1) *Let E be a finite extension of a field F. Then E is normal over F if and only if E is the splitting field of some polynomial in $F[t]$.*

11.1 Normal and separable extensions

Proof. First of all assume that E is normal over F. Since $(E : F)$ is finite, we can write $E = F(x_1, x_2, \ldots, x_k)$. Let $f_i = \text{Irr}_F(x_i)$. Now f_i has the root x_i in E and so by normality of the extension all its roots are in E. Put $f = f_1 f_2 \ldots f_k \in F[t]$. Then all the roots of f belong to E and also they generate the field E. Hence E is the splitting field of f.

The converse is harder to prove. Suppose that E is the splitting field of some $f \in F[t]$, and denote the roots of f by $a_1, a_2, \ldots, a_r$, so that $E = F(a_1, a_2, \ldots, a_r)$. Let g be an irreducible polynomial over F with a root b in E. Further let K be the splitting field of g over E. Then $F \subseteq E \subseteq K$. Let $b^* \in K$ be another root of g. Our task is to show that $b^* \in E$.

Since $g = \text{Irr}_F(b) = \text{Irr}_F(b^*)$, there is an isomorphism $\theta_0 : F(b) \to F(b^*)$ such that $\theta_0(b) = b^*$ and $\theta_0|_F$ is the identity map: here we have applied the basic result (10.3.2). Put $g_1 = \text{Irr}_{F(b)}(a_1)$ and note that g_1 divides f over $F(b)$ since $f(a_1) = 0$. Now consider $g_1^* = \theta_0(g_1) \in F(b^*)[t]$. Then g_1^* divides $\theta_0(f) = f$ over $F(b^*)$. This demonstrates that the roots of g_1^* are among $a_1, a_2, \ldots, a_r$.

Let a_{i_1} be any root of g_1^*. By (10.3.2) once again, there is an isomorphism $\theta_1 : F(b, a_1) \to F(b^*, a_{i_1})$ such that $\theta_1(a_1) = a_{i_1}$ and $\theta_1|_{F(b_1)} = \theta_0$. Next write $g_2 = \text{Irr}_{F(b,a_1)}(a_2)$ and $g_2^* = \theta_1(g_2)$. The roots of g_2^* are among $a_1, a_2, \ldots, a_r$, by the argument used above. Let a_{i_2} be any root of g_2^*. Now extend θ_1 to an isomorphism $\theta_2 : F(b, a_1, a_2) \to F(b^*, a_{i_1}, a_{i_2})$ such that $\theta_2(a_2) = a_{i_2}$ and $\theta_2|_{F(b,a_1)} = \theta_1$.

After r applications of this argument we will have an isomorphism

$$\theta : F(b, a_1, a_2, \ldots, a_r) \to F(b^*, a_{i_1}, a_{i_2}, \ldots, a_{i_r})$$

with the properties $\theta(a_j) = a_{i_j}$, $\theta(b) = b^*$ and $\theta|_F = $ the identity map. But $b \in E = F(a_1, a_2, \ldots, a_r)$ by hypothesis, so $b^* = \theta(b) \in F(a_{i_1}, a_{i_2}, \ldots, a_{i_r}) \subseteq E$, as required. □

Separable polynomials. Contrary to what one might first think, it is possible for an irreducible polynomial to have repeated roots. This phenomenon is called *inseparability*.

Example (11.1.3) Let p be a prime and let f be the polynomial $t^p - x$ in $\mathbb{Z}_p\{x\}[t]$: here x and t are distinct indeterminates and $\mathbb{Z}_p\{x\}$ is the field of rational functions in x over $\mathbb{Z}_p$. Then f is irreducible over $\mathbb{Z}_p[x]$ by Eisenstein's Criterion (7.4.9) since x is clearly an irreducible element of $\mathbb{Z}_p[x]$. Gauss's Lemma (7.3.7) shows that f is irreducible over $\mathbb{Z}_p\{x\}$. Let a be a root of f in its splitting field. Then $f = t^p - a^p = (t-a)^p$ since $\binom{p}{i} \equiv 0 \pmod{p}$ if $0 < i < p$. It follows that f has all its roots equal to a.

Let f be an irreducible polynomial over a field F. Then f is said to be *separable* if all its roots are different, i.e., f is a product of distinct linear factors over its splitting field. The example above shows that $t^p - x$ is not separable over $\mathbb{Z}_p\{x\}$, a field with prime characteristic. The criterion which follows shows that the phenomenon of inseparability can only occur for fields with prime characteristic.

(11.1.2) *Let f be an irreducible polynomial over a field F.*

(i) *If $\operatorname{char}(F) = 0$, then f is always separable.*

(ii) *If $\operatorname{char}(F) = p > 0$, then f is inseparable if and only if $f = g(t^p)$ for some irreducible polynomial g over F.*

Proof. There is no loss in supposing f to be monic. Assume first that $\operatorname{char}(F) = 0$ and let a be a root of f in its splitting field. If a has multiplicity greater than 1, then (7.4.7) shows that $t - a \mid f'$ where f' is the derivative of f. Thus $f'(a) = 0$. Writing $f = a_0 + a_1 t + \cdots + a_n t^n$, we have $f' = a_1 + 2a_2 t + \cdots + n a_n t^{n-1}$. But $f = \operatorname{Irr}_F(a)$, so f divides f'. Since $\deg(f') < \deg((f))$, this can only mean that $f' = 0$, i.e., $i a_i = 0$ for all $i > 0$ and so $a_i = 0$. But then f is constant, which is impossible. Therefore a is not a repeated root and f is separable.

Now assume that $\operatorname{char}(F) = p > 0$ and again let a be a multiple root of f. Arguing as before, we conclude that $i a_i = 0$ for $i > 0$. In this case all we can conclude is that $a_i = 0$ if p does not divide i. Hence

$$f = a_0 + a_p t^p + a_{2p} t^{2p} + \cdots + a_{rp} t^{rp}$$

where rp is the largest positive multiple of p not exceeding n. It follows that $f = g(t^p)$ where $g = a_0 + a_p t + \cdots + a_{rp} t^r$. Note that if g were reducible, then so would f be.

Conversely, assume that $f = g(t^p)$ where $g = \sum_{i=0}^{r} a_i t^i \in F[t]$. We show that f is inseparable. Let b_i be a root of $t^p - a_i$ in the splitting field E of the polynomial $(t^p - a_1)(t^p - a_2) \ldots (t^p - a_r)$. Then $a_i = b_i^p$ and hence

$$f = \sum_{i=0}^{r} b_i^p t^{ip} = \left(\sum_{i=0}^{r} b_i t^i \right)^p.$$

From this it follows that every root of f has multiplicity at least p. Hence f is inseparable. □

Separable extensions. Let E be an extension of a field F. An element x of E is said to be *separable over F* if x is algebraic and if its multiplicity as a root of $\operatorname{Irr}_F(x)$ is 1. If x is algebraic but inseparable, then the final arguments of the proof of (11.1.2) shows that its irreducible polynomial is a prime power of a polynomial, so that *all* its roots have multiplicity greater then 1. Therefore $x \in E$ is separable over F if and only if $\operatorname{Irr}_F(x)$ is a separable polynomial.

If every element of E is separable over F, then E is called a *separable extension* of F. Finally, a field F is said to be *perfect* if every algebraic extension of F is separable. Since any irreducible polynomial over a field of characteristic 0 is separable, all fields of characteristic 0 are perfect. There is a simple criterion for a field of prime characteristic to be perfect.

11.1 Normal and separable extensions

(11.1.3) *Let F be a field of prime characteristic p. Then F is perfect if and only if $F = F^p$ where F^p is the subfield $\{a^p \mid a \in F\}$.*

Proof. In the first place F^p is a subfield of F since $(a \pm b)^p = a^p \pm b^p$, $(a^{-1})^p = (a^p)^{-1}$ and $(ab)^p = a^p b^p$. Now assume that $F = F^p$. If $f \in F[t]$ is irreducible but inseparable, then $f = g(t^p)$ for some $g \in F[t]$ by (11.1.2). Let $g = \sum_{i=0}^{r} a_i t^i$; then $a_i = b_i^p$ for some $b_i \in F$ since $F = F^p$. Therefore

$$f = \sum_{i=0}^{r} a_i t^{pi} = \sum_{i=0}^{r} b_i^p t^{pi} = \left(\sum_{i=0}^{r} b_i t^i\right)^p,$$

which is impossible since f is irreducible. Thus f is separable. This shows that if E is an algebraic extension of F, then it is separable. Hence F is a perfect field.

Conversely, assume that $F \neq F^p$ and choose $a \in F - F^p$. Consider the polynomial $f = t^p - a$. First we claim that f is irreducible over F. Indeed suppose this is false, so that $f = gh$ where g and h in $F[t]$ are monic with smaller degrees than f. Now $f = t^p - a = (t - b)^p$ where b is a root of f in its splitting field, and it follows that $g = (t - b)^i$ and $h = (t - b)^j$ where $i + j = p$ and $0 < i, j < p$. Since $\gcd\{i, p\} = 1$, we may write $1 = iu + pv$ for suitable integers u, v. It follows that $b = (b^i)^u (b^p)^v = (b^i)^u a^v \in F$ since $b^i \in F$, and so $a = b^p \in F^p$, a contradiction. Thus f is irreducible and by (11.1.2) it is inseparable. It follows that F cannot be a perfect field. □

Corollary (11.1.4) *Every finite field is perfect.*

Proof. Let F be a field of order p^m with p a prime. Every element f of F satisfies the equation $t^{p^m} - t = 0$ by (10.3.1). Hence $F = F^p$ and F is perfect. □

It is desirable to have a usable criterion for a finite extension of prime characteristic to be separable.

(11.1.5) *Let E be a finite extension of a field F with prime characteristic p. Then E is separable over F if and only if $E = F(E^p)$.*

Proof. Assume first that E is separable over F and let $a \in E$. Writing $f = \mathrm{Irr}_{F(a^p)}(a)$, we observe that f divides $t^p - a^p = (t - a)^p$. Since f is a separable polynomial, it follows that $f = t - a$ and thus $a \in F(a^p) \subseteq F(E^p)$.

Conversely, assume that $E = F(E^p)$ and let $x \in E$; we need to prove that $f = \mathrm{Irr}_F(x)$ is separable over F. If this is false, then $f = g(t^p)$ for some $g \in F[t]$, with say $g = \sum_{i=0}^{k} a_i t^i$. Since $a_0 + a_1 x^p + \cdots + a_k x^{kp} = 0$, the field elements $1, x^p, \ldots, x^{kp}$ are linearly dependent over F. On the other hand, $k < kp = \deg(f) = (F(x) : F)$, so that $1, x, \ldots, x^k$ must be linearly independent over F. Extend $\{1, x, \ldots, x^k\}$ to an F-basis of E, say $\{y_1, y_2, \ldots, y_n\}$, using (8.2.5).

We have $E = Fy_1 + Fy_2 + \cdots + Fy_n$ and thus $E^p \subseteq Fy_1^p + Fy_2^p + \cdots + Fy_n^p$. Therefore $E = F(E^p) = Fy_1^p + Fy_2^p + \cdots + Fy_n^p$. It now follows that $y_1^p, y_2^p, \ldots, y_n^p$ are linearly independent since $n = (E : F)$. This shows that $1, x^p, \ldots, x^{kp}$ are linearly independent, a contradiction. □

Corollary (11.1.6) *Let $E = F(a_1, a_2, \ldots, a_k)$ be an extension of a field F where each a_i is separable over F. Then E is separable over F.*

Proof. We may assume that $\mathrm{char}(F) = p > 0$. Since a_i is separable over F, we have $a_i \in F(a_i^p)$, as in the first paragraph of the preceding proof. Hence $a_i \in F(E^p)$ and $E = F(E^p)$. Therefore E is separable over F by (11.1.5). □

Notice the consequence of the last result: *the splitting field of a separable polynomial is a separable extension.*

We conclude this section by addressing a natural question which may already have occurred to the reader: when is a finite extension E of F a simple extension, i.e., when is $E = F(x)$ for some x? An important result on this problem is:

(11.1.7) (Theorem of the Primitive Element) *Let E be a finite separable extension of a field F. Then there is an element a such that $E = F(a)$.*

Proof. The proof is easy when E is finite. For then $E - \{0\}$ is a cyclic group by (10.3.6), generated by a, say. Then $E = \{0, 1, a, \ldots, a^{q-1}\}$ where $q = |E|$, and $E = F(a)$.

From now on we assume E to be infinite. Since $(E : F)$ is finite, we have that $E = F(u_1, u_2, \ldots, u_n)$ for some u_i in E. The proof proceeds by induction on n. If $n > 2$, then $F(u_1, u_2, \ldots, u_{n-1}) = F(v)$ for some v, by induction hypothesis, and hence $E = F(v, u_n) = F(a)$ for some a by the case $n = 2$. Thus we have only to deal with the case $n = 2$. From now on write

$$E = F(u, v).$$

Next we introduce the polynomials $f = \mathrm{Irr}_F(u)$ and $g = \mathrm{Irr}_F(v)$; these are separable polynomials since E is separable over F. Let the roots of f and g in a splitting field containing F be $u = x_1, x_2, \ldots, x_m$ and $v = y_1, y_2, \ldots, y_n$ respectively. Here all the x_i are different, as are all the y_j. From this we may conclude that for $j \neq 1$ there is at most one element z_{ij} in F such that

$$u + z_{ij} v = x_i + z_{ij} y_j,$$

namely $z_{ij} = (x_i - u)(v - y_j)^{-1}$. Since F is infinite, it is possible to choose an element z in F which is different from each z_{ij}; then $u + zv \neq x_i + zy_j$ for all $(i, j) \neq (1, 1)$.

With this choice of z, put $a = u + zv \in E$. We will argue that $E = F(a)$. Since $g(v) = 0 = f(u) = f(a - zv)$, the element v is a common root of the polynomials g and $f(a - zt) \in F(a)[t]$. Now these polynomials have no other common roots. For if

y_j were one, then $a - zy_j = x_i$ for some i, which implies that $u + zv = a = x_i + zy_j$; this is contrary to the choice of z. It follows that $t - v$ is the unique (monic) gcd of g and $f(a - zt)$ in $E[t]$. Now the gcd of these polynomials must actually lie in the smaller ring $F(a)[t]$: for the gcd can be computed by using the Euclidean Algorithm, which is valid for $F(a)[t]$ since it depends only on the Division Algorithm. Therefore $v \in F(a)$ and $u = a - zv \in F(a)$. Finally $E = F(u, v) = F(a)$. □

Since an algebraic number field is by definition a finite extension of $\mathbb{Q}$, we deduce:

Corollary (11.1.8) *If E is an algebraic number field, then $E = \mathbb{Q}(a)$ for some a in E.*

Exercises (11.1)

1. Which of the following field extensions are normal?
   (a) $\mathbb{Q}(3^{1/3})$ of $\mathbb{Q}$; (b) $\mathbb{Q}(3^{1/3}, e^{2\pi i/3})$ of $\mathbb{Q}$; (c) $\mathbb{R}$ of $\mathbb{Q}$; (d) $\mathbb{C}$ of $\mathbb{R}$.

2. Let $F \subseteq K \subseteq E$ be field extensions with all degrees finite. If E is normal over F, then it is normal over K, but K need not be normal over F.

3. Let $f \in F[t]$ where $\text{char}(F) = p > 0$, and assume that f is monic with degree p^n and all its roots are equal in its splitting field. Prove that $f = t^{p^n} - a$ for some $a \in F$.

4. Let E be a finite extension of a field F of characteristic $p > 0$ such that $(E : F)$ is not divisible by p. Show that E is separable over F.

5. Let $F \subseteq K \subseteq E$ be field extensions with all degrees finite and E separable over F. Prove that E is separable over K.

6. Let $F \subseteq K \subseteq E$ be field extensions with all degrees finite. If E is separable over K and K is separable over F, show that E is separable over F.

11.2 Automorphisms of field extensions

Fields, just like groups, possess automorphisms and these play a crucial role in field theory. An *automorphism of a field F* is defined to be a bijective ring homomorphism $\alpha : F \rightarrow F$; thus $\alpha(x+y) = \alpha(x) + \alpha(y)$ and $\alpha(xy) = \alpha(x)\alpha(y)$. The automorphisms of a field are easily seen to form a group with respect to functional composition. If E is a field extension of F, we interested in *automorphisms of E over F*, i.e., automorphisms of E whose restriction to F is the identity function. For example, complex conjugation is an automorphism of $\mathbb{C}$ over $\mathbb{R}$. The set of automorphisms of E over F is a subgroup of the group of all automorphisms of F and is denoted by

$$\text{Gal}(E/F):$$

this is the *Galois*[1] *group* of E over F.

---
[1] Évariste Galois (1811–1831)

Suppose that $E = F(a)$ is a simple algebraic extension of F with degree n. Then every element of E has the form $x = \sum_{i=0}^{n-1} c_i a^i$ where $c_i \in F$ and thus $\alpha(x) = \sum_{i=0}^{n-1} c_i \alpha(a)^i$, where $\alpha \in \text{Gal}(E/F)$. If b is any root of the polynomial $f = \text{Irr}_F(a)$, then $0 = \alpha(f(b)) = f(\alpha(b))$, so that $\alpha(b)$ is also a root of f in E. Thus each α in $\text{Gal}(E/F)$ gives rise to a permutation $\pi(\alpha)$ of X, the set of all distinct roots of f in E. What is more, the mapping

$$\pi : \text{Gal}(E/F) \to \text{Sym}(X)$$

is evidently a group homomorphism, i.e., α is a permutation representation of the Galois group on X.

Now π is in fact a faithful permutation representation of $\text{Gal}(E/F)$ on X. For, if $\pi(\alpha)$ is the identity permutation, then $\alpha(a) = a$ and hence α is the identity automorphism. For this reason it is often useful to think of the elements of $\text{Gal}(E/F)$ as permutations of the set of roots X.

Next let b be any element of X. Then $F \subseteq F(b) \subseteq E = F(a)$, and also

$$(F(b) : F) = \deg(f) = (F(a) : F)$$

by (10.1.4) since $f = \text{Irr}_F(b)$. It follows that $F(b) = F(a) = E$ by (10.1.7). Since $\text{Irr}_F(a) = f = \text{Irr}_F(b)$, we may apply (10.3.2) to produce an automorphism α of E over F such that $\alpha(a) = b$. Since $\alpha \in \text{Gal}(E/F)$, we have proved that the group $\text{Gal}(E/F)$ acts *transitively* on the set X. Finally, if α in $\text{Gal}(E/F)$ fixes some b in X, then α must equal the identity since $E = F(b)$. This shows that $\text{Gal}(E/F)$ acts *regularly* on X and it follows from (5.2.2) that $|X| = |\text{Gal}(E/F)|$.

These conclusions are summed up in the following fundamental result.

(11.2.1) *Let $E = F(a)$ be a simple algebraic extension of a field F. Then $\text{Gal}(E/F)$ acts regularly on the set X of distinct roots of $\text{Irr}_F(a)$ in E. Therefore*

$$|\text{Gal}(E/F)| = |X| \leq (E : F).$$

An extension of a subfield F which is finite, separable and normal is said to be *Galois* over F. For such extensions we have:

Corollary (11.2.2) *If E is a Galois extension of a field F with degree n, then $\text{Gal}(E/F)$ is isomorphic with a regular subgroup of S_n and*

$$|\text{Gal}(E/F)| = n = (E : F).$$

For by (11.1.7) $E = F(a)$ for some $a \in E$. Also $\text{Irr}_F(a)$ has n distinct roots in E by normality and separability.

The Galois group of a polynomial. Suppose that f is a non-constant polynomial over a field F and let E be the splitting field of f: recall from (10.3.4) that this field is unique. Then the *Galois group of the polynomial f* is defined to be

$$\text{Gal}(f) = \text{Gal}(E/F).$$

This is always a finite group by (11.2.1). The basic properties of the Galois group are given in the next result.

(11.2.3) *Let f be a non-constant polynomial of degree n over a field F. Then:*

(i) $\operatorname{Gal}(f)$ *is isomorphic with a permutation group on the set of all roots of f; thus* $|\operatorname{Gal}(f)|$ *divides* $n!$;

(ii) *if all the roots of f are distinct, then f is irreducible if and only if $\operatorname{Gal}(f)$ acts transitively on the set of roots of f.*

Proof. Let E denote the splitting field of f, so that $\operatorname{Gal}(f) = \operatorname{Gal}(E/F)$. If a is a root of f in E, then $f(\alpha(a)) = \alpha(f(a)) = 0$, so that $\alpha(a)$ is also a root of f. If α fixes every root of f, then α is the identity automorphism since E is generated by F and the roots of f. Hence $\operatorname{Gal}(f)$ is isomorphic with a permutation group on the set of distinct roots of f. If there are r such roots, then $r \leq n$ and $|\operatorname{Gal}(f)| \mid r! \mid n!$, so that $|\operatorname{Gal}(f)| \mid n!$.

Next we assume that all the roots of f are different. Let f be irreducible. If a and b are roots of f, then $\operatorname{Irr}_F(a) = f = \operatorname{Irr}_F(b)$, and by (10.3.2) there exists $\alpha \in \operatorname{Gal}(f)$ such that $\alpha(a) = b$. It follows that $\operatorname{Gal}(f)$ acts transitively on the set of roots of f.

Conversely, suppose that $\operatorname{Gal}(f)$ acts transitively on the roots of f, yet f is reducible. Then we can write $f = g_1 g_2 \ldots g_k$ where $g_i \in F[t]$ is irreducible and $k \geq 2$. Let a_1, a_2 be roots of g_1, g_2 respectively. By transitivity there exists $\alpha \in \operatorname{Gal}(f)$ such that $\alpha(a_1) = a_2$. But $0 = \alpha(g_1(a)) = g_1(\alpha(a)) = g_1(a_2)$. Hence $g_2 = \operatorname{Irr}_F(a_2)$ divides g_1, so that $g_1 = g_2$. Therefore g_1^2 divides f and the roots of f cannot all be different, a contradiction which shows that f is irreducible. □

Corollary (11.2.4) *Let f be a separable polynomial of degree n over a field F and let E be its splitting field. Then $|\operatorname{Gal}(f)| = (E : F)$ and $|\operatorname{Gal}(f)|$ is divisible by n.*

Proof. Note that E is separable and hence Galois over F by (11.1.6). Hence $|\operatorname{Gal}(E/F)| = |\operatorname{Gal}(f)| = (E : F)$ by (11.2.2). Further f is irreducible by definition, so $\operatorname{Gal}(f)$ acts transitively on the n roots of f; therefore n divides $|\operatorname{Gal}(f)|$ by (5.2.2). □

We shall now look at some polynomials whose Galois groups can be readily computed.

Example (11.2.1) Let $f = t^3 - 2 \in \mathbb{Q}[t]$. Then $\operatorname{Gal}(f) \simeq S_3$.

To see this let E denote the splitting field of f; thus E is Galois over $\mathbb{Q}$. Then $E = \mathbb{Q}(2^{1/3}, e^{2\pi i/3})$ and one can easily check that $(E : \mathbb{Q}) = 6$, so that $|\operatorname{Gal}(f)| = 6$. Since $\operatorname{Gal}(f)$ is isomorphic with a subgroup of S_3, it follows that $\operatorname{Gal}(f) \simeq S_3$.

In fact it is not difficult to write down the six elements of the group $\operatorname{Gal}(f)$. Put $a = 2^{1/3}$ and $\omega = e^{2\pi i/3}$; then $E = \mathbb{Q}(a, \omega)$. Since $E = \mathbb{Q}(a)(\omega)$ and $t^3 - 2$ is the irreducible polynomial of both a and $a\omega$ over $\mathbb{Q}(\omega)$, there is an automorphism α of E

over $\mathbb{Q}$ such that $\alpha(a) = a\omega, \alpha(\omega) = \omega$. Clearly α has order 3. Of course $\alpha^2(a) = a\omega^2$ and $\alpha^2(\omega) = \omega$. It is easy to identify an automorphism β such that $\beta(a) = a$ and $\beta(\omega) = \omega^2$; for β is just complex conjugation. Two more automorphisms of order 2 are formed by composition: $\gamma = \alpha\beta$ and $\delta = \alpha^2\beta$. It is quickly seen that γ maps ω to ω^2 and a to $a\omega$, while δ maps ω to ω^2 and a to $a\omega^2$. Thus the elements of the Galois group $\mathrm{Gal}(f)$ are $1, \alpha, \alpha^2, \beta, \gamma, \delta$.

Example (11.2.2) Let p be a prime and put $f = t^p - 1 \in \mathbb{Q}[t]$. Then $\mathrm{Gal}(f) \simeq U(\mathbb{Z}_p)$, a cyclic group of order $p - 1$.

Put $a = e^{2\pi i/p}$, a primitive p-th root of unity; then the roots of f are $1, a, a^2, \ldots, a^{p-1}$ and its splitting field is $E = \mathbb{Q}(a)$. Now $f = (t-1)(1 + t + t^2 + \cdots + t^{p-1})$ and the second factor is $\mathbb{Q}$-irreducible by Example (7.4.6). Hence the irreducible polynomial of a is $1 + t + t^2 + \cdots + t^{p-1}$ and $|\mathrm{Gal}(f)| = (E : F) = p - 1$.

To show that $\mathrm{Gal}(f)$ is cyclic, we construct a group isomorphism

$$\theta : U(\mathbb{Z}_p) \to \mathrm{Gal}(f).$$

If $1 \leq i < p$, define $\theta(i + p\mathbb{Z})$ to be θ_i where $\theta_i(a) = a^i$ and θ_i is trivial on $\mathbb{Q}$; this is an automorphism by (10.3.2). Obviously θ_i is the identity only if $i = 1$, so θ is injective. Since $U(\mathbb{Z}_p)$ and $\mathrm{Gal}(f)$ both have order $p - 1$, they are isomorphic.

Conjugacy in field extensions. Let E be an extension of a field F. Two elements a and b of E are said to be *conjugate over F* if $\alpha(a) = b$ for some $\alpha \in \mathrm{Gal}(E/F)$. In normal extensions conjugacy amounts to the elements having the same irreducible polynomial, as the next result shows.

(11.2.5) *Let E be a finite normal extension of a field F. Then two elements a and b of E are conjugate over F if and only if they have the same irreducible polynomial.*

Proof. If a and b have the same irreducible polynomial, (10.3.2) shows that there is a field isomorphism $\theta : F(a) \to F(b)$ such that $\theta(a) = b$ and θ is the identity map on F. Now by (11.1.1) E is the splitting field of some polynomial over F, and hence over $F(a)$. Consequently (10.3.3) can be applied to extend θ to an isomorphism $\alpha : E \to E$ such that θ is the restriction of α to $F(a)$. Hence $\alpha \in \mathrm{Gal}(E/F)$ and $\alpha(a) = b$, which shows that a and b are conjugate over F.

To prove the converse, suppose that $b = \alpha(a)$ where $a, b \in E$ and $\alpha \in \mathrm{Gal}(E/F)$. Put $f = \mathrm{Irr}_F(a)$ and $g = \mathrm{Irr}_F(b)$. Then $0 = \alpha(f(a)) = f(\alpha(a)) = f(b)$. Therefore g divides f and it follows that $f = g$ since f and g are monic and irreducible. □

The next result is of critical importance in Galois theory: it asserts that the only elements of an extension that are fixed by every automorphism are the elements of the base field.

(11.2.6) *Let E be a Galois extension of a field F and let $a \in E$. Then $\alpha(a) = a$ for all automorphisms α of E over F if and only if $a \in F$.*

Proof. Assume that $\alpha(a) = a$ for all $\alpha \in \text{Gal}(E/F)$ and put $f = \text{Irr}_F(a)$. Since E is normal over F, all the roots of f are in E. If b is any such root, there exists α in $\text{Gal}(E/F)$ such that $\alpha(a) = b$. Hence $b = a$ and the roots of f are all equal. But f is separable since E is separable over F. Therefore $f = t - a$, which shows that a belongs to F. □

Roots of unity. At this point we postpone further development of the theory of Galois extensions until the next section and concentrate on roots of unity. Let F be a field and n a positive integer. A root a of the polynomial $t^n - 1 \in F[t]$ is called an *n-th root of unity over F*; thus $a^n = 1$. If $a^m \neq 1$ for all proper divisors m of n, then a is said to be a *primitive n-th root of unity*. Now if $\text{char}(F) = p$ divides n, there are no primitive n-th roots of unity over F: for then $t^n - 1 = (t^{n/p} - 1)^p$ and every n-th root of unity has order at most n/p. However we will show next that if $\text{char}(F)$ does not divide n, then primitive n-th roots of unity over F always exist.

(11.2.7) *Let F be a field whose characteristic does not divide the positive integer n and let E be the splitting field of $t^n - 1$ over F. Then:*

  (i) *primitive n-th roots of unity exist in E; furthermore these form a cyclic subgroup of order n.*

  (ii) *$\text{Gal}(E/F) \simeq U(\mathbb{Z}_n)$, so that the Galois group is abelian with order dividing $\phi(n)$.*

Proof. (i) Set $f = t^n - 1$, so that $f' = nt^{n-1}$. Since $\text{char}(F)$ does not divide n, the polynomials f and f' are relatively prime. It follows via (7.4.7) that f has n distinct roots in its splitting field E, namely the n-th roots of unity. Clearly these roots form a subgroup of $U(E)$, say H, which has order n, and by (10.3.6) H is cyclic, say $H = \langle x \rangle$. Here x has order n and thus it is a primitive n-th root of unity.

(ii) Let a be a primitive n-th root of unity in E. Then the roots of $t^n - 1$ are $a^i, i = 0, 1, \ldots, n - 1$, and $E = F(a)$. If $\alpha \in \text{Gal}(E/F)$, then α is completely determined by $\alpha(a) = a^i$ where $0 \leq i < n$ and i is relatively prime to n. Further the mapping $\alpha \mapsto i + n\mathbb{Z}$ is an injective homomorphism from the Galois group into $U(\mathbb{Z}_n)$. By Lagrange's Theorem $|\text{Gal}(E/F)|$ divides $|U(\mathbb{Z}_n)| = \phi(n)$. □

Corollary (11.2.8) *The number of primitive n-th roots of unity over a field whose characteristic does not divide n is $\phi(n)$, where ϕ is Euler's function.*

For if a is a fixed primitive n-th root of unity, the primitive n-th roots of unity are just the powers a^i where $1 \leq i < n$ and i is relatively prime to n.

Cyclotomic polynomials. Assume that F is a field whose characteristic does not divide the positive integer n and denote the primitive n-th roots of unity over F by

$a_1, a_2, \ldots, a_{\phi(n)}$. The *cyclotomic polynomial of order n over F* is defined to be

$$\Phi_n = \prod_{i=1}^{\phi(n)} (t - a_i),$$

which is a monic polynomial of degree $\phi(n)$. Since every n-th root of unity is a primitive d-th root of unity for some divisor d of n, we have immediately that

$$t^n - 1 = \prod_{d \mid n} \Phi_d.$$

This leads to the formula

$$\Phi_n = \frac{t^n - 1}{\prod_{d \mid\mid n} \Phi_d},$$

where $d \mid\mid n$ means that d is a proper divisor of n. Using this formula, we can compute Φ_n recursively, i.e., if we know Φ_d for all proper divisors d of n, then we can calculate Φ_n. The formula also tells us that $\Phi_n \in F[t]$. For $\Phi_1 = t - 1 \in F[t]$ and if $\Phi_d \in F[t]$ for all proper divisors d of n, then $\Phi_n \in F[t]$.

Example (11.2.3) We have seen that $\Phi_1 = t - 1$. Hence $\Phi_2 = \frac{t^2-1}{t-1} = t + 1$, $\Phi_3 = \frac{t^3-1}{t-1} = t^2 + t + 1$, and

$$\Phi_4 = \frac{t^4 - 1}{(t-1)(t+1)} = t^2 + 1.$$

There is in fact an explicit formula for Φ_n. This involves the *Möbius function* μ, which is well-known from number theory. It is defined by the rules:

$$\mu(1) = 1, \quad \mu(p_1 p_2 \ldots p_k) = (-1)^k,$$

if $p_1, p_2, \ldots, p_k$ are distinct primes, and

$$\mu(n) = 0$$

if n is divisible by the square of a prime. Here is the formula for Φ_n.

(11.2.9) *The cyclotomic polynomial of order n over any field whose characteristic does not divide n is given by*

$$\Phi_n = \prod_{d \mid n} (t^d - 1)^{\mu(n/d)}.$$

Proof. First we need to establish an auxiliary property of the Möbius function,

$$\sum_{d \mid n} \mu(d) = \begin{cases} 1 & \text{if } n = 1 \\ 0 & \text{if } n > 1. \end{cases}$$

This is obvious if $n = 1$, so assume that $n > 1$ and write $n = p_1^{e_1} p_2^{e_2} \ldots p_k^{e_k}$ where the p_i are distinct primes. If d is a square-free divisor of n, then d has the form $p_{i_1} p_{i_2} \ldots p_{i_r}$ where $1 \leq i_1 < i_2 < \cdots < i_r \leq n$, which corresponds to the term $(-1)^r t_{i_1} t_{i_2} \ldots t_{i_r}$ in the product $(1-t_1)(1-t_2)\ldots(1-t_n)$; note also that $\mu(d) = (-1)^r$. Therefore we obtain the identity

$$(1 - t_1)(1 - t_2)\ldots(1 - t_n) = \sum \mu(p_{i_1} p_{i_2} \ldots p_{i_r}) t_{i_1} t_{i_2} \ldots t_{i_r},$$

where the sum is over all i_j satisfying $1 \leq i_1 < i_2 < \cdots < i_r \leq n$. Set all $t_i = 1$ to get $\sum \mu(p_{i_1}, p_{i_2}, \ldots p_{i_r}) = 0$. Since $\mu(d) = 0$ if d is not square-free, we may rewrite the last equation as $\sum_{d|n} \mu(d) = 0$.

We are now in a position to establish the formula for Φ_n. Write

$$\Psi_n = \prod_{e|n}(t^e - 1)^{\mu(n/e)}.$$

Then $\Psi_1 = t - 1 = \Phi_1$. Assume that $\Psi_d = \Phi_d$ for all $d < n$. Then by definition of Ψ_d, we have

$$\prod_{d|n} \Psi_d = \prod_{d|n} \prod_{e|d}(t^e - 1)^{\mu(d/e)} = \prod_{f|n}(t^f - 1)^{\sum_{f|d|n} \mu(d/f)}.$$

Next for a fixed f dividing d we have

$$\sum_{f|d|n} \mu(d/f) = \sum_{\frac{d}{f}|\frac{n}{f}} \mu(d/f),$$

which equals 1 or 0 according as $f = n$ or $f < n$. It therefore follows that

$$\prod_{d|n} \Psi_d = t^n - 1 = \prod_{d|n} \Phi_d.$$

Since $\Psi_d = \Phi_d$ if $d < n$, cancellation yields $\Psi_n = \Phi_n$ and the proof is complete. □

Example (11.2.4) Use the formula of (11.2.9) to compute the cyclotomic polynomial of order 12 over $\mathbb{Q}$.

The formula yields

$$\Phi_{12} = (t - 1)^{\mu(12)}(t^2 - 1)^{\mu(6)}(t^3 - 1)^{\mu(4)}(t^4 - 1)^{\mu(3)}(t^6 - 1)^{\mu(2)}(t^{12} - 1)^{\mu(1)},$$

which equals

$$(t^2 - 1)(t^4 - 1)^{-1}(t^6 - 1)^{-1}(t^{12} - 1) = t^4 - t^2 + 1$$

since $\mu(12) = \mu(4) = 0$, $\mu(2) = \mu(3) = -1$ and $\mu(6) = \mu(1) = 1$.

Example (11.2.5) If p is a prime, then $\Phi_p = 1 + t + t^2 + \cdots + t^{p-1}$.

For $\Phi_p = (t-1)^{\mu(p)}(t^p - 1)^{\mu(1)} = \frac{t^p - 1}{t - 1} = 1 + t + t^2 + \cdots + t^{p-1}$, since $\mu(p) = -1$.

Since we are interested in computing the Galois group of a cyclotomic polynomial over $\mathbb{Q}$, it is important to know whether Φ_n is irreducible. This is certainly true when n is prime by Example (7.4.6) and Example (11.2.5). The general result is:

(11.2.10) *The cyclotomic polynomial Φ_n is irreducible over $\mathbb{Q}$ for all integers n.*

Proof. Assume that Φ_n is reducible over $\mathbb{Q}$; then Gauss's Lemma (7.3.7) tells us that it must be reducible over $\mathbb{Z}$. Since Φ_n is monic, it has a monic divisor f which is irreducible over $\mathbb{Z}$. Choose a root of f, say a, so that f is the irreducible polynomial of a over $\mathbb{Q}$. Now a is a primitive n-th root of unity, so, if p is any prime not dividing n, then a^p is also a primitive n-th root of unity and is thus a root of Φ_n. Hence a^p is a root of some monic $\mathbb{Q}$-irreducible divisor g of Φ_n in $\mathbb{Z}[t]$. Of course g is the irreducible polynomial of a^p.

Suppose first that $f \neq g$. Thus $t^n - 1 = fgh$ for some $h \in \mathbb{Z}[t]$ since f and g are distinct $\mathbb{Z}$-irreducible divisors of $t^n - 1$. Also $g(a^p) = 0$ implies that f divides $g(t^p)$ and thus $g(t^p) = fk$ where $k \in \mathbb{Z}[t]$. The canonical homomorphism from $\mathbb{Z}$ to $\mathbb{Z}_p$ induces a homomorphism from $\mathbb{Z}[t]$ to $\mathbb{Z}_p[t]$; let $\bar{f}, \bar{g}, \ldots$ denote images of $f, g, \ldots$ under this homomorphism. Then $\bar{f}\bar{k} = \bar{g}(t^p) = (\bar{g}(t))^p$ since $x^p \equiv x \pmod{p}$ for all integers x. Now $\mathbb{Z}_p[t]$ is a PID and hence a UFD. Since $\bar{f}\bar{k} = \bar{g}^p$, the polynomials $\bar{f}$ and $\bar{g}$ have a common irreducible divisor in $\mathbb{Z}_p[t]$. This means that $\bar{f}\bar{g}\bar{h} \in \mathbb{Z}_p[t]$ is divisible by the square of this irreducible factor and hence $t^n - 1 \in \mathbb{Z}_p[t]$ has a multiple root in its splitting field. However $(t^n - 1)' = nt^{n-1}$ is relatively prime to $t^n - 1$ in $\mathbb{Z}_p[t]$ since p does not divide n. This is a contradiction by (7.4.7). It follows that $f = g$.

We have now proved that a^p is a root of f for all primes p not dividing n. Hence a^m is a root of f for $1 \leq m < n$ and $\gcd\{m, n\} = 1$. Thus $\deg(f) \geq \phi(n) = \deg(\Phi_n)$, which shows that $f = \Phi_n$ and Φ_n is irreducible. □

We are now able to compute the Galois group of a cyclotomic polynomial.

(11.2.11) *If n is a positive integer, the Galois group of Φ_n over $\mathbb{Q}$ is isomorphic with $U(\mathbb{Z}_n)$, an abelian group of order $\phi(n)$.*

Proof. Let E denote the splitting field of Φ_n over $\mathbb{Q}$ and let a be a primitive n-th root of unity in E. The roots of Φ_n are a^i where $i = 1, 2, \ldots, n-1$ and $\gcd\{i, n\} = 1$. Hence $E = \mathbb{Q}(a)$ and Φ_n is the irreducible polynomial of a. Thus $|\operatorname{Gal}(E/F)| = \deg(\Phi_n) = \phi_n$. If $1 \leq i < n$ and i is relatively prime to n, there is an automorphism α_i of E over $\mathbb{Q}$ such that $\alpha_i(a) = a^i$ since a and a^i have the same irreducible polynomial. Moreover the map $i + n\mathbb{Z} \mapsto \alpha_i$ is easily seen to be an injective group homomorphism from $U(\mathbb{Z})$ to $\operatorname{Gal}(E/F)$. Both groups have order $\phi(n)$, so they are isomorphic. □

The splitting field of $\Phi_n \in \mathbb{Q}[t]$ is called a *cyclotomic number field*. Thus the Galois group of a cyclotomic number field is abelian.

Exercises (11.2)

1. Give an example E of a finite simple extension of a field F such that $\mathrm{Gal}(E/F) = 1$, but $E \neq F$.

2. If $E = \mathbb{Q}(\sqrt{5})$, find $\mathrm{Gal}(E/F)$.

3. If $E = \mathbb{Q}(\sqrt{2}, \sqrt{3})$, find $\mathrm{Gal}(E/F)$.

4. Find the Galois groups of the following polynomials in $\mathbb{Q}[t]$:
   (a) $t^2 + 1$; (b) $t^3 - 4$; (c) $t^3 - 2t + 4$.

5. Let $f \in F[t]$ and suppose that $f = f_1 f_2 \ldots f_k$ where the f_i are polynomials over the field F. Prove that $\mathrm{Gal}(f)$ is isomorphic with a subgroup of the direct product $\mathrm{Gal}(f_1) \times \mathrm{Gal}(f_2) \times \cdots \times \mathrm{Gal}(f_k)$.

6. Prove that the Galois group of $\mathrm{GF}(p^m)$ over $\mathrm{GF}(p)$ is a cyclic group of order m and that it is generated by the automorphism $a \mapsto a^p$.

7. Give an example to show that $\mathrm{Gal}(\Phi_n)$ need not be cyclic.

8. Let p be a prime not dividing the positive integer n. Show that if Φ_n is irreducible over $\mathrm{GF}(p)$, then $\phi(n)$ is the smallest positive integer m such that $p^m \equiv 1 \pmod{n}$.

9. Show that Φ_5 is reducible over $\mathrm{GF}(11)$ and find an explicit factorization of it in terms of irreducibles.

11.3 The Fundamental Theorem of Galois Theory

Armed with the techniques of the last two sections, we can now approach the famous theorem of the title. First some terminology: let E be a Galois extension of a field F. By an *intermediate field* we shall mean a subfield S such that $F \subseteq S \subseteq E$. If H is a subgroup of $\mathrm{Gal}(E/F)$, the *fixed field of H*

$$\mathrm{Fix}(H)$$

is the set of elements of E which are fixed by every element of H. It is quickly verified that $\mathrm{Fix}(H)$ is a subfield and $F \subseteq \mathrm{Fix}(H) \subseteq E$, i.e., $\mathrm{Fix}(H)$ is an intermediate field.

(11.3.1) *Let E be a Galois extension of a field F. Let S denote an intermediate field and let H denote a subgroup of the Galois group $G = \mathrm{Gal}(E/F)$. Then:*

(i) *the mappings $H \mapsto \mathrm{Fix}(H)$ and $S \mapsto \mathrm{Gal}(E/S)$ are mutually inverse, inclusion reversing bijections;*

(ii) $(E : \mathrm{Fix}(H)) = |H|$ *and* $(\mathrm{Fix}(H) : F) = |G : H|$;

(iii) $(E:S) = |\text{Gal}(E/S)|$ and $(S:F) = |G:\text{Gal}(E/S)|$.

Thus the theorem asserts the existence of a bijection from subfields between E and F to subgroups of the Galois group G; further the bijection reverses set inclusions. Such a bijection is called a *Galois correspondence*. Thus the Fundamental Theorem allows us to translate a problem about subfields into one about subgroups, which sometimes makes the problem easier to solve.

Proof of (11.3.1). (i) In the first place $\text{Fix}(\text{Gal}(E/S)) = S$ by (11.2.6). To show that we have mutually inverse bijections we still have to prove that $\text{Gal}(E/\text{Fix}(H)) = H$. By the Theorem of the Primitive Element (11.1.7), $E = F(a)$ for some a in E. Define a polynomial f in $E[t]$ by

$$f = \prod_{\alpha \in H}(t - \alpha(a)).$$

Note that all the roots of f are distinct: for $\alpha_1(a) = \alpha_2(a)$ implies that $\alpha_1 = \alpha_2$ since $E = F(a)$. Hence $\deg(f) = |H|$. Also elements of H permute the roots of f, so that $\alpha(f) = f$ for all $\alpha \in H$. Therefore the coefficients of f lie in $K = \text{Fix}(H)$. In addition $f(a) = 0$, so $\text{Irr}_K(a)$ divides f, and since $E = K(a)$, it follows that

$$(E:K) = \deg(\text{Irr}_K(a)) \leq \deg(f) = |H|.$$

Hence $|\text{Gal}(E/K)| \leq |H|$. But clearly $H \leq \text{Gal}(E/K)$, so that $H = \text{Gal}(E/K)$, as required.

(ii) Since E is Galois over $\text{Fix}(H)$, we have $(E:\text{Fix}(H)) = |\text{Gal}(E/\text{Fix}(H))| = |H|$ by (i). The second statement follows from

$$(E:\text{Fix}(H)) \cdot (\text{Fix}(H):F) = (E:F) = |G|.$$

(iii) The first statement is obvious. For the second statement we have

$$(E:S)(S:F) = (E:F)$$

and $(E:S) = \text{Gal}(E/S)$, while $(E:F) = |G|$. The result now follows. □

Normal extensions and normal subgroups. If E is a Galois extension of a field F, intermediate subfields which are normal over F must surely correspond to subgroups of $\text{Gal}(E/F)$ which are in some way special. In fact these are exactly the normal subgroups of $\text{Gal}(E/F)$. To prove this a simple lemma about Galois groups of conjugate subfields is called for. If $\alpha \in \text{Gal}(E/F)$ and $F \subseteq S \subseteq E$, write $\alpha(S) = \{\alpha(a) \mid a \in S\}$. Clearly $\alpha(S)$ is a subfield and $F \subseteq \alpha(S) \subseteq E$: here $\alpha(S)$ is called a *conjugate* of S.

(11.3.2) *Let E be an extension of a field F and let S be an intermediate field. If $\alpha \in \text{Gal}(E/F)$, then $\text{Gal}(E/\alpha(S)) = \alpha\,\text{Gal}(E/S)\alpha^{-1}$.*

11.3 The Fundamental Theorem of Galois Theory

Proof. Let $\beta \in \mathrm{Gal}(E/F)$. Then $\beta \in \mathrm{Gal}(E/\alpha(S))$ if and only if $\beta(\alpha(a)) = \alpha(a)$, i.e., $\alpha^{-1}\beta\alpha(a) = a$ for all $a \in S$, or equivalently $\alpha^{-1}\beta\alpha \in \mathrm{Gal}(E/S)$. Hence $\beta \in \mathrm{Gal}(E/\alpha(S))$ if and only if $\beta \in \alpha\,\mathrm{Gal}(E/S)\alpha^{-1}$. □

The connection between normal extensions and normal subgroups can now be made.

(11.3.3) *Let E be a Galois extension of a field F and let S be an intermediate field. Then the following statements about S are equivalent:*

(i) *S is normal over F;*

(ii) *$\alpha(S) = S$ for all $\alpha \in \mathrm{Gal}(E/F)$;*

(iii) *$\mathrm{Gal}(E/S) \triangleleft \mathrm{Gal}(E/F)$.*

Proof. (i) implies (ii). Let $a \in S$ and write $f = \mathrm{Irr}_F(a)$. Since S is normal over F and f has a root in S, all the roots of f are in S. If $\alpha \in \mathrm{Gal}(E/F)$, then $\alpha(a)$ is also a root of f since $f(\alpha(a)) = \alpha(f(a)) = 0$. Therefore $\alpha(a) \in S$ and $\alpha(S) \subseteq S$. By the same argument $\alpha^{-1}(S) \subseteq S$, so that $S \subseteq \alpha(S)$ and $\alpha(S) = S$.

(ii) implies (iii). Suppose that $\alpha \in \mathrm{Gal}(E/F)$. By (11.3.2)

$$\alpha\,\mathrm{Gal}(E/S)\alpha^{-1} = \mathrm{Gal}(E/\alpha(S)) = \mathrm{Gal}(E/S),$$

which shows that $\mathrm{Gal}(E/S) \triangleleft \mathrm{Gal}(E/F)$.

(iii) implies (i). Starting with $\mathrm{Gal}(E/S) \triangleleft \mathrm{Gal}(E/F)$, we have for any $\alpha \in \mathrm{Gal}(E/F)$ that

$$\mathrm{Gal}(E/S) = \alpha\,\mathrm{Gal}(E/S)\alpha^{-1} = \mathrm{Gal}(E/\alpha(S))$$

by (11.3.2). Apply the function Fix to $\mathrm{Gal}(E/S) = \mathrm{Gal}(E/\alpha(S))$ to obtain $S = \alpha(S)$ by the Fundamental Theorem of Galois Theory. Next let f in $F[t]$ be irreducible with a root a in S, and suppose b is another root of f. Then $b \in E$ since E is normal over F. Because $\mathrm{Irr}_F(a) = f = \mathrm{Irr}_F(b)$, there is an α in $\mathrm{Gal}(E/F)$ such that $\alpha(a) = b$. Therefore $b \in \alpha(S) = S$, from which it follows that S is normal over F. □

Corollary (11.3.4) *If E is a Galois extension of a field F and S is an intermediate field which is normal over F, then*

$$\mathrm{Gal}(S/F) \simeq \mathrm{Gal}(E/F)/\mathrm{Gal}(E/S).$$

Proof. Let $\alpha \in \mathrm{Gal}(E/F)$; then $\alpha(S) = S$ by (11.3.3), and thus $\alpha|_S \in \mathrm{Gal}(S/F)$. What is more, the restriction map $\alpha \mapsto \alpha|_S$ is a homomorphism from $\mathrm{Gal}(E/F)$ to $\mathrm{Gal}(S/F)$ with kernel equal to $\mathrm{Gal}(E/S)$. The First Isomorphism Theorem then tells us that $\mathrm{Gal}(E/F)/\mathrm{Gal}(E/S)$ is isomorphic with a subgroup of $\mathrm{Gal}(S/F)$. But in addition we have

$$|\mathrm{Gal}(E/F)/\mathrm{Gal}(E/S)| = (E:F)/(E:S) = (S:F) = |\mathrm{Gal}(S/F)|$$

since S is Galois over F. Therefore $\mathrm{Gal}(E/F)/\mathrm{Gal}(E/S) \simeq \mathrm{Gal}(S/F)$. □

Example (11.3.1) Let E denote the splitting field of $t^3 - 2 \in \mathbb{Q}[t]$. Thus $E = \mathbb{Q}(a, \omega)$ where $a = 2^{1/3}$ and $\omega = e^{2\pi i/3}$. We have seen that $(E:\mathbb{Q}) = 6$ and $G = \mathrm{Gal}(E/F) \simeq S_3$. Now G has exactly six subgroups, which are displayed in the Hasse diagram below.

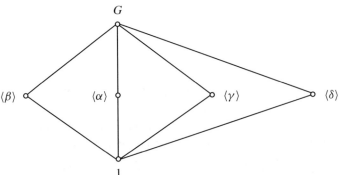

Here $\alpha(a) = a\omega$ and $\alpha(\omega) = \omega$; $\beta(a) = a$ and $\beta(\omega) = \omega^2$; $\gamma(a) = a\omega$ and $\gamma(\omega) = \omega^2$; $\delta(a) = a\omega^2$ and $\delta(\omega) = \omega^2$ – see Example (11.2.1). Each subgroup H corresponds to its fixed field $\mathrm{Fix}(H)$ under the Galois correspondence. For example, $\mathrm{Fix}(\langle\alpha\rangle) = \mathbb{Q}(\omega)$ and $\mathrm{Fix}(\langle\beta\rangle) = \mathbb{Q}(a)$. The normal subgroups of G are 1, $\langle\alpha\rangle$ and G; the three corresponding normal extensions are E, $\mathbb{Q}(\omega)$ and $\mathbb{Q}$.

Allowing for inclusion reversion, we may display the six subfields of E in the Hasse diagram below.

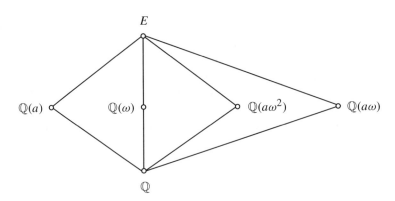

Since every subgroup of an abelian group is normal, we deduce at once from (11.3.3):

(11.3.5) *If E is a Galois extension of a field F and $\mathrm{Gal}(E/F)$ is abelian, then every intermediate field is normal over F.*

For example, by (11.2.11) the Galois group of the cyclotomic polynomial $\Phi_n \in \mathbb{Q}[t]$ is abelian. Therefore *every subfield of a cyclotomic number field is normal over $\mathbb{Q}$*.

As an impressive demonstration of the power of Galois theory, we shall prove the *Fundamental Theorem of Algebra*, which was mentioned in 7.4. All known proofs of this theorem employ some analysis. Here we use only the *Intermediate Value Theorem*: if f is a continuous function assuming values a and b, then f assumes all values between a and b. In fact this result is only required for polynomial functions.

(11.3.6) *Let f be a non-constant polynomial over $\mathbb{C}$. Then f is the product of linear factors over $\mathbb{C}$.*

Proof. First note that $f\bar{f}$ has real coefficients; therefore we can replace f by this polynomial, so that there is no loss in assuming that f has real coefficients. We can also assume that $\deg(f) > 1$. Let E be the splitting field of f over $\mathbb{C}$. Then E is the splitting field of $(t^2 + 1)f$ over $\mathbb{R}$. Hence E is Galois over $\mathbb{R}$ since the characteristic is 0. Put $G = \mathrm{Gal}(E/\mathbb{R})$. Then $|G| = (E : \mathbb{R}) = (E : \mathbb{C}) \cdot (\mathbb{C} : \mathbb{R}) = 2(E : \mathbb{C})$, and we may conclude that $|G|$ is even.

Let H be a Sylow 2-subgroup of G and put $F = \mathrm{Fix}(H)$. Then $\mathbb{R} \subseteq F \subseteq E$ and $(F : \mathbb{R}) = |G : H|$ is odd. Let $a \in F$ and set $g = \mathrm{Irr}_{\mathbb{R}}(a)$. Since $\deg(g) = (\mathbb{R}(a) : \mathbb{R})$, which divides $(F : \mathbb{R})$, we conclude that $\deg(g)$ is odd. Hence $g(x) > 0$ for large positive x and $g(x) < 0$ for large negative x. This is our opportunity to apply the Intermediate Value Theorem and conclude that $g(x) = 0$ for some real number x. But g is irreducible over $\mathbb{R}$, so $\deg(g) = 1$; hence $a \in \mathbb{R}$ and $F = \mathbb{R}$. This implies that $H = G$ and G is a 2-group.

Let $G_0 = \mathrm{Gal}(E/\mathbb{C}) \leq G$; then G_0 is a 2-group. Now $G_0 = 1$ implies that $(E : \mathbb{R}) = 2$ and hence $E = \mathbb{C}$. Therefore $G_0 \neq 1$ and there exists a maximal (proper) subgroup M of G_0; by (9.2.7) $M \triangleleft G_0$ and $|G_0 : M| = 2$. Now put $S = \mathrm{Fix}(M)$. By (11.3.1) we have

$$(S : \mathbb{C}) = \frac{|\mathrm{Gal}(E/\mathbb{C})|}{|\mathrm{Gal}(E/S)|} = \frac{|G_0|}{|M|} = 2.$$

Hence any s in S has irreducible polynomial over $\mathbb{C}$ of degree at most 2: say it is $t^2 + at + b$ where $a, b \in \mathbb{C}$. By the quadratic formula $s = -\frac{1}{2}(a \pm \sqrt{a^2 - 4b}) \in \mathbb{C}$, and it follows that $S = \mathbb{C}$, a contradiction. $\square$

Constructing regular n-gons. We return to the last of the ruler and compass problems discussed in 10.2, which was left unresolved. The problem is to construct a regular n-gon of side 1 unit using ruler and compass only.

To see what is involved here, let θ_n be the angle between lines joining the centroid to neighboring vertices of the n-gon; of course $\theta_n = \frac{2\pi}{n}$. By elementary geometry, if

d is the distance from the centroid C to a vertex, then $d \sin \frac{1}{2}\theta_n = \frac{1}{2}$. Hence

$$d = \frac{1}{2 \sin\left(\frac{1}{2}\theta_n\right)} = \frac{1}{\sqrt{2(1 - \cos\theta_n)}}.$$

It follows via (10.2.1) that the regular n-gon is constructible by ruler and compass if and only if $\cos\theta_n$ is constructible from the set $\{(0,0), (1,0)\}$.

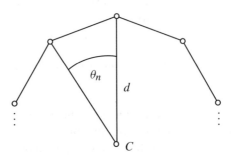

The definitive result can now be proved.

(11.3.7) *A regular n-gon of side 1 can be constructed by ruler and compass if and only if n has the form $2^k p_1 p_2 \ldots p_k$ where $k \geq 0$ and the p_i are distinct Fermat primes, i.e., of the form $2^{2^t} + 1$.*

Proof. Assume that the regular n-gon is constructible, so that $\cos\theta_n$ is constructible. By (10.2.2) $(\mathbb{Q}(\cos\theta_n) : \mathbb{Q})$ must be a power of 2. Now put $c = e^{2\pi i/n}$, a primitive n-th root of unity. Then $\cos\theta_n = \frac{1}{2}(c + c^{-1})$, so that $\mathbb{Q}(\cos\theta_n) \subseteq \mathbb{Q}(c)$. Since $c + c^{-1} = 2\cos\theta_n$, we have $c^2 - 2c\cos\theta_n + 1 = 0$. Hence $(\mathbb{Q}(c) : \mathbb{Q}(\cos\theta_n)) = 2$ and $(\mathbb{Q}(c) : \mathbb{Q}) = 2^d$ for some d. Recall from (11.2.10) that $\mathrm{Irr}_{\mathbb{Q}}(c) = \Phi_n$, which has degree $\phi(n)$. Writing $n = 2^k p_1^{e_1} \ldots p_r^{e_r}$ with distinct odd primes p_i and $e_i > 0$, we have $\phi(n) = 2^{k-1}(p_1^{e_1} - p_1^{e_1-1}) \ldots (p_r^{e_r} - p_r^{e_r-1})$ by (2.3.8). This must equal 2^d. Hence $e_i = 1$ and $p_i - 1$ is a power of 2. Since $2^s + 1$ can only be prime if s is a power of 2 (see Exercise (2.2.11)), each p_i is a Fermat prime.

Conversely, assume that n has the form indicated. Then $\mathbb{Q}(c)$ is Galois over $\mathbb{Q}$ and $(\mathbb{Q}(c) : \mathbb{Q}) = \phi(n)$, which is a power of 2 by the formula for $\phi(n)$. Hence $G = \mathrm{Gal}(\mathbb{Q}(c) : \mathbb{Q})$ is a finite 2-group. Therefore a composition series of G has all its factors of order 2 and by the Fundamental Theorem of Galois Theory there is a chain of subfields

$$\mathbb{Q} = F_0 \subset F_1 \subset \cdots \subset F_\ell = \mathbb{Q}(c)$$

such that F_{i+1} is Galois over F_i and $(F_{i+1} : F_i) = 2$.

We argue by induction on i that every real number in F_i is constructible. Let x in $F_{i+1} - F_i$ be real. Then $\mathrm{Irr}_{F_i}(x) = t^2 + at + b$ where $a, b \in F_i$. Thus $x^2 + ax + b = 0$ and $\left(x + \frac{1}{2}a\right)^2 = \frac{1}{4}a^2 - b > 0$ since x is real. Put $x' = x + \frac{1}{2}a$; then $x'^2 \in F_i$. By induction hypothesis x'^2 is constructible and (10.2.1) shows that x' is constructible, whence so is x. □

Example (11.3.2) A regular n-gon is constructible for $n = 3, 4, 5, 6$, but not for $n = 7$.

In fact, the only known Fermat primes are $3, 5, 17, 257 = 2^{2^3} + 1$ and $65,537 = 2^{2^4} + 1$. Since 7 is not a Fermat prime, it is impossible to construct a regular 7-gon using ruler and compass.

Exercises (11.3)

1. For each of the following polynomials over $\mathbb{Q}$ display the lattice of subgroups of the Galois group and the corresponding lattice of subfields of the splitting field:
    (a) $t^2 - 5$; (b) $t^4 - 5$; (c) $(t^2 + 1)(t^2 + 3)$.

2. Determine the normal subfields of the splitting fields in Exercise (11.3.1).

3. Use the Fundamental Theorem of Galois Theory and Exercise (11.2.6) to prove that $GF(p^m)$ has exactly one subfield of order p^d for each divisor d of n and no subfields of other orders – see also Exercise (10.3.2).

4. Let $E = \mathbb{Q}(\sqrt{2}, \sqrt{3})$. Find all the subgroups of the Galois group and hence all subfields of E.

5. Find all finite fields with exactly two subfields.

6. Let E be a Galois extension of a field F and let p^k divide $(E : F)$ where p is a prime. Prove that there is an intermediate field S such that $(E : S) = p^k$.

7. If E is a Galois extension of a field F and there is exactly one proper intermediate field, what can be said about $\mathrm{Gal}(E/F)$?

8. If E is a Galois extension of F and $(E : F)$ is the square of a prime, then each intermediate field is normal over F.

9. A regular 2^k-gon of side 1 is constructible.

10. For which values of n in the range 10 to 20 can a regular n-gon of side 1 be constructed?

11. Show that if a is a real number such that $(\mathbb{Q}(a) : \mathbb{Q})$ is a power of 2 and $\mathbb{Q}(a)$ is normal over $\mathbb{Q}$, then a is constructible from the points $(0, 0)$ and $(1, 0)$.

12. Let p be a prime and let $f = t^p - t - a \in F[t]$ where $F = GF(p)$. Denote by E the splitting field of f over F.
    (a) If x is a root of f in E, show that the set of all roots of f is $\{x + b \mid b \in F\}$, and that $E = F(x)$.
    (b) Prove that f is irreducible over F if and only if $a \neq 0$.
    (c) Prove that $\mathrm{Gal}(f)$ has order 1 or p.

11.4 Solvability of equations by radicals

One of the oldest parts of algebra is concerned with the problem of solving equations of the form $f(t) = 0$ where f is a non-constant polynomial over $\mathbb{Q}$ or $\mathbb{R}$. The object is to find a formula giving the solutions of the equation, which involves the coefficients of f, square roots, cube roots etc. The most familiar cases are when $\deg(f) \leq 2$; if the degree is 1, we are just solving a single linear equation. If the degree is 2, there is the familiar formula for the solutions of a quadratic equation. For equations of degree 3 and 4 the problem is harder, but methods of solution had been found by the 16th Century. Thus if $\deg(f) \leq 4$, there are explicit formulas for the roots of $f(t) = 0$, which in fact involve radicals of the form $\sqrt[k]{}$ for $k \leq 4$.

The problem of finding formulas for the solutions of equations of degree 5 and higher is one that fascinated mathematicians for hundreds of years. An enormous amount of ingenuity was expended in vain efforts to solve the general equation of the fifth degree. It was only with the work of Abel, Galois and Ruffini[2] in the early 19th Century that it became clear why all efforts had been in vain. It is a fact that the solvability of a polynomial equation is inextricably linked to the solvability of the Galois group of the polynomial. The symmetric group S_n is solvable for $n < 5$, but it is insolvable for $n \geq 5$. This explains why early researchers were able to solve the general equation of degree n only for $n \leq 4$. Without the aid of group theory it is impossible to understand this failure. Our aim here is explain why the solvability of the Galois group governs the solvability of a polynomial equation.

Radical extensions. Let E be an extension of a field F. Then E is called a *radical extension* of F if there is a chain of subfields

$$F = E_0 \subseteq E_1 \subseteq E_2 \subseteq \cdots \subseteq E_m = E$$

such that $E_{i+1} = E_i(a_{i+1})$ where a_{i+1} has irreducible polynomial over E_i of the form $t^{n_{i+1}} - b_i$. It is natural to refer to a_{i+1} as a *radical* and write $a_{i+1} = \sqrt[n_{i+1}]{b_i}$, but here one has to keep in mind that a_{i+1} may not be uniquely determined by b_i. Since

$$E = F\left(\sqrt[n_1]{b_1}, \sqrt[n_2]{b_2}, \ldots, \sqrt[n_m]{b_m}\right),$$

elements of E are expressible as polynomial functions of the radicals $\sqrt[n_i]{b_i}$.

Next let f be a non-constant polynomial over F with splitting field K. Then f, or the equation $f = 0$, is said to be *solvable by radicals* if K is contained in some radical extension of F. This means that the roots of f are obtained by forming a finite sequence of successive radicals, starting with elements of F. This definition gives a precise expression of our intuitive idea of what it means for a polynomial equation to be solvable by radicals.

To make progress with the problem of describing the radical extensions we must have a better understanding of polynomials of the form $t^n - a$.

---

[2] Paulo Ruffini (1765–1822).

(11.4.1) *Let F be a field and n a positive integer. Assume that F contains a primitive n-th root of unity. Then for any a in F the group $\mathrm{Gal}_F(t^n - a)$ is cyclic with order dividing n.*

Proof. Let z be a primitive n-th root of unity in F and denote by b a root of $f = t^n - a$ in its splitting field E. Then the roots of f are bz^i, $i = 0, 1, \ldots, n - 1$. If $\alpha \in \mathrm{Gal}(f) = \mathrm{Gal}(E/F)$, then $\alpha(b) = bz^{i(\alpha)}$ for some $i(\alpha)$ and α is completely determined by $i(\alpha)$: this is because $\alpha|_F$ is the identity map and $E = F(b)$ since $z \in F$. Now the assignment $\alpha \mapsto i(\alpha) + n\mathbb{Z}$ is an injective homomorphism from $\mathrm{Gal}(f)$ to $\mathbb{Z}_n$: for $\alpha\beta(b) = \alpha(bz^{i(\beta)}) = \alpha(b)z^{i(\beta)} = bz^{i(\alpha)+i(\beta)}$ and thus $i(\alpha\beta) \equiv i(\alpha) + i(\beta)$ (mod n). It follows that $\mathrm{Gal}(f)$ is isomorphic with a subgroup of $\mathbb{Z}_n$, whence it is a cyclic group with order dividing n. □

The principal theorem is now within reach.

(11.4.2) (Galois) *Let f be a non-constant polynomial over a field F of characteristic 0. If f is solvable by radicals, then $\mathrm{Gal}(f)$ is a solvable group.*

Proof. Let E denote the splitting field of f. Then by definition $E \subseteq R$ where R is a radical extension of F. Hence there are subfields R_i such that

$$F = R_0 \subseteq R_1 \subseteq \cdots \subseteq R_m = R$$

where $R_{i+1} = R_i(a_{i+1})$ and $\mathrm{Irr}_{R_i}(a_{i+1}) = t^{n_{i+1}} - b_i$, with $b_i \in R_i$. It follows that $(R_{i+1} : R_i) = n_{i+1}$ and hence $n = (R : F) = n_1 n_2 \ldots n_m$.

Let K and L be the splitting fields of the polynomial $t^n - 1$ over F and R respectively. Setting $L_i = K(R_i)$, we then have a chain of subfields

$$K = L_0 \subseteq L_1 \subseteq \cdots \subseteq L_m = L.$$

Notice that all the roots of $t^{n_{i+1}} - b_i$ are in L_{i+1}: for $a_{i+1} \in L_{i+1}$ and K contains all primitive n_{i+1}-th roots of unity since n_{i+1} divides n. Therefore L_{i+1} is the splitting field of $t^{n_{i+1}} - b_i$ over L_i and consequently L_{i+1} is normal over L_i. Since $\mathrm{char}(F) = 0$, the field L_{i+1} is separable and hence Galois over L_i. All the relevant subfields are displayed in the Hasse diagram on the next page.

Now set $G_i = \mathrm{Gal}(L/L_i)$. By (11.3.3) $G_{i+1} \triangleleft G_i$ and we have the series of subgroups $1 = G_m \triangleleft G_{m-1} \triangleleft \cdots \triangleleft G_1 \triangleleft G_0 = \mathrm{Gal}(L/K)$. Also by (11.3.4)

$$G_i/G_{i+1} = \mathrm{Gal}(L/L_i)/\mathrm{Gal}(L/L_{i+1}) \simeq \mathrm{Gal}(L_{i+1}/L_i),$$

which last is a cyclic group by (11.4.1) since L_{i+1} is the splitting field of $t^{n_{i+1}} - b_i$ over L_i. Therefore $\mathrm{Gal}(L/K)$ is a solvable group.

Next K is normal over F, being a splitting field, and thus $\mathrm{Gal}(L/K) \triangleleft \mathrm{Gal}(L/F)$. Also

$$\mathrm{Gal}(L/F)/\mathrm{Gal}(L/K) \simeq \mathrm{Gal}(K/F),$$

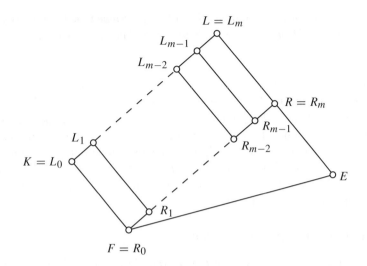

which is abelian by (11.2.7). It follows that $\mathrm{Gal}(L/F)$ is solvable. Finally, E is normal over F, so that $\mathrm{Gal}(L/E) \triangleleft \mathrm{Gal}(L/F)$ and

$$\mathrm{Gal}(f) = \mathrm{Gal}(E/F) \simeq \mathrm{Gal}(L/F)/\mathrm{Gal}(L/E).$$

But a quotient group of a solvable group is also solvable by (9.2.2). Therefore $\mathrm{Gal}(f)$ is solvable as claimed. □

It can be shown – although we shall not do so here – that the converse of (11.4.2) is valid: see [1] for a detailed proof. Thus there is the following definitive result.

(11.4.3) *Let f be a non-constant polynomial over a field of characteristic 0. Then f is solvable by radicals if and only if $\mathrm{Gal}(f)$ is a solvable group.*

Let $n = \deg(f)$. Then $\mathrm{Gal}(f)$ is isomorphic with a subgroup of the symmetric group S_n by (11.2.3). Now if $n \leq 4$, then S_n, and hence $\mathrm{Gal}(f)$, is solvable. Therefore by (11.4.3) *every polynomial with degree 4 or less is solvable by radicals.*

On the other hand, when $n \geq 5$, the symmetric group S_n is not solvable since A_n is simple by (9.1.7). Thus one is led to suspect that not every polynomial equation of degree 5 is solvable by radicals. Actual examples of polynomials that are not solvable by radicals are furnished by the next result.

(11.4.4) *Let $f \in \mathbb{Q}[t]$ be an irreducible polynomial of prime degree p and assume that f has exactly two complex roots. Then $\mathrm{Gal}(f) \simeq S_p$ and hence f is not solvable by radicals if $p \geq 5$.*

Proof. Label the roots of f in its splitting field $a_1, a_2, \ldots, a_p$; these are all different since f is separable. Two of these roots are complex conjugates, say $\bar{a}_1 = a_2$, while the others are all real. We can think of Gal(f) as a group of permutations of the set of roots $\{a_1, a_2, \ldots, a_p\}$ and Gal(f) acts transitively since f is irreducible. Therefore p divides $|\text{Gal}(f)|$ by (5.2.2), and Cauchy's Theorem (5.3.9) shows that there is an element of order p in Gal(f). Hence Gal(f) contains a p-cycle, say $\pi = (a_1, a_{i_2}, \ldots, a_{i_p})$. Replacing π by a suitable power, we may assume that $i_2 = 2$. Now relabel the roots $a_3, a_4, \ldots, a_p$ so that $\pi = (a_1, a_2, \ldots, a_p)$.

Complex conjugation, i.e., $\sigma = (a_1 a_2)$, is an element of Gal(f) with order 2. Conjugation of σ by powers of π shows that Gal(f) also contains all adjacent transpositions $(a_i a_{i+1})$, for $i = 1, 2, \ldots, p-1$. But any permutation is expressible as a product of adjacent transpositions – see Exercise (3.1.4) – and therefore Gal(f) = S_n. □

Example (11.4.1) The polynomial $t^5 - 6t + 3 \in \mathbb{Q}[t]$ is not solvable by radicals.

In the first place the polynomial is irreducible over $\mathbb{Q}$ by Eisenstein's Criterion and Gauss's Lemma. In addition calculus tells us that the graph of the function $t^5 - 6t + 3$ crosses the t-axis exactly three times, so that there are three real roots and two complex ones. Thus Gal($t^5 - 6t + 3$) $\simeq S_5$ and the result follows.

Example (11.4.2) The polynomial $f = t^5 + 8t^3 - t^2 + 12t - 2$ is solvable by radicals.

Here the situation is different since f factorizes as $(t^2 + 2)(t^3 + 6t - 1)$. Therefore its Gal(f) is isomorphic with a subgroup of

$$\text{Gal}(t^2 + 2) \times \text{Gal}(t^3 + 6t - 1)$$

by Exercise (11.2.5). The latter is a solvable group. Hence by (9.2.2) Gal(f) is solvable and f is solvable by radicals.

Symmetric functions. As the final topic of the chapter, we present an account of the elementary theory of symmetric functions and explore its relationship with Galois theory. Let F be any field and put $E = F\{x_1, x_2, \ldots, x_n\}$, the field of rational functions over F in distinct indeterminates $x_1, x_2, \ldots, x_n$. By a *symmetric function* in E we mean a g such that

$$g(x_{\pi(1)}, x_{\pi(2)}, \ldots, x_{\pi(n)}) = g(x_1, x_2, \ldots, x_n)$$

for all $\pi \in S_n$. Thus g is unaffected by any permutation of the indeterminates $x_1, x_2, \ldots, x_n$. It is easy to verify that the elementary functions form a subfield of E.

Next consider the polynomial

$$f = (t - x_1)(t - x_2) \ldots (t - x_n) \in E[t]$$

where t is another indeterminate. Then expansion shows that

$$f = t^n - s_1 t^{n-1} + s_2 t^{n-2} - \cdots + (-1)^n s_n$$

where $s_1 = \sum_{i=1}^n x_i$, $s_2 = \sum_{i<j=1}^n x_i x_j$, and in general

$$s_j = \sum_{i_1 < i_2 < \cdots < i_j = 1}^n x_{i_1} x_{i_2} \ldots x_{i_j},$$

the sum being over all tuples $(i_1, i_2, \ldots, i_j)$ such that $1 \le i_1 < i_2 < \cdots < i_j \le n$. It is obvious that these s_j are symmetric functions: they are known as the *elementary symmetric functions* in $x_1, x_2, \ldots, x_n$. For example, when $n = 3$, there are three elementary symmetric functions,

$$x_1 + x_2 + x_3, \quad x_1 x_2 + x_2 x_3 + x_1 x_3, \quad x_1 x_2 x_3.$$

Put $S = F(s_1, s_2, \ldots, s_n)$, which is a subfield of E. Then $f \in S[t]$ and E is generated by S and the roots of f, i.e., $x_1, x_2, \ldots, x_n$. Hence E is the splitting field of f over S. Since all the roots of f are distinct, (11.1.6) shows that E is separable and hence Galois over S. Therefore $\mathrm{Gal}(f) = \mathrm{Gal}(E/S)$ has order $(E : S)$. We now proceed to determine the Galois group of f over S. With the same notation the definitive result is:

(11.4.5) $\mathrm{Gal}(f) \simeq S_n$.

Proof. Since $\mathrm{Gal}(f)$ permutes the roots $x_1, x_2, \ldots, x_n$ faithfully, we can identify it with a subgroup of S_n. Let $\pi \in S_n$ and define $\alpha_\pi : E \to E$ by the rule

$$\alpha_\pi(g(x_1, x_2, \ldots, x_n)) = g(x_{\pi(1)}, x_{\pi(2)}, \ldots, x_{\pi(n)}):$$

then α_π is evidently an automorphism of E. Since α_π fixes all the elementary symmetric functions, it fixes every element of $S = F(s_1, s_2, \ldots, s_n)$ and therefore $\alpha_\pi \in \mathrm{Gal}(E/S) = \mathrm{Gal}(f)$. Finally, all the α_π are different. Therefore $\mathrm{Gal}(f) = S_n$. □

From this we quickly deduce a famous theorem.

Corollary (11.4.6) (The Symmetric Function Theorem) *If F is any field and $s_1, s_2, \ldots, s_n$ are the elementary symmetric functions in indeterminates $x_1, x_2, \ldots, x_n$, then $F(s_1, s_2, \ldots, s_n)$ is the field of all symmetric functions in $x_1, x_2, \ldots, x_n$. Also the symmetric polynomials form a subring which is generated by F and the $s_1, s_2, \ldots, s_n$.*

Proof. Let $S = F(s_1, s_2, \ldots, s_n) \subseteq E = F\{x_1, x_2, \ldots, x_n\}$. Since $\mathrm{Gal}(E/S) \simeq S_n$, the subfield of symmetric functions in E is by definition $\mathrm{Fix}(\mathrm{Gal}(E/S))$, i.e., S by the Fundamental Theorem of Galois Theory. The statement about polynomials is left as an exercise. □

Generic polynomials. Let F be an arbitrary field and write K for the rational function field in indeterminates $x_1, x_2, \ldots, x_n$ over F. The polynomial

$$f = t^n - x_1 t^{n-1} + x_2 t^{n-2} - \cdots + (-1)^n x_n$$

is called a *generic polynomial*. The point to note is that we can obtain from f any monic polynomial of degree n in $F[t]$ by replacing $x_1, x_2, \ldots, x_n$ by suitable elements of F. It is therefore not surprising that the Galois group of f over K is as large as it could be.

(11.4.7) *With the above notation,* $\mathrm{Gal}(f) \simeq S_n$.

Proof. Let $u_1, u_2, \ldots, u_n$ be the roots of f in its splitting field E over K. Then $f = (t - u_1)(t - u_2) \ldots (t - u_n)$ and thus $x_i = s_i(u_1, u_2, \ldots, u_n)$ where s_i is the i-th elementary symmetric function in n indeterminates $y_1, y_2, \ldots, y_n$, all of which are different from $x_1, x_2, \ldots, x_n, t$.

The assignment $x_i \mapsto s_i$ determines a ring homomorphism

$$\phi_0 : F\{x_1, x_2, \ldots, x_n\} \to F\{y_1, y_2, \ldots, y_n\};$$

observe here that $f(s_1, \ldots, s_n) = 0$ implies that $f(x_1, \ldots, x_n) = 0$ because $x_i = s_i(u_1, \ldots, u_n)$. So ϕ_0 is actually an isomorphism from $K = F\{x_1, x_2, \ldots, x_n\}$ to $L = F(s_1, s_2, \ldots, s_n) \subseteq F\{y_1, y_2, \ldots, y_n\}$. Set $f^* = \phi_0(f)$; thus

$$f^* = t^n - s_1 t^{n-1} + s_2 t^{n-2} - \cdots + (-1)^n s_n = (t - y_1)(t - y_2) \ldots (t - y_n),$$

by definition of the elementary symmetric functions s_i.

By (10.3.2) we may extend ϕ_0 to an isomorphism ϕ from E, the splitting field of f over K, to the splitting field of f^* over L. Therefore ϕ induces a group isomorphism from $\mathrm{Gal}(f)$ to $\mathrm{Gal}(f^*)$. But we know that $\mathrm{Gal}(f^*) \simeq S_n$ by (11.4.5). Hence $\mathrm{Gal}(f) \simeq S_n$. □

Corollary (11.4.8) (Abel, Ruffini) *If F is a field of characteristic 0, the generic polynomial $t^n - x_1 t^{n-1} + x_2 t^{n-2} - \cdots + (-1)^n x_n$ is insolvable by radicals if $n \geq 5$.*

Thus, as one would expect, there is no general formula for the roots of the generic polynomial of degree $n \geq 5$ in terms of its coefficients $x_1, x_2, \ldots, x_n$.

Exercises (11.4)

1. Let $F \subseteq K \subseteq E$ be field extensions with K radical over F and E radical over K. Prove that E is radical over F.

2. Let $F \subseteq K \subseteq E$ be field extensions with E radical and Galois over F. Prove that E is radical over K.

3. The polynomial $t^5 - 3t + 2$ in $\mathbb{Q}[t]$ is solvable by radicals.

4. If p is a prime larger than 11, then $t^5 - pt + p$ in $\mathbb{Q}[t]$ is not solvable by radicals.

5. If $f \in F[t]$ is solvable by radicals and $g \mid f$ in $F[t]$, then g is solvable by radicals.

6. Let $f = f_1 f_2$ where $f_1, f_2 \in F[t]$ and F has characteristic 0. If f_1 and f_2 are solvable by radicals, then so is f. Deduce that every non-constant reducible polynomial of degree < 6 over $\mathbb{Q}$ is solvable by radicals.

7. Let F be a field of characteristic 0 and let $f \in F[t]$ be non-constant with splitting field E. Prove that there is a unique smallest intermediate field S such that S is normal over F and f is solvable by radicals over S. [Hint: show first that there is a unique maximum solvable normal subgroup in any finite group].

8. For every integer $n \geq 5$ there is a polynomial of degree n over $\mathbb{Q}$ which is insolvable by radicals.

9. Let G be any finite group. Prove that there is a Galois extension E of some algebraic number field F such that $\text{Gal}(E/F) \simeq G$. [You may assume there is an algebraic number field whose Galois group over $\mathbb{Q}$ is isomorphic with S_n]. (Remark: the general problem of whether every finite group is the Galois group of some algebraic number field over $\mathbb{Q}$ is still open; it is known to be true for solvable groups.)

10. Write each of the following symmetric polynomials as polynomials in the elementary symmetric functions s_1, s_2, s_3 in x_1, x_2, x_3.
    (a) $x_1^2 + x_2^2 + x_3^2$;
    (b) $x_1^2 x_2 + x_1 x_2^2 + x_2^2 x_3 + x_2 x_3^2 + x_1^2 x_3 + x_1 x_3^2$;
    (c) $x_1^3 + x_2^3 + x_3^3$.

Chapter 12
Further topics

The final chapter contains additional material some of which is not ordinarily part of an undergraduate algebra course. With the exception of coding theory, the topics are closely connected with previous chapters. Readers with the time and inclination are encouraged to read at least some of these sections.

12.1 Zorn's Lemma and its applications

The background to Zorn's Lemma lies in the kind of set theory we are using. So far we have been functioning — quite naively — in what is called the *Gödel–Bernays Theory*. In this the primitive, or undefined, notions are *class, membership* and *equality*. On the basis of these concepts and the accompanying axioms, the usual elementary properties of sets can be derived. In addition we have made extensive use of the Well Ordering Law for $\mathbb{Z}$, and its corollary the Principle of Mathematical Induction – see 2.1.

However, the set theory just described does not provide a satisfactory basis for dealing with infinite sets. For example, suppose that H is a subgroup of infinite index in a group G. We might wish to form a left transversal to H in G. Now this would involve making a simultaneous choice of one element from each of the infinitely many left cosets of H. That such a choice is possible is asserted by the well-known *Axiom of Choice*. However this axiom is known to be independent of the Gödel–Bernays axioms. Thus, if we want to be able to form left transversals in infinite groups, we must assume the Axiom of Choice or else something equivalent to it. For many purposes in algebra the most useful set theoretic axiom is what has become known as *Zorn's Lemma*. Despite the name, this is an axiom that must be assumed and not a lemma.

Zorn's Lemma. *Let $(S, \preceq)$ be a non-empty partially ordered set with the property that every chain in S has an upper bound in S. Then S contains at least one maximal element.*

Here the terminology calls for some explanation. Recall from 1.2 that a *chain* in a partially ordered set S is a subset C which is linearly ordered by the partial order $\preceq$. An *upper bound* for C is an element s of S such that $c \preceq s$ is valid for all c in C. Finally, a *maximal element* of S is an element m such that $m \preceq s$ and $s \in S$ imply that $s = m$. Note that in general a partially ordered set may contain several maximal elements or none at all.

We now demonstrate how Zorn's Lemma can be used to prove some crucial theorems in algebra.

Existence of a basis in a vector space. It was shown in Chapter 8 that every finitely generated non-zero vector space has a basis – see (8.2.6). Zorn's Lemma can be used to extend this fundamental result to infinitely generated vector spaces.

(12.1.1) *Every non-zero vector space has a basis.*

Proof. Let V be a non-zero vector space over a field F and define $\mathcal{S}$ to be the set of all linearly independent subsets of V. Then $\mathcal{S}$ is non-empty since it contains the singleton set $\{v\}$ where v is any non-zero vector in V. Further, the inclusion relation is a partial order on $\mathcal{S}$, so $(\mathcal{S}, \subseteq)$ is a partially ordered set. To apply Zorn's Lemma, we need to verify that every chain in $\mathcal{S}$ has an upper bound.

Let $\mathcal{C}$ be a chain in $\mathcal{S}$. There is an obvious candidate for an upper bound, namely the union $U = \bigcup_{X \in \mathcal{C}} X$. Notice that U is linearly independent: for any relation of linear dependence in U will involve a *finite* number of elements of $\mathcal{S}$ and so the relation will hold in some $X \in \mathcal{C}$. Here it is vital that $\mathcal{C}$ be linearly ordered. Thus $U \in \mathcal{C}$ and obviously U is an upper bound for $\mathcal{C}$.

It is now possible to apply Zorn's Lemma to obtain a maximal element in $\mathcal{S}$, say B. By definition B is linearly independent: to show that B is a basis we must prove that B generates V. Assume this to be false and let v be a vector in V which is not expressible as a linear combination of vectors in B; then in particular $v \notin B$ and hence $B \subset \{v\} \cup B = B'$. By maximality of B, the set B' does not belong to $\mathcal{S}$ and thus it is linearly dependent. So there is a linear relation

$$a_1 u_1 + a_2 u_2 + \cdots + a_m u_m + cv = 0$$

where $u_i \in B$ and c and the a_i belong to F and not all of them are 0. Now if $c = 0$, then $a_1 u_1 + a_2 u_2 + \cdots + a_m u_m = 0$, so that $a_1 = a_2 = \cdots = a_m = 0$ since $u_1, u_2, \ldots, u_m$ are linearly dependent. Therefore $c \neq 0$ and we can solve the equation for v, obtaining

$$v = (-c^{-1} a_1) u_1 + (-c^{-1} a_2) v_2 + \cdots + (-c^{-1} a_m) u_m,$$

which contradicts the choice of v. Hence B generates V. □

It can also be shown that any two bases of a vector space have the same cardinal, so that it is possible to define the dimension of an infinitely generated vector space to be this cardinal.

Maximal ideals in rings. Recall that a maximal ideal I of a ring R is a largest proper ideal. If R is commutative and has an identity, then by (6.3.7) this is equivalent to I being an ideal such that R/I is a field. Maximal ideals were used in 7.4 to construct fields, but only in circumstances where it was clear that they existed, for example when the ascending chain on ideals held. Zorn's Lemma can be used to produce maximal ideals under more general circumstances.

(12.1.2) *Any ring with identity has at least one maximal ideal.*

Proof. Let $\mathcal{S}$ denote the set of all proper ideals of the ring R. Now the zero ideal is proper since it does not contain 1_R. Thus $\mathcal{S}$ is not empty. Of course $\mathcal{S}$ is partially ordered by inclusion. Let $\mathcal{C}$ be a chain in $\mathcal{S}$ and define U to be $\bigcup_{I \in \mathcal{C}} I$. It is easily seen that U is an ideal. If $U = R$, then 1_R belongs to some I in $\mathcal{C}$, from which it follows that $R = RI \subseteq I$ and $I = R$. By this contradiction we may infer that $U \neq R$, so that $U \in \mathcal{S}$. Now we can apply Zorn's Lemma to get a maximal element of S, i.e., a maximal ideal of R. □

An immediate consequence of (12.1.2) is:

(12.1.3) *If R is a commutative ring with identity, there is a quotient ring of R which is a field.*

On the other hand, not all rings have maximal ideals.

Example (12.1.1) There exist non-zero commutative rings without maximal ideals.

To see why, we start with the additive abelian group $\mathbb{Q}$ and turn it into a ring by declaring all products to be 0. Then $\mathbb{Q}$ becomes a commutative ring in which every subgroup is an ideal. But $\mathbb{Q}$ cannot have a maximal subgroup: for if S were one, $\mathbb{Q}/S$ would be a group without proper non-trivial subgroups and so $|\mathbb{Q}/S| = p$, a prime. But this is impossible since $\mathbb{Q} = p\mathbb{Q}$. It follows that the ring $\mathbb{Q}$ has no maximal ideals.

The existence of algebraic closures. Another important application of Zorn's Lemma is to show that for every field F there is a largest algebraic extension, its *algebraic closure*. The construction of such a largest extension is the kind of task to which Zorn's Lemma is well-suited.

Let E be a field extension of F, with $F \subseteq E$. Then E is called an *algebraic closure of F* if the following two conditions hold:

(i) E is algebraic over F;

(ii) every irreducible polynomial in $E[t]$ has degree 1.

Notice that the second requirement amounts to saying that if K is an algebraic extension of E, then $K = E$. A field which equals its algebraic closure is called an *algebraically closed field*. For example, the complex field $\mathbb{C}$ is algebraically closed by the Fundamental Theorem of Algebra – see (11.3.6).

Our objective is to prove the following theorem:

(12.1.4) *Every field has an algebraic closure.*

Proof. Let F be an arbitrary field. The first step is to choose a set which is large enough to accommodate the algebraic closure. To avoid getting embroiled in technicalities, we shall be somewhat casual about this: but in fact what is needed is a set S with cardinal greater than $\aleph_0|F|$. In particular $|F| < |S|$ and so there exists an injection $\alpha : F \to S$. Use the map α to turn $\text{Im}(\alpha)$ into a field, by defining

$$\alpha(x) + \alpha(y) = \alpha(x+y) \quad \text{and} \quad \alpha(x)\alpha(y) = \alpha(xy)$$

where $x, y \in F$. Clearly $\text{Im}(\alpha)$ is a field isomorphic with F. Thus, replacing F by $\text{Im}(\alpha)$, we may assume that $F \subseteq S$.

To apply Zorn's Lemma we need to introduce a suitable partially ordered set. Let $\mathcal{K}$ denote the set of all subsets E such that $F \subseteq E \subseteq S$ and the field operations of F may be extended to E in such a way that E becomes a field which is algebraic over F. Quite obviously $F \in \mathcal{K}$, so that $\mathcal{K}$ is not empty. A partial order $\preceq$ on $\mathcal{K}$ is defined as follows: if $E_1, E_2 \in \mathcal{K}$, then $E_1 \preceq E_2$ means that $E_1 \subseteq E_2$ and the field operations of E_2 are consistent with those of E_1. Thus E_1 is actually a subfield of E_2. It is quite easy to see that $\preceq$ is a partial order on $\mathcal{K}$. Thus we have our partially ordered set $(\mathcal{K}, \preceq)$.

Next the union U of a chain $\mathcal{C}$ in $\mathcal{K}$ is itself in $\mathcal{K}$. For by the definition of $\preceq$ the field operations of all the members of $\mathcal{C}$ are consistent and so they may be combined to give the field operations of U. It follows that $U \in \mathcal{K}$ and clearly U is an upper bound for $\mathcal{C}$ in $\mathcal{K}$. Zorn's Lemma may now be applied to yield a maximal element of $\mathcal{K}$, say E.

By definition E is algebraic over F. What needs to be established is that any irreducible polynomial f in $E[t]$ has degree 1. Suppose that in fact $\deg(f) > 1$. Put $E' = E[t]/(f)$, an algebraic extension of E and hence of F by (10.1.8). If we write $E_0 = \{a + (f) \mid a \in E\}$, then $E_0 \subseteq E'$ and there is an isomorphism $\beta : E_0 \to E$ given by $\beta(a + (f)) = a$. It is at this point that the largeness of the set S is important. One can show without too much trouble that $|E' - E_0| < |S - E|$, using the inequalities $|E| \leq \aleph_0|F|$ and $|E[t]| < |S|$. Accepting this fact, we may form an injective map $\beta_1 : E' - E_0 \mapsto S - E$. Combine β_1 with $\beta : E_0 \to E$ to produce an injection $\gamma : E' \to S$. Thus $\gamma(a + (f)) = a$ for a in E.

Now use the map γ to make $J = \text{Im}(\gamma)$ into a field, by defining $\gamma(x_1) + \gamma(x_2)$ to be $\gamma(x_1 + x_2)$ and $\gamma(x_1)\gamma(x_2)$ to be $\gamma(x_1 x_2)$. Then $\gamma : E' \mapsto J$ is an isomorphism of fields and $\gamma(E_0) = E$. Since E' is algebraic over E_0, it follows that J is algebraic over E, and therefore $J \in \mathcal{K}$. However $E \neq J$ since $E_0 \neq E'$, which contradicts the maximality of E and completes the proof. □

While some details in the above proof may seem tricky, the reader should appreciate the simplicity of the essential idea, that of building a largest algebraic extension using Zorn's Lemma. It can be shown, although we shall not do it here, that *every field has a unique algebraic closure up to isomorphism.*

For example, the algebraic closure of $\mathbb{R}$ is $\mathbb{C}$, while that of $\mathbb{Q}$ is the field of all algebraic numbers. Another example of interest is the algebraic closure of the Galois field GF(p), which is an algebraically closed field of prime characteristic p.

As a final example of the use of Zorn's Lemma, we prove a result on cardinal numbers which was stated without proof in 1.4.

(12.1.5) (The Law of Trichotomy) *If A and B are sets, then exactly one of the following must hold,*
$$|A| < |B|, \quad |A| = |B|, \quad |B| < |A|.$$

Proof. Because of the Cantor–Bernstein Theorem (1.4.2), it is enough to show that either $|A| \leq |B|$ or $|B| \leq |A|$ holds. Clearly A and B can be assumed non-empty.

Consider the set $\mathcal{F}$ of all pairs (X, α) where $X \subseteq A$ and $\alpha : X \to B$ is an injective function. A partial order $\preceq$ on $\mathcal{F}$ is defined by $(X, \alpha) \preceq (X', \alpha')$ if $X \subseteq X'$ and $\alpha'|_X = \alpha$. It is obvious that $\mathcal{F}$ is not empty. Now let $\mathcal{C} = \{(X_i, \alpha_i) \mid i \in I\}$ be a chain in $\mathcal{F}$. Put $U = \bigcup_{i \in I} X_i$ and define $\alpha : U \to B$ by extending the α_i, which are consistent functions, to U. Then (U, α) is an upper bound for $\mathcal{C}$ in $\mathcal{F}$.

Now apply Zorn's Lemma to obtain a maximal element (X, α) of $\mathcal{F}$. We claim that either $X = A$ or $\text{Im}(\alpha) = B$. Indeed suppose that both statements are false, and let $a \in A - X$ and $b \in B - \text{Im}(\alpha)$. Put $Y = X \cup \{a\}$ and define $\beta : Y \to B$ by $\beta(a) = b$ and $\beta|_X = \alpha$. Then β is injective since $b \notin \text{Im}(\alpha)$, and clearly $(\alpha, X) \preceq (\beta, Y)$, which is a contradiction. Therefore either $X = A$ and hence $|A| \leq |B|$ by definition of the linear ordering of cardinals, or else $\text{Im } \alpha = B$. In the latter case for each b in B choose an a_b in A such that $\alpha(a_b) = b$: then the map $b \mapsto a_b$ is an injection from B to A and thus $|B| \leq |A|$. □

Other axioms equivalent to Zorn's Lemma. We record three other axioms which are logically equivalent to Zorn's Lemma.

(I) The Axiom of Well-Ordering.
Every non-empty set can be well-ordered.

Recall from 1.2 that a *well-order* on a set is a linear order such that each non-empty subset has a first element. (Compare the Axiom of Well-Ordering with the Well-Ordering Law, which implies that $\leq$ is a well-order on $\mathbb{N}$).

(II) The Principle of Transfinite Induction.
Let S be a well-ordered set and let T be a non-empty subset of S. Suppose that $t \in T$ whenever it is true that $x \in T$ for all x in S such that $x < t$. Then T must equal S.

This result, which is the basis for the method of proof by transfinite induction, should be compared with the Principal of Mathematical Induction (2.1.1).

(III) The Axiom of Choice.
Let $\{S_i \mid i \in I\}$ be a non-empty set whose elements S_i are non-empty sets. Then there is at least one choice function $\alpha : I \to \bigcup_{i \in I} S_i$, i.e., a function α such that $\alpha(i) \in S_i$. Informally we may express this by saying that it is possible to choose an element simultaneously from every set S_i.

For a clear account of the equivalence of these axioms see [5].

Exercises (12.1)

1. Let R be a commutative ring and let $0 \neq r \in R$. Prove that there is an ideal I which is maximal subject to not containing r. Then prove that I is *irreducible*, i.e., it is not the intersection of two larger ideals.

2. Deduce from Exercise (12.1.1) that every proper ideal of R is an intersection of irreducible ideals.

3. Let G be a non-trivial finitely generated group. Prove that G has at least one maximal subgroup. Deduce that $\varphi(G) \neq G$ where $\varphi(G)$ is the Frattini subgroup of G.

4. Let G be a group and let X be a subset and g an element of G such that $g \notin X$. Prove that there is a subgroup H which is maximal subject to $X \subseteq H$ and $g \notin H$.

5. Generalize (9.2.8) by showing that in an arbitrary group G the Frattini subgroup $\varphi(G)$ consists of all non-generators. [Hint: let $g \in \varphi(G)$ and assume that $G = \langle g, X \rangle$ but $G \neq \langle X \rangle$. Apply Exercise (12.1.4)].

6. Let G be an arbitrary group and p a prime. Show that G has a *maximal p-subgroup*, i.e., a subgroup which is maximal subject to every element having order a power of a prime p. Then prove that a maximal p-subgroup of a finite group is simply a Sylow p-subgroup.

7. Let P be a prime ideal of R, a commutative ring with identity. Prove that there is a largest prime ideal Q containing P. Also show that Q is a maximal ideal.

12.2 More on roots of polynomials

In this section we plan to complete certain topics begun but left unfinished in Chapter 11. In particular, the concept of the *discriminant* of a polynomial is introduced and this is applied to the Galois groups of polynomials of degree ≤ 4.

The discriminant of a polynomial. Let f be a non-constant monic polynomial in t over a field F and let $n = \deg(f)$. Let the roots of f in its splitting field E be $a_1, a_2, \ldots, a_n$ and define

$$\Delta = \prod_{i<j=1}^{n} (a_i - a_j),$$

which is an element of E. Note that Δ depends on the order of the roots, i.e., it is only determined up to its sign. Also f has all its roots distinct if and only if $\Delta \neq 0$: let us assume this to be the case. Thus E is Galois over F.

If $\alpha \in \operatorname{Gal}(f) = \operatorname{Gal}(E/F)$, then α permutes the roots $a_1, a_2, \ldots, a_n$, and $\alpha(\Delta) = \pm \Delta$. Indeed $\alpha(\Delta) = \Delta$ precisely when α produces an *even* permutation of the a_i's. So in any event α fixes

$$D = \Delta^2.$$

12.2 More on roots of polynomials

Here D is called the *discriminant* of f: it is independent of the order of the roots of f. Since D is fixed by every automorphism α and E is Galois over F, it follows from (11.2.6) that D must belong to F. The question arises as to how D is related to the coefficients of the original polynomial f.

(12.2.1) *Let f be a non-constant polynomial over a field F. Then the discriminant D of f is expressible as a polynomial in the coefficients of f.*

Proof. It may be assumed that f has distinct roots $a_1, a_2, \ldots, a_n$ since otherwise $D = 0$. Then

$$f = (t - a_1)(t - a_2) \ldots (t - a_n) = t^n - s_1 t^{n-1} + s_2 t^{n-2} - \cdots + (-1)^n s_n$$

where $s_1, s_2, \ldots, s_n$ are the elementary symmetric functions of degree $1, 2, \ldots, n$ in $a_1, a_2, \ldots, a_n$, (see 11.4). Now $D = \prod_{i<j=1}^{n}(a_i - a_j)^2$ is obviously a symmetric function of $a_1, a_2, \ldots, a_n$. By the Symmetric Function Theorem (11.4.6), D is expressible as a polynomial in $s_1, s_2, \ldots, s_n$, which are polynomials in the coefficients of f. □

We now examine the discriminant of polynomial of degrees 2 and 3.

Example (12.2.1) Let $f = t^2 + ut + v$. If the roots of f are a_1 and a_2, then $\Delta = a_1 - a_2$ and $D = (a_1 - a_2)^2$. This can be rewritten in the form $D = (a_1 + a_2)^2 - 4a_1 a_2$. Now clearly $u = -(a_1 + a_2)$ and $v = a_1 a_2$, so we arrive at the familiar formula for the discriminant of the quadratic $t^2 + ut + v$,

$$D = u^2 - 4v.$$

Example (12.2.2) Consider a general cubic polynomial $f = t^3 + ut^2 + vt + w$ and let a_1, a_2, a_3 be its roots. Then $D = (a_1 - a_2)^2 (a_2 - a_3)^2 (a_1 - a_3)^2$. Also $u = -(a_1 + a_2 + a_3)$, $v = a_1 a_2 + a_2 a_3 + a_1 a_3$ and $w = -a_1 a_2 a_3$. By a rather laborious calculation we can expand D and write it in terms of the elements u, v, w. What emerges is the formula

$$D = u^2 v^2 - 4v^3 - 4u^3 w - 27 w^2 + 18 uvw.$$

This expression can be simplified by making a judicious change of variable. Put $t' = t + \frac{1}{3}u$, so that $t = t' - \frac{1}{3}u$. On substituting for t in $f = t^3 + ut^2 + vt + w$, we find that $f = t'^3 + pt' + q$ where $p = v - \frac{1}{3}u^2$ and $q = w + \frac{2}{27}u^3 - \frac{1}{3}uv$. This shows that no generality is lost by assuming that f has no term in t^2 and

$$f = t^3 + pt + q.$$

The formula for the discriminant now reduces to the more manageable expression

$$D = -4p^3 - 27q^2.$$

Next we will relate the discriminant to the Galois group of a polynomial.

(12.2.2) *Let F be a field whose characteristic is not 2 and let f be a monic polynomial in $F[t]$ with distinct roots $a_1, a_2, \ldots, a_n$. Write $\Delta = \prod_{i<j=1}^{n}(a_i - a_j)$. If $G = \mathrm{Gal}(f)$ is identified with a subgroup of S_n, then $\mathrm{Fix}(G \cap A_n) = F(\Delta)$.*

Proof. Let $H = G \cap A_n$ and note that $H \triangleleft G$ and $|G : H| \le 2$. If E is the splitting field of f, then $F \subseteq F(\Delta) \subseteq \mathrm{Fix}(H) \subseteq E$ since elements of H, being even permutations, fix Δ. Now E is Galois over F, so we have

$$(F(\Delta) : F) \le (\mathrm{Fix}(H) : F) = |G : H| \le 2.$$

If $H = G$, then $F = F(\Delta) = \mathrm{Fix}(H)$ and $\Delta \in F$. The statement is therefore true in this case.

Now suppose that $|G : H| = 2$, and let $\alpha \in G - H$. Then $\alpha(\Delta) = -\Delta$ since α is odd. Since char$(F) \ne 2$, we have $\Delta \ne -\Delta$ and hence $\Delta \notin F$. Thus $(F(\Delta) : F) = 2$ and $\mathrm{Fix}(H) = F(\Delta)$. □

Corollary (12.2.3) *With the above notation, $\mathrm{Gal}(f) \le A_n$ if and only if $\Delta \in F$.*

We now apply these ideas to investigate the Galois groups of polynomials of low degree.

Galois groups of polynomials of degree ≤ 4. Let F be a field such that char$(F) \ne 2$.

(i) Consider a quadratic $f = t^2 + ut + v \in F[t]$. Then $|\mathrm{Gal}(f)| = 1$ or 2. By (12.2.3) $|\mathrm{Gal}(f)| = 1$ precisely when $\Delta \in F$, i.e., $\sqrt{u^2 - 4v} \in F$. This is the familiar condition for f to have both its roots in F. Of course $|\mathrm{Gal}(f)| = 2$ if and only if $\Delta \notin F$, which is the irreducible case.

(ii) Next let f be the cubic $t^3 + pt + q \in F[t]$. We saw that

$$\Delta = \sqrt{D} = \sqrt{-4p^3 - 27q^2}.$$

If f is reducible over F, it must have a quadratic factor f_1 and clearly $\mathrm{Gal}(f) = \mathrm{Gal}(f_1)$, which has order 1 or 2. Thus we can assume f is irreducible. We know from (11.2.2) that $\mathrm{Gal}(f) \le S_3$, and that $|\mathrm{Gal}(f)|$ is divisible by 3 since it acts transitively on the roots of f. Hence $\mathrm{Gal}(f) = A_3$ or S_3. By (12.2.3) $\mathrm{Gal}(f) = A_3$ if and only if $\Delta \in F$; otherwise $\mathrm{Gal}(f) = S_3$.

(iii) Finally, let f be a monic polynomial of degree 4 in $F[t]$. If f is reducible, and $f = f_1 f_2$ with $\deg(f_i) \le 3$, then $\mathrm{Gal}(f)$ is isomorphic with a subgroup of $\mathrm{Gal}(f_1) \times \mathrm{Gal}(f_2)$, (see Exercise (11.2.5)). The structure of $\mathrm{Gal}(f_i)$ is known from (i) and (ii). So assume that f is irreducible. Then $\mathrm{Gal}(f) \le S_4$ and 4 divides $|\mathrm{Gal}(f)|$. Now the subgroups of S_4 whose orders are divisible by 4 are $\mathbb{Z}_4$, V (the Klein 4-group), Dih(8), A_4 and S_4; thus $\mathrm{Gal}(f)$ must be one of these. In fact all five cases can occur, although we shall not attempt to prove this here.

Explicit formulas for the roots of cubic and quartic equations over $\mathbb{R}$ were found in the early 16th century by Scipione del Ferro (1465–1526), Gerolamo Cardano (1501–1576), Niccolo Tartaglia (1499–1526) and Lodovico Ferrari (1522–1565). An interesting account of their discoveries and of the mathematical life of the times can be found in [15].

Exercises (12.2)

1. Find the Galois groups of the following quadratic polynomials over $\mathbb{Q}$:
$t^2 - 5t + 6$; $t^2 + 5t + 1$, $(t+1)^2$.

2. Find the Galois group of the following cubic polynomials over $\mathbb{Q}$:
$t^3 + 4t^2 + 2t - 7$; $t^3 - t - 1$; $t^3 - 3t + 1$; $t^3 + 6t^2 + 11t + 5$.

3. Let f be a cubic polynomial over $\mathbb{Q}$ and let D be its discriminant. Show that f has three real roots if and only if $D \geq 0$. Apply this to the polynomial $t^3 + pt + q$.

4. Let f be an irreducible quartic polynomial over $\mathbb{Q}$ with exactly two real roots. Show that $\mathrm{Gal}(f) \simeq \mathrm{Dih}(8)$ or S_4.

5. (*How to solve cubic equations*). Let $f = t^3 + pt + q \in \mathbb{R}[t]$. The following procedure, due essentially to Scipione del Ferro, will produce a root of f.
   (a) Suppose that $t = u - v$ is a root of f and show that $(u^3 - v^3) + (p - 3uv)(u - v) = -q$;
   (b) find a root of the form $u - v$ by solving $\begin{cases} u^3 - v^3 = -q \\ uv = \frac{p}{3} \end{cases}$ for u and v.

6. The procedure of Exercise (12.2.5) yields one root $u - v$ of $f = t^3 + pt + q$. Show that the other two roots of f are $\omega u - \omega^2 v$ and $\omega^2 u - \omega v$ where $\omega = e^{2\pi i/3}$. (These are known as *Cardano's formulas*.)

7. Use the methods of the last two exercises to find all the roots of the polynomial $t^3 + 3t + 1$.

8. Solve the cubic equation $t^3 + 3t^2 + 6t + 3 = 0$ by first transforming it to one of the form $t'^3 + pt' + q = 0$.

12.3 Generators and relations for groups

When groups first entered the mathematical arena towards the close of the 18th century, they were exclusively permutation groups and they were studied in connection with the theory of equations. A hundred years later groups arose from a different source, geometry, and usually these groups were most naturally described by listing a set of *generators* and a set of *defining relations* which the generators had to satisfy. A very simple example is where there is just one generator x and a single defining relation

$x^n = 1$ where n is a positive integer. Intuitively one would expect these to determine a cyclic group of order n.

As another example, suppose that a group has two generators x and y subject to the three relations $x^2 = 1 = y^2$ and $xy = yx$. Now the Klein 4-group fits this description, with $x = (12)(34)$ and $y = (13)(24)$. Thus it seems reasonable that any group with these generators and relations should be a Klein 4-group. Of course this cannot be substantiated until we have explained exactly what is meant by a group with given sets of generators and defining relations. Even when the generators are subject to no defining relations, a precise definition is still lacking: this is the important case of a *free group*. Thus our first task is to define free groups.

Free groups. A free group is best defined in terms of a certain *mapping property*. Let F be a group, X a non-empty set and $\sigma : X \to F$ a function. Then F, or more precisely the pair (F, σ), is said to be *free on X* if, for each function α from X to a group G there is a *unique* homomorphism $\beta : F \to G$ such that $\beta\sigma = \alpha$, i.e., the triangle diagram below *commutes*:

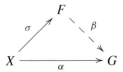

The terminology "commutes" is intended to signify that if one starts at X and proceeds around the triangle to G in either direction, composing functions, the functions obtained are identical, i.e., $\beta\sigma = \alpha$.

First a comment about this definition. *The function $\sigma : X \to F$ is necessarily injective*. For suppose that $\sigma(x_1) = \sigma(x_2)$ and $x_1 \neq x_2$. Let G be any group with two or more elements and choose any function $\alpha : X \to G$ such that $\alpha(x_1) \neq \alpha(x_2)$. Then $\beta\sigma(x_1) = \beta\sigma(x_2)$ and so $\alpha(x_1) = \alpha(x_2)$, a contradiction. Thus we could replace X by the set $\operatorname{Im}(\alpha)$, which has the same cardinality, and take X to be a subset of F with σ the inclusion map. What the mapping property then asserts is that every mapping from X to a group G can be extended to a unique homomorphism from F to G.

At first sight the definition of a free group may seem abstract, but patience is called for. Soon a concrete description of free groups will be given. In the meantime our first order of business must be to show that free groups actually exist, a fact that is not obvious from the definition.

(12.3.1) *Let X be any non-empty set. Then there exists a group F and a function $\sigma : X \to F$ such that (F, σ) is free on X and F is generated by $\operatorname{Im}(\sigma)$.*

Proof. Roughly speaking, the idea of the proof is to construct F by forming "words" in X which are combined in a formal manner by juxtaposition, while at the same time allowing for cancellation of segments like xx^{-1} or $x^{-1}x$ where $x \in X$.

12.3 Generators and relations for groups

The first step is to choose a set disjoint from X with the same cardinality. Since the purpose of this move is to accommodate inverses of elements of X, it is appropriate to take the set to be $X^{-1} = \{x^{-1} \mid x \in X\}$. But keep in mind that x^{-1} is merely a symbol at this point and does not denote an inverse. By *a word in* X is meant any finite sequence w of elements of the set $X \cup X^{-1}$, written for convenience in the form

$$w = x_1^{q_1} x_2^{q_2} \ldots x_r^{q_r},$$

where $q_i = \pm 1$, $x_i^1 = x_i \in X$ and $r \geq 0$. The case $r = 0$, when the sequence is empty, is the *empty word*, which is written 1. Two words are to be considered *equal* if they have the same entries in each position, i.e., they look exactly alike.

The *product* of two words $w = x_1^{q_1} \ldots x_r^{q_r}$ and $v = y_1^{p_1} \ldots y_s^{p_s}$ is formed in the obvious way by juxtaposition, i.e.,

$$wv = x_1^{q_1} \ldots x_r^{q_r} y_1^{p_1} \ldots y_s^{p_s},$$

with the convention that $w1 = w = 1w$. This is clearly an associative binary operation on the set S of all words in X. The *inverse* of the word w is defined to be

$$w^{-1} = x_r^{-q_r} \ldots x_1^{-q_1},$$

with the convention that $1^{-1} = 1$. So far S, together with the product operation, is a semigroup with an identity element, i.e., a monoid. We now introduce a device which permits the cancellation of segments of a word with the form xx^{-1} or $x^{-1}x$. Once this is done, instead of a monoid, we will have a group.

A relation $\sim$ on the set S is defined in the following way: $w \sim v$ means that it is possible to go from w to v by means of a finite sequence of operations of the following types:

(a) insertion of xx^{-1} or $x^{-1}x$ as consecutive symbols of a word where $x \in X$;

(b) deletion of any such sequences from a word.

For example, $xyy^{-1}z \sim t^{-1}txz$ where $x, y, z, t \in X$. It is easy to check that the relation $\sim$ is an equivalence relation on S. Let F denote the set of all equivalence classes $[w]$ where $w \in S$. Our aim is to make F into a group, which will turn out to be a free group on the set X.

If $w \sim w'$ and $v \sim v'$, then it is readily seen that $wv \sim w'v'$. It is therefore meaningful to define the *product* of the equivalence classes $[w]$ and $[v]$ by the rule

$$[w][v] = [wv].$$

It follows from this that $[w][1] = [w] = [1][w]$ for all words w. Also $[w][w^{-1}] = [ww^{-1}] = [1]$ since ww^{-1} is plainly equivalent to 1. Finally, we verify the associative law:

$$([u][v])[w] = [uv][w] = [(uv)w] = [u(vw)] = [u][vw] = [u]([v][w]).$$

Consequently F is a group in which $[1]$ is the identity element and $[w^{-1}]$ is the inverse of $[w]$. Furthermore F is generated by the subset $\overline{X} = \{[x] \mid x \in X\}$; for if $w = x_1^{q_1} x_2^{q_2} \ldots x_r^{q_r}$, then

$$[w] = [x_1]^{q_1} [x_2]^{q_2} \ldots [x_r]^{q_r} \in \langle \overline{X} \rangle.$$

It remains to prove that in F we have a free group on the set X. For this purpose we define a function $\sigma : X \to F$ by the rule $\sigma(x) = [x]$; thus $\mathrm{Im}(\sigma) = \overline{X}$ and this subset generates F. Next let $\alpha : X \to G$ be a map from X into some other group G. To show that (F, σ) is free on X we need to produce a homomorphism $\beta : F \to G$ such that $\beta \sigma = \alpha$. There is only one reasonable candidate here: define β by the rule

$$\beta \left(\left[x_1^{q_1} \ldots x_r^{q_r} \right] \right) = \alpha(x_1)^{q_1} \ldots \alpha(x_r)^{q_r}.$$

The first thing to note is that β is well-defined: for any other element in the equivalence class $\left[x_1^{q_1} \ldots x_r^{q_r} \right]$ differs from $x_1^{q_1} \ldots x_r^{q_r}$ only by segments of the form xx^{-1} or $x^{-1}x$, $(x \in X)$, and these will contribute to the image under β merely $\alpha(x)\alpha(x)^{-1}$ or $\alpha(x)^{-1}\alpha(x)$, i.e., the identity. It is a simple direct check that β is a homomorphism. Notice also that $\beta\sigma(x) = \beta([x]) = \alpha(x)$, so that $\beta\sigma = \alpha$.

Finally, we have to establish the uniqueness of β. If $\beta' : F \to G$ is another homomorphism for which $\beta'\sigma = \alpha$, then $\beta\sigma = \beta'\sigma$ and thus β and β' agree on $\mathrm{Im}(\sigma)$. But $\mathrm{Im}(\sigma)$ generates the group F, from which it follows that $\beta = \beta'$. Therefore (F, σ) is free on X. □

Reduced words. Now that we know free groups exist, we would like to find a convenient form for their elements. Let F be the free group on the set X just constructed. A word in X is called *reduced* if it contains no pairs of consecutive symbols xx^{-1} or $x^{-1}x$ where $x \in X$. The empty word is considered to be reduced. Now if w is any word, we can delete subsequences xx^{-1} and $x^{-1}x$ from w until a reduced word is obtained. Thus each equivalence class $[w]$ contains at least one reduced word. However the really important point to establish is that there is a unique reduced word in each equivalence class.

(12.3.2) *Each equivalence class of words on X contains a unique reduced word.*

Proof. There are likely to be many different ways to cancel segments xx^{-1} or $x^{-1}x$ from a word. For this reason a direct approach to proving uniqueness would be complicated. An indirect argument will be used which avoids this difficulty.

Let R denote the set of all reduced words in X. The idea of the proof is to introduce a permutation representation of the free group F on the set R. Let $u \in X \cup X^{-1}$: then a function $u' : R \to R$ is determined by the rule

$$u'(x_1^{q_r} \ldots x_r^{q_r}) = \begin{cases} u x_1^{q_1} \ldots x_r^{q_r} & \text{if } u \neq x_1^{-q_1} \\ x_2^{q_2} \ldots x_r^{q_r} & \text{if } u = x_1^{-q_1} \end{cases}$$

Here of course $x_1^{q_1} \ldots x_r^{q_r}$ is a reduced word, and one has to observe that after applying the function u' the word is still reduced. Next u' is a permutation of R since its inverse is the function $(u^{-1})'$. Now let $\alpha : X \to \mathrm{Sym}(R)$ be defined by the assignment $u \mapsto u'$. By the mapping property of the free group F there is a homomorphism $\beta : F \to \mathrm{Sym}(R)$ such that $\beta\sigma = \alpha$: notice that this implies $\beta([x]) = x'$.

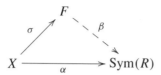

Now suppose that v and w are two equivalent reduced words; we will prove that $v = w$. Certainly $[v] = [w]$, so $\beta([v]) = \beta([w])$. If $v = x_1^{q_1} \ldots x_r^{q_r}$, then $[v] = [x_1^{q_1}] \ldots [x_r^{q_r}]$ and we have

$$\beta([v]) = \beta([x_1])^{q_1} \ldots \beta([x_r])^{q_r} = \beta\sigma(x_1)^{q_1} \ldots \beta\sigma(x_r)^{q_r}.$$

Hence

$$\beta([v]) = \alpha(x_1)^{q_1} \ldots \alpha(x_r)^{q_r} = (x_1^{q_1})' \ldots (x_r^{q_r})'.$$

Applying the function $\beta([v])$ to the empty word 1, which is reduced, we obtain $x_1^{q_1} \ldots x_r^{q_r} = v$ since this word is reduced. Similarly $\beta([w])$ sends the empty word to w. Therefore $v = w$. □

Normal form. The argument used in the proof of (12.3.2) is a subtle one, which it is well worth rereading. But the main point is to appreciate the significance of (12.3.2): it provides us with a unique way of representing the elements of the constructed free group F on the set X. Each element of F has the form $[w]$ where w is a uniquely determined reduced word, say $w = x_1^{q_1} \ldots x_r^{q_r}$ where of course $q_i = \pm 1$, $r \geq 0$. No consecutive terms xx^{-1} or $x^{-1}x$ occur in w. Now $[w] = [x_1]^{q_1} \ldots [x_r]^{q_r}$; on combining consecutive terms of this product which involve the same x_i, we conclude that the element $[w]$ may be uniquely written in the form

$$[w] = [x_1]^{\ell_1} [x_2]^{\ell_2} \ldots [x_s]^{\ell_s},$$

where $s \geq 0$, ℓ_i is a non-zero integer and $x_i \neq x_{i+1}$: strictly speaking we have relabelled the x_i's here. To simplify the notation let us drop the distinction between x and $[x]$, so that now $X \subseteq F$. Then every element w of F has the unique form

$$w = x_1^{\ell_1} x_2^{\ell_2} \ldots x_s^{\ell_s},$$

where $s \geq 0$, $\ell_i \neq 0$ and $x_i \neq x_{i+1}$. This is called the *normal form* of w.

For example, if $X = \{x\}$, each element of F has the unique normal form x^ℓ, where $\ell \in \mathbb{Z}$. Thus $F = \langle x \rangle$ is an infinite cyclic group.

In fact the existence of a normal form is characteristic of free groups in the following sense.

(12.3.3) *Let X be a subset of a group G and suppose that each element g of G can be uniquely written in the form $g = x_1^{\ell_1} x_2^{\ell_2} \ldots x_s^{\ell_s}$ where $x_i \in X$, $s \geq 0$, $\ell_i \neq 0$, and $x_i \neq x_{i+1}$. Then G is free on X.*

Proof. Let F be the free group of equivalence classes of words in the set X and let $\sigma : X \to F$ be the associated injection; thus $\sigma(x) = [x]$. Apply the mapping property with the inclusion map $\alpha : X \to G$, i.e., $\alpha(x) = x$ for all $x \in X$. Then there is a homomorphism $\beta : F \to G$ such that $\beta\sigma = \alpha$. Since $X = \text{Im}(\alpha)$ generates G, it follows that $\text{Im}(\beta) = G$ and β is surjective. Finally, the uniqueness of the normal form guarantees that β is injective. For if $\beta([x_1]^{\ell_1} \ldots [x_r]^{\ell_r}) = 1$ with $r > 0$, $x_i \neq x_{i+1}$, $\ell_i \neq 0$, then $(\beta\sigma(x_1))^{\ell_1} \ldots (\beta\sigma(x_r))^{\ell_r} = 1$, and hence $x_1^{\ell_1} \ldots x_r^{\ell_r} = 1$, a contradiction. Therefore β is an isomorphism and $F \simeq G$, so that G is free on X. □

So far we have worked with a particular free group on a set X, the group constructed from equivalence classes of words in X. But in fact all free groups on the same set are isomorphic, a fact which allows us to deal only with free groups of words. This follows from the next result.

(12.3.4) *Let F_1 be free on X_1 and F_2 free on X_2 where $|X_1| = |X_2|$. Then $F_1 \simeq F_2$.*

Proof. Let $\sigma_1 : X_1 \to F_1$ and $\sigma_2 : X_2 \to F_2$ be the respective injections for the free groups F_1 and F_2, and let $\alpha : X_1 \to X_2$ be a bijection, which exists since $|X_1| = |X_2|$. Then by the mapping property there are commutative diagrams

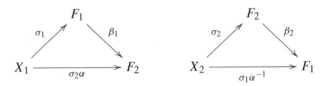

in which β_1 and β_2 are homomorphisms. Thus $\beta_1 \sigma_1 = \sigma_2 \alpha$ and $\beta_2 \sigma_2 = \sigma_1 \alpha^{-1}$. Hence $\beta_2 \beta_1 \sigma_1 = \beta_2 \sigma_2 \alpha = \sigma_1 \alpha^{-1} \alpha = \sigma_1$ and consequently the diagram

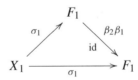

commutes. But the identity map on F_1 will also make this diagram commute and therefore $\beta_2 \beta_1$ must equal the identity map by the uniqueness requirement. In a similar fashion we see that $\beta_1 \beta_2$ is the identity on F_2, so that $\beta_1 : F_1 \to F_2$ is an isomorphism. □

Examples of free groups. At this point the reader may still regard free groups as mysterious objects, despite our success in constructing them. It is time to remedy this by exhibiting some real life examples.

Example (12.3.1) Consider the functions α and β on the set $\mathbb{C} \cup \{\infty\}$ defined by the rules $\alpha(x) = x + 2$ and $\beta(x) = \frac{x}{2x+1}$. Here the symbol ∞ is required to satisfy the formal rules $\frac{1}{\infty} = 0$, $\frac{1}{0} = \infty$, $\frac{\infty}{\infty} = 1$, $2\infty = \infty$, $\infty + 1 = \infty$. Thus $\alpha(\infty) = \infty$ and $\beta(\infty) = \frac{1}{2}$. The first thing to notice is that α and β are bijections since they have inverses given by $\alpha^{-1}(x) = x - 2$ and $\beta^{-1}(x) = \frac{x}{1-2x}$. Of course this can be checked by computing the composites $\alpha\alpha^{-1}$, $\alpha^{-1}\alpha$, $\beta\beta^{-1}$, $\beta^{-1}\beta$. Define F to be the group $\langle \alpha, \beta \rangle$, which is a subgroup of the symmetric group on the set $\mathbb{C} \cup \{\infty\}$. We are going to prove that F is a free group on $\{\alpha, \beta\}$. To accomplish this it is enough to show that no non-trivial reduced word in α, β can equal 1: for then each element of F has a unique normal form and (12.3.3) can be applied.

Since direct calculations with the functions α and β would be tedious, we adopt a geometric approach. Observe that each non-trivial power of α maps the interior of the unit circle in the complex plane to its exterior. Also a non-trivial power of β maps the exterior of the unit circle to its interior with $(0, 0)$ removed: the truth of the last statement is best seen from the equation $\beta\left(\frac{1}{x}\right) = \frac{1}{x+2}$. It follows from this observation that no mapping of the form $\alpha^{\ell_1}\beta^{m_1} \ldots \alpha^{\ell_r}\beta^{m_r}$ can be trivial unless all the l_i and m_i are 0.

Example (12.3.2) An even more concrete example of a free group is provided by the matrices

$$A = \begin{pmatrix} 1 & 2 \\ 0 & 1 \end{pmatrix} \quad \text{and} \quad B = \begin{pmatrix} 1 & 0 \\ 2 & 1 \end{pmatrix}$$

These generate a subgroup F_1 of $\text{GL}_2(\mathbb{Z})$ which is free on $\{A, B\}$.

To see why this is true, first consider a matrix

$$U = \begin{bmatrix} a & b \\ c & d \end{bmatrix} \in \text{GL}_2(\mathbb{C}).$$

There is a corresponding permutation $\theta(U)$ of $\mathbb{C} \cup \{\infty\}$ defined by

$$\theta(U) : x \mapsto \frac{ax+b}{cx+d}.$$

This is called a *linear fractional transformation*. It is easy to verify that $\theta(UV) = \theta(U)\theta(V)$, so that $\theta : \text{GL}_2(\mathbb{C}) \to \text{Sym}(\mathbb{C}\cup\{\infty\})$ is a homomorphism. Thus the linear fractional transformations form a subgroup $\text{Im}(\theta)$. Now $\theta(A) = \alpha$ and $\theta(B) = \beta$. Hence, if some non-trivial reduced word in A and B were to equal the identity matrix, the corresponding word in α and β would equal the identity permutation, which is impossible by Example (12.3.1). Therefore F_1 is free on $\{A, B\}$ by (12.3.3).

Next we will use normal form to obtain some structural information about free groups.

(12.3.5) *Let F be a free group on a set X. Then*

(i) *each non-trivial element of F has infinite order;*

(ii) *if F is not infinite cyclic, i.e. $|X| > 1$, then $Z(F) = 1$.*

Proof. (i) Let $1 \neq f \in F$ and suppose that $f = x_1^{\ell_1} x_2^{\ell_2} \ldots x_s^{\ell_s}$ is the normal form. Now if $x_1 = x_s$, we can replace f by the conjugate $x_s^{\ell_s} f x_s^{-\ell_s} = x_1^{\ell_1+\ell_s} x_2^{\ell_2} \ldots x_{s-1}^{\ell_{s-1}}$, which has the same order as f. For this reason there is nothing to be lost in assuming that $x_1 \neq x_s$. Let m be a positive integer; then

$$f^m = (x_1^{\ell_1} \ldots x_s^{\ell_s})(x_1^{\ell_1} \ldots x_s^{\ell_s}) \ldots (x_1^{\ell_1} \ldots x_s^{\ell_s}),$$

with m factors, which is in normal form since $x_1 \neq x_s$. It follows that $f^m \neq 1$ and hence f has infinite order.

(ii) Assume that $1 \neq f \in Z(F)$ and let $f = x_1^{\ell_1} x_2^{\ell_2} \ldots x_s^{\ell_s}$ be the normal form of f. By conjugating f as in (i), we may assume that $x_1 \neq x_s$. Then $fx_1 = x_1^{\ell_1} x_2^{\ell_2} \ldots x_s^{\ell_s} x_1$ and $x_1 f = x_1^{\ell_1+1} x_2^{\ell_2} \ldots x_s^{\ell_s}$ are both in normal form, except that $x_1^{\ell_1+1}$ could be trivial; but in any event $fx_1 \neq x_1 f$ and so $f \notin Z(G)$. □

Generators and relations. The next result shows why free groups are worth studying: they occupy a key position in group theory since their quotients account for all groups.

(12.3.6) *Let G be a group and X a set of generators for G. If F is a free group on X, then there is a surjective homomorphism $\theta : F \to G$ and hence $G \simeq F/\operatorname{Ker}(\theta)$. Thus every group is isomorphic with a quotient of a free group.*

Proof. Let (F, σ) be free on X. The existence of the homomorphism θ follows on applying the mapping property of the free group F to the diagram

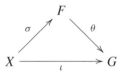

where ι is the inclusion map. Then $x = \iota(x) = \theta\sigma(x) \in \operatorname{Im}(\theta)$ for all x in X. Hence $G = \operatorname{Im}(\theta)$. □

We are finally ready to define a group with a given set of generators and defining relations. Let X be a non-empty set and let F be the free group on X with $X \subseteq F$. Suppose that R is a subset of F and define N to be the *normal closure* of R in F, i.e., N is the subgroup generated by all conjugates in F of elements of R. Let

$$G = F/N.$$

12.3 Generators and relations for groups

Certainly the group G has as generators the elements xN where $x \in X$; also $r(xN) = r(x)N = N = 1_G$ for all $r \in R$. Hence the relations $r = 1$ hold in G. Here of course $r(xN)$ is the element of G obtained from r by replacing each x by xN. Then G is called *the group with generators X and defining relations $r = 1$, where $r \in R$*: in symbols

$$G = \langle X \mid r = 1, r \in R \rangle.$$

Elements of R are called *defining relators* and the group may also be written

$$G = \langle X \mid R \rangle.$$

The pair $\langle X, R \rangle$ is called a *presentation* of G. An element w in the normal subgroup N is a *relator*; it is expressible as a product of conjugates of defining relators and their inverses. The relator w is said to be a *consequence* of the defining relators in R. Finally, a presentation $\langle X, R \rangle$ is called *finite* if X and R are both finite.

Our first concern is to prove that every group can be defined by a presentation, which is the next result.

(12.3.7) *Every group has a presentation.*

Proof. Start with any group G and choose a set X of generators for it, for example $X = G$ will do. Let F be a free group on X. Then by (12.3.6) there is a surjective homomorphism $\theta : F \to G$ and $G \simeq F/\mathrm{Ker}(\theta)$. Next choose a subset R of $\mathrm{Ker}(\theta)$ whose normal closure in F is $\mathrm{Ker}(\theta)$ – for example we could take R to be $\mathrm{Ker}(\theta)$. Then we have $G \simeq F/\mathrm{Ker}(\theta) = \langle X \mid R \rangle$, which is a presentation of G. □

In the proof just given there are many possible choices for X and R, so a group has many presentations. This is one reason why it is notoriously difficult to extract information about the structure of a group from a given presentation. Another and deeper reason for this difficulty arises from the *insolvability of the word problem*. Roughly speaking this asserts that it is impossible to write a computer program which can decide if a word in the generators of a group which is given by a finite presentation equals the identity element. (See [13] for a very readable account of the word problem). One consequence of this failure is that we will have to exploit special features of a group presentation if we hope to derive information about the group structure from it.

Despite the difficulties inherent in working with presentations of groups, there is one very useful tool available to us.

(12.3.8) (Von Dyck's[1] Theorem) *Let G and H be groups with presentations $\langle X \mid R \rangle$ and $\langle Y \mid S \rangle$ respectively. Assume that $\alpha : X \to Y$ is a surjective map such that $\alpha(x_1)^{\ell_1} \alpha(x_2)^{\ell_2} \ldots \alpha(x_k)^{\ell_k}$ is a relation of H whenever $x_1^{\ell_1} x_2^{\ell_2} \ldots x_k^{\ell_k} \in R$. Then there is a surjective homomorphism $\theta : G \to H$ such that $\theta|_X = \alpha$.*

---
[1] Walter von Dyck (1856–1934).

Proof. Let F be a free group on X; then $G = F/N$ where N is the normal closure of R in F. By the mapping property of free groups there is a homomorphism $\theta_0 : F \to H$ such that $\theta_0|_X = \alpha$. Now $\theta_0(r) = 1$ for all $r \in R$ and thus $\theta_0(a) = 1$ for all a in N, the normal closure of R in F. Hence θ_0 induces a homomorphism $\theta : G \to H$ such that $\theta(fN) = \theta_0(f)$. Finally $Y \subseteq \text{Im}(\theta_0)$ since α is surjective, so θ_0, and hence θ, is surjective. □

We will shortly show how Von Dyck's Theorem can be used to obtain information about groups given by a presentation, but first it will be used to establish:

(12.3.8) *Every finite group has a finite presentation.*

Proof. Let $G = \{g_1, g_2, \ldots, g_n\}$, where $g_1 = 1$, be a finite group of order n. Then $g_i g_j = g_{v(i,j)}$ and $g_i^{-1} = g_{u(i)}$ where $u(i), v(i, j) \in \{1, 2, \ldots, n\}$. Now let $\bar{G}$ be the group with generators $\bar{g}_1, \bar{g}_2, \ldots, \bar{g}_n$ and defining relations $\bar{g}_i \bar{g}_j = \bar{g}_{v(i,j)}$, $\bar{g}_i^{-1} = \bar{g}_{u(i)}$, where $i, j = 1, 2, \ldots, n$. Plainly $\bar{G}$ has a finite presentation. Apply Von Dyck's Theorem to $\bar{G}$ and G where α is the assignment $\bar{g}_i \mapsto g_i$, noting that each defining relator of $\bar{G}$ is a relator of G. It follows that there is a surjective homomorphism $\theta : \bar{G} \to G$ such that $\theta(\bar{g}_i) = g_i$. Now every element $\bar{g}$ of $\bar{G}$ is expressible as a product of $\bar{g}_i$'s and their inverses; therefore, from the defining relators, we see that $\bar{g}$ equals some $\bar{g}_k$. It follows that $|\bar{G}| \leq n$. But $G \simeq \bar{G}/\text{Ker}(\theta)$ and hence $n = |G| \leq |\bar{G}| \leq n$. Thus $|G| = |\bar{G}|$, so that $\text{Ker}(\theta) = 1$ and $G \simeq \bar{G}$. □

Next we consider some explicit examples of groups given by a finite presentation.

Example (12.3.3) Let $G = \langle x \mid x^n \rangle$ where n is a positive integer.

The free group F on $\{x\}$ is generated by x: thus F is infinite cyclic and $G = F/\langle x^n \rangle \simeq \mathbb{Z}_n$, a cyclic group of order n, as one would expect.

Example (12.3.4) Let $G = \langle x, y \mid xy = yx, x^2 = 1 = y^2 \rangle$.

Since $xy = yx$, the group G is abelian; also every element of G has the form $x^i y^j$ where $i, j \in \{0, 1\}$ because $x^2 = 1 = y^2$; hence $|G| \leq 4$. On the other hand, the Klein 4-group V is generated by the permutations $a = (12)(34)$ and $b = (13)(24)$, and furthermore $ab = ba$ and $a^2 = 1 = b^2$. Hence Von Dyck's Theorem can be applied to yield a surjective homomorphism $\theta : G \to V$ such that $\theta(x) = a$ and $\theta(y) = b$. Thus $G/\text{Ker}(\theta) \simeq V$. Since $|G| \leq 4 = |V|$, it follows that $\text{Ker}(\theta) = 1$ and θ is an isomorphism. Therefore G is a Klein 4-group.

A more challenging example of a presentation is

Example (12.3.5) Let $G = \langle x, y \mid x^2 = y^3 = (xy)^2 = 1 \rangle$.

Our first move is to find an upper bound for $|G|$. Let $H = \langle y \rangle$; this is a subgroup of order 1 or 3. Write $\mathcal{S} = \{H, xH\}$; we will argue that $\mathcal{S}$ is the set of *all* left cosets of H in G. To establish this it is sufficient to show that $x\mathcal{S} = \mathcal{S} = y\mathcal{S}$ since it will then follow that $\mathcal{S}$ contains every left coset of H. Certainly $x\mathcal{S} = \mathcal{S}$ since $x^2 = 1$.

Next $xyxy = (xy)^2 = 1$ and hence $yx = xy^2$ since $y^{-1} = y^2$. It follows that $yxH = xy^2H = xH$ and thus $y8 = 8$. Since $|H| \leq 3$ and $|G : H| = |8| \leq 2$, we deduce that $|G| \leq 6$.

Next observe that the symmetric group S_3 is generated by the permutations $a = (12)(3)$ and $b = (123)$, and that $a^2 = b^3 = (ab)^2 = 1$ since $ab = (1)(23)$. By Von Dyck's theorem there is a surjective homomorphism $\theta : G \to S_3$. Since $|G| \leq 6$, it follows that θ is an isomorphism and $G \simeq S_3$.

The method of the last two examples can be useful when a finite group is given by a presentation. The procedure is to choose a subgroup for whose order one has an upper bound, and then by coset enumeration to find an upper bound for the index. This gives an upper bound for the order of the group. The challenge is then to identify the group by comparing it with a known group for which the defining relations hold.

Exercises (12.3)

1. Let F be a free group on a set X. Prove that an element f of F belongs to the derived subgroup F' if and only if the sum of the exponents of x in f is 0 for every x in X.

2. If F is a free group, then F/F' is a direct product of infinite cyclic groups.

3. Let G be the subgroup of $GL_2(\mathbb{C})$ generated by $\begin{bmatrix} 1 & a \\ 0 & 1 \end{bmatrix}$ and $\begin{bmatrix} 1 & 0 \\ a & 1 \end{bmatrix}$ where $|a| \geq 2$. Prove that G is a free group.

4. (*The projective property of free groups*). Let there be given groups and homomorphisms $\alpha : F \to H$ and $\beta : G \to H$ where F is a free group and β is surjective. Show that there is a homomorphism $\gamma : F \to G$ such that $\beta\gamma = \alpha$, i.e., the triangle

commutes.

5. Let G be a group with a normal subgroup N such that G/N is a free group. Prove that there is free subgroup H such that $G = HN$ and $H \cap N = 1$.

6. Let H be a subgroup with finite index in a free group F. If $1 \neq K \leq F$, prove that $H \cap K \neq 1$.

In the next three exercises identify the groups with the given presentations.

7. $\langle x, y \mid x^2 = 1 = y^4, xy = yx \rangle$.

8. $\langle x, y \mid x^3 = (xy)^2 = y^3 = 1 \rangle$.

9. $\langle x, y \mid x^2 = (xy)^2 = y^5 = 1 \rangle$.

10. Let G be a group which has a presentation with n generators and r defining relators. Prove that if $r < n$, then G is infinite. [Hint: consider the abelian group G/G' and use Exercise (12.3.2)].

11. Let F_1 and F_2 be free groups on finite sets X_1 and X_2 respectively. If $F_1 \simeq F_2$, prove that $|X_1| = |X_2|$. (In fact this result is true for free groups on infinite sets. Thus a free group is determined up to isomorphism by the cardinality of the set on which it is free). [Hint: consider F_i/F_i^2 as a vector space over GF(2)].

12.4 An introduction to error correcting codes

This is the age of information technology in which every day vast amounts of data are transmitted electronically over great distances. The data are generally in the form of bit strings, i.e., sequences of 0's and 1's. Inevitably errors occur from time to time during the process of transmission, so that the message received may differ from the one transmitted. An *error correcting code* allows the detection and correction of erroneous messages. The essential idea here is that the transmitted codewords should not be too close to one another, i.e., they should not agree in too many entries. For this makes it more likely that an error can be detected and the original message recovered. Over the last fifty years an entire mathematical theory of error-correcting codes has evolved.

Fundamental concepts. Let Q be a finite set with q elements; this is called the *alphabet*. A *word w of length n over Q* is just an n-tuple of elements of Q, written for convenience

$$w = (w_1 w_2 \ldots w_n), \quad w_i \in Q.$$

The set of all words of length n over Q is called *n-dimensional Hamming[2] space* and is denoted by

$$H_n(q).$$

This is the set of all conceivable messages of length n that could be sent. Notice that $|H_n(q)| = q^n$. If Q is a finite field, Hamming space is an n-dimensional vector space over Q. In practice Q is usually the field of 2 elements, when Hamming n-space is the set of all bit strings of length n.

It will be important to have a measure of how far apart two words are: the natural measure to use is the number of entries in which the words differ. If v and w belong to $H_n(q)$, the *distance* between v and w is defined to be

$$d(v, w) = |\{i \mid v_i \neq w_i\}|,$$

---
[2] Richard Wesley Hamming (1915–1998).

i.e., the number of positions where v and w have different entries. The *weight* of a word v is its distance from the zero word,

$$wt(v) = d(v, 0).$$

So $wt(v)$ is just the number of non-zero entries of v. Clearly, $d(u, v)$ is the smallest number of errors that could have been made if the word u is transmitted and it is received wrongly as v.

The basic properties of the distance function are given in the following result.

(12.4.1) *Let* $u, v, w \in H_n(q)$. *Then:*

(i) $d(v, w) \geq 0$ *and* $d(v, w) = 0$ *if only if* $v = w$;

(ii) $d(v, w) = d(w, v)$;

(iii) $d(u, w) \leq d(u, v) + d(v, w)$.

These three properties assert that the function $d : H_n(q) \times H_n(q) \to \mathbb{N}$ *is a metric on the Hamming space* $H_n(q)$.

Proof of (12.4.1). Statements (i) and (ii) are obviously true. To prove (iii) note that u can be changed to v by $d(u, v)$ entry changes and v can then be changed to w by $d(v, w)$ changes. Thus u can be changed to w by $d(u, v) + d(v, w)$ entry changes. Therefore $d(u, w) \leq d(u, v) + d(v, w)$. □

Codes. A *code of length* n over an alphabet Q with q elements, or more briefly a q-*ary code of length* n, is simply a subset C of $H_n(Q)$ which has at least two elements. The elements of C are called *codewords*: it is codewords which are transmitted in an actual message.

A code C is said to be e-*error detecting* if $c_1, c_2 \in C$ and $d(c_1, c_2) \leq e$ always imply that $c_1 = c_2$. So the distance between distinct codewords is always greater than e. Equivalently, a codeword cannot be transmitted and received as a different codeword if e or fewer errors have occurred. In this sense the code C is able to detect up to e errors.

Next a q-ary code of length n is called e-*error correcting* if for every w in $H_n(q)$ there is at most one codeword c such that $d(w, c) \leq e$. This means that if a codeword c is received as a different word w and at most e errors have occurred, it is possible to recover the original codeword by looking at all words v in $H_n(q)$ such that $d(w, v) \leq e$: exactly one of these is a codeword and it must be the transmitted codeword c. Clearly a code which is e-error correcting is e-error detecting.

An important parameter of a code is the minimum distance between distinct codewords; this is called the *minimum distance* of the code. The following result is basic.

(12.4.2) *Let C be a code with minimum distance d. Then:*

(i) *C is e-error detecting if and only if $d \geq e + 1$;*

(ii) *C is e-error correcting if and only if $d \geq 2e + 1$.*

Proof. (i) Suppose that $d \geq e + 1$. If c_1, c_2 are distinct codewords, then $d(c_1, c_2) \geq d \geq e+1$. Hence C is e-error detecting. Conversely, assume that $d \leq e$. By definition of d there exist $c_1 \neq c_2$ in C such that $d(c_1, c_2) = d \leq e$, so that C is not e-error detecting.

(ii) Assume that C is not e-error correcting, so that there is a word w and codewords $c_1 \neq c_2$ such that $d(c_1, w) \leq e$ and $d(w, c_2) \leq e$. Then

$$d \leq d(c_1, c_2) \leq d(c_1, w) + d(w, c_2) \leq 2e$$

by (12.4.1). Hence $d < 2e + 1$.

Conversely suppose that $d < 2e + 1$ and let c_1 and c_2 be codewords at distance d apart. Put $f = \left[\frac{1}{2}d\right]$, i.e., the greatest integer $\leq \frac{1}{2}d$; thus $f \leq \frac{1}{2}d \leq e$. We claim that $d - f \leq e$. This is true when d is even since $d - f = d - \frac{1}{2}d = \frac{1}{2}d \leq e$. If d is odd, then $f = \frac{d-1}{2}$ and $d - f = \frac{d+1}{2} < e + 1$; therefore $d - f \leq e$. Next we can pass from c_1 to c_2 by changing exactly d entries. Let w be the word obtained from c_1 after the first f entry changes. Then $d(c_1, w) = f \leq e$, while $d(c_2, w) = d - f \leq e$. Therefore C is not e-error correcting. □

Corollary (12.4.3) *If a code has minimum distance d, its maximum error detection capacity is $d - 1$ and its maximum error correction capacity is $\left[\frac{d-1}{2}\right]$.*

Example (12.4.1) Consider the binary code C of length 5 with the three codewords

$$c_1 = (1\,0\,0\,1\,0), \quad c_2 = (0\,1\,1\,0\,0), \quad c_3 = (1\,0\,1\,0\,1).$$

Clearly the minimum distance of C is 3. Hence C is 2-error detecting and 1-error correcting. For example, suppose that c_2 is transmitted and is received as $w = (1\,1\,0\,0\,0)$, so two entry errors have occurred. The error can be detected since $w \notin C$. But C is not 2-error detecting since if $v = (1\,1\,1\,0\,0)$, then $d(c_2, v) = 1$ and $d(c_3, v) = 2$. Thus if v is received and up to two errors occurred, we cannot tell whether c_2 or c_3 was the transmitted codeword.

Bounds for the size of a code. It is evident from (12.4.2) that for a code to have good error correcting capability it must have large minimum distance. But the price to be paid for this is that fewer codewords are available. An interesting question is: what is the maximum size of a q-ary code with given length n and minimum distance d, or equivalently with maximum error correcting capacity e? We will begin with a lower bound, which guarantees the existence of a code of a certain size.

12.4 An introduction to error correcting codes

(12.4.4) (The Varshamov–Gilbert lower bound) *Let n, q, d be positive integers with $d \leq n$. Then there is a q-ary code C of length n and minimum distance d in which the number of codewords is at least*

$$\frac{q^n}{\sum_{i=0}^{d-1} \binom{n}{i}(q-1)^i}.$$

Before embarking on the proof we introduce the important concept of the *r-ball with center w*,

$$B_r(w).$$

This is the set of all words in $H_n(q)$ at distance r or less from w. Thus a code C is *e-error correcting* if and only if the e-balls $B_e(c)$ with c in C are pairwise disjoint.

Proof of (12.4.4). The first step is to establish a basic formula for the size of an r-ball,

$$|B_r(w)| = \sum_{i=0}^{r} \binom{n}{i}(q-1)^i.$$

To construct a word in $B_r(w)$ we alter at most r entries of w. Choose the i entries to be altered in $\binom{n}{i}$ ways and then replace them by different elements of Q in $(q-1)^i$ ways. This gives a count of $\binom{n}{i}(q-1)^i$ words at distance i from w; the formula now follows at once.

To start the construction choose any q-ary code C_0 of length n with minimum distance d; for example, C_0 might consist of the zero word and a single word of weight d. If the union of the $B_{d-1}(c)$ with $c \in C_0$ is not $H_n(q)$, there is a word w whose distance from every word in C_0 is at least d. Let $C_1 = C_0 \cup \{w\}$; this is a larger code than C_0 which has the same minimum distance d. Repeat the procedure for C_1 and then as often as possible. Eventually a code C with minimum distance d will be obtained which cannot be enlarged; when this occurs, we will have

$$H_n(q) = \bigcup_{c \in C} B_{d-1}(c).$$

Therefore

$$q^n = |H_n(q)| = \left| \bigcup_{c \in C} B_{d-1}(c) \right| \leq |C| \cdot |B_{d-1}(c)|$$

for any c in C. Hence $|C| \geq q^n / |B_{d-1}(c)|$ and the bound has been established. □

Next we give an upper bound for the size of an e-error correcting code.

(12.4.5) (The Hamming upper bound) *Let C be any q-ary code of length n which is e-error correcting. Then*

$$|C| \leq \frac{q^n}{\sum_{i=0}^{e} \binom{n}{i}(q-1)^i}.$$

Proof. Since C is e-error correcting, the e-balls $B_e(c)$ for $c \in C$ are pairwise disjoint. Hence

$$\left| \bigcup_{c \in C} B_e(c) \right| = |C| \cdot |B_e(c)| \leq |H_n(q)| = q^n,$$

for any c in C. Therefore $|C| \leq q^n/|B_e(c)|$, as required. □

A q-ary code C for which the Hamming upper bound is attained is called a *perfect code*. In this case

$$|C| = \frac{q^n}{\sum_{i=0}^{e} \binom{n}{i}(q-1)^i},$$

and clearly this happens precisely when $H_n(q)$ is the union of the disjoint balls $B_e(c)$, $c \in C$, i.e., every word lies at distance $\leq e$ from exactly one codeword. Perfect codes are desirable since the number of codewords is maximized for the given error correcting capacity; however they are relatively rare.

Example (12.4.2) (The binary repetition code) A very simple example of a perfect code is the binary code C of length $2e+1$ with just two codewords,

$$c_0 = (0, 0, \ldots, 0) \quad \text{and} \quad c_1 = (1, 1, \ldots, 1).$$

Clearly C has minimum distance $d = 2e+1$ and its maximum error correction capacity is e by (12.4.3). A word w belongs $B_e(c_0)$ if more of its entries are 0 than 1; otherwise $w \in B_e(c_1)$. Thus $B_e(c_0) \cap B_e(c_1) = \emptyset$ and $B_e(c_0) \cup B_e(c_1) = H_{2e+1}(2)$.

Linear codes. Let Q denote GF(q), the field of q elements, where q is now a prime power. Then Hamming space $H_n(q)$ is the n-dimensional vector space Q_n of all n-row vectors over Q. A q-ary code C of length n is called *linear* if it is a subspace of $H_n(Q)$. Linear codes form an important class of codes; they have the advantage that they can be specified by giving a basis instead of listing all the codewords. Linear codes can also be described by matrices, as will be seen in the sequel.

A computational advantage of linear codes is indicated by:

(12.4.6) *The minimum distance of a linear code equals the minimal weight of a non-zero codeword.*

Proof. Let C be a linear code. If $c_1, c_2 \in C$, then $d(c_1, c_2) = wt(c_1 - c_2)$ and $c_1 - c_2 \in C$. Hence the minimum distance equals the minimum weight. □

A point to keep in mind here is that in order to compute the minimum distance of a code C one must compute $\binom{|C|}{2}$ distances, whereas to compute the minimum weight of C only the distances from the zero word need be found, so that at most $|C|$ computations are necessary.

As for codes in general, it is desirable to have linear codes with large minimum distance and as many codewords as possible. There is a version of the Varshamov–Gilbert lower bound for linear codes.

(12.4.7) *Let d and n be positive integers with $d \leq n$ and let q be a prime power. Then there is a linear q-ary code C of length n and minimum distance d for which the number of codewords is at least*

$$\frac{q^n}{\sum_{i=0}^{d-1} \binom{n}{i}(q-1)^i}.$$

Proof. We refer to the proof of (12.4.4). To start the construction choose a linear q-ary code C_0 of length n and minimum distance d; for example, the subspace generated by a single word of weight d will suffice. If $\bigcup_{c \in C_0} B_{d-1}(c) \neq H_n(q)$, choose a word w in $H_n(q)$ which belongs to no $B_{d-1}(c)$ with c in C_0. Thus $w \notin C_0$. Define C_1 to be the subspace generated by C_0 and w. We claim that C_1 still has minimum distance d. To prove this it is sufficient to show that $wt(c') \geq d$ for any c' in $C_1 - C_0$; this is because of (12.4.6). Now we may write $c' = c_0 + aw$ where $c_0 \in C_0$ and $0 \neq a \in Q$. Then

$$wt(c') = wt(c_0 + aw) = wt(-a^{-1}c_0 - w) = d(-a^{-1}c_0, w) \geq d$$

by choice of w since $-a^{-1}c_0 \in C_0$. Note also that $\dim(C_0) < \dim(C_1)$.

Repeat the argument above for C_1, and then as often as possible. After at most n steps we arrive at a subspace C with minimum distance d such that $\bigcup_{c \in C} B_{d-1}(c) = H_n(q)$. It now follows that $|C| \cdot |B_{d-1}(c)| \geq q^n$ for any c in C, which gives the bound. □

Example (12.4.3) Let $q = 2$, $d = 3$ and $n = 31$. According to (12.4.7) there is a linear binary code C of length 31 with minimum distance 3 such that

$$|C| \geq \frac{2^{31}}{1 + 31 + \binom{31}{2}} = 4,320,892.652.$$

However C is a subspace of $H_{31}(2)$, so its order must be a power of 2. Hence $|C| \geq 2^{23} = 8,388,608$. In fact there is a larger code of this type with 2^{26} codewords, a so-called Hamming code – see Example (12.4.7) below.

Generator matrix and check matrix of a linear code. Let C be a linear q-ary code of length n and let k be the dimension of C as a subspace of $H_n(q)$. Thus $|C| = q^k$. Choose an ordered basis $c_1, c_2, \ldots, c_k$ for C and write

$$G = \begin{bmatrix} c_1 \\ c_2 \\ \vdots \\ c_k \end{bmatrix}.$$

This $k \times n$ matrix over $Q = GF(q)$ is called a *generator matrix* for C. If c is any codeword, then $c = a_1 c_1 + \cdots + a_k c_k$ for suitable $a_i \in Q$. Thus $c = aG$ where

$a = (a_1, \ldots, a_k) \in H_k(q)$. Hence each codeword is uniquely expressible in the form aG with $a \in H_k(q)$. It follows that the code C *is the row space of the matrix* G, i.e., the subspace of $H_n(q)$ generated by all the rows of G. Note that the rank of G is k since its rows are linearly independent.

Recall from 8.1 that the *null space* N of G consists of all n-column vectors x^T such that $Gx^T = 0$: here of course $x \in H_n(q)$. Next choose an ordered basis for N and use the transposes of its elements to form the rows of a matrix H. This is called a *check matrix* for C. Since G has rank k, we can apply (8.3.8) to obtain $\dim(N) = n - k$, so that H is an $(n-k) \times n$ matrix over Q. Since the columns of H^T belong to N, the null space of G, we obtain the important equation

$$GH^T = 0.$$

It should be kept in mind that G and H depend on choices of bases for C and N.

At this point the following result about matrices is relevant.

(12.4.8) *Let G and H be $k \times n$ and $(n-k) \times n$ matrices respectively over $Q = \mathrm{GF}(q)$, each having linearly independent rows. Then the following statements are equivalent:*

(i) $GH^T = 0$;

(ii) *row space*$(G) = \{x \in H_n(q) \mid xH^T = 0\}$;

(iii) *row space*$(H) = \{x \in H_n(q) \mid xG^T = 0\}$.

Proof. Let $S = \{x \in H_n(q) \mid xH^T = 0\}$; then $x \in S$ if and only if $0 = (xH^T)^T = Hx^T$, i.e., x^T belongs to null space(H). This implies that S is a subspace and $\dim(S) = n - (n-k) = k$. Now assume that $GH^T = 0$. Then $x \in$ row space(G) implies that $xH^T = 0$, so that $x \in S$ and row space$(G) \subseteq S$. But $\dim(\text{row space}(G)) = k = \dim(S)$, whence we obtain $S =$ row space(G). Thus (i) implies (ii). It is clear that (ii) implies (i), and thus (i) and (ii) are equivalent.

Next we observe that $GH^T = 0$ if and only if $HG^T = 0$, by applying the transpose. Thus the roles of G and H are interchangeable, which means that (i) and (iii) are equivalent. □

Let us return now to the discussion of the linear code C with generator matrix G and check matrix H. From (12.4.8) we conclude that

$$C = \text{row space}(G) = \{w \in H_n(q) \mid wH^T = 0\}.$$

So the check matrix H provides a convenient way of checking whether a given word w is a codeword. At this point we have proved half of the next result.

(12.4.9) (i) *If C is a linear q-ary code with generator matrix G and check matrix H, then $GH^T = 0$ and $C = \{w \in H_n(q) \mid wH^T = 0\}$.*

12.4 An introduction to error correcting codes

(ii) *If G and H are $k \times n$ and $(n - k) \times n$ matrices respectively over $\mathrm{GF}(q)$ with linearly independent rows and if $GH^T = 0$, then $C = \{w \in H_n(q) \mid wH^T = 0\}$ is a linear q-ary code of length n and dimension k with generator matrix G and check matrix H.*

Proof. Only (ii) requires comment. Clearly C is a linear q-ary code of length n. By (12.4.8) C is the row space of G. Hence $\dim(C) = k$ and G is a generator matrix for C. Finally, the null space of G consists of all w in $H_n(q)$ such that $Gw^T = 0$, i.e., $wG^T = 0$; this is the row space of H by (12.4.8). Hence G and H are corresponding generator and check matrices for C. □

On the basis of (12.4.9) we show how to construct a linear q-ary code of length n and dimension $n - \ell$ with check matrix equal to a given $\ell \times n$ matrix H over $\mathrm{GF}(q)$ of rank ℓ. Define $C = \{x \in H_n(q) \mid xH^T = 0\}$; this is a linear q-ary code. Pass from H to its reduced row echelon form $H' = [1_\ell \mid A]$ where A is an $\ell \times (n - \ell)$ matrix: note that interchanges of columns, i.e., of word entries, may be necessary to achieve this and $H' = EHF$ for some non-singular E and F. Writing G' for $[-A^T \mid 1_{n-\ell}]$, we have

$$G'H'^T = \begin{bmatrix} -A^T \mid 1_{n-\ell} \end{bmatrix} \begin{bmatrix} 1_\ell \\ A^T \end{bmatrix} = 0.$$

Hence $0 = G'H'^T = (G'F^T)H^T E^T$, so $(G'F^T)H^T = 0$ since E^T is non-singular. Put $G = G'F^T$; thus $GH^T = 0$ and by (12.4.9) G is a generator matrix and H a check matrix for C. Note that if no column interchanges are needed to go from H to H', then $F = 1$ and $G = G'$.

Example (12.4.4) Consider the matrix

$$H = \begin{bmatrix} 1 & 1 & 1 & 0 & 1 & 0 & 0 \\ 1 & 1 & 0 & 1 & 0 & 1 & 0 \\ 1 & 0 & 1 & 1 & 0 & 0 & 1 \end{bmatrix}$$

over $\mathrm{GF}(2)$. Here $n = 7$ and $\ell = 3$. The rank of H is 3, so it determines a linear binary code C of dimension $7 - 3 = 4$. Put H in reduced row echelon form,

$$H' = \begin{bmatrix} 1 & 0 & 0 & 0 & 1 & 1 & 1 \\ 0 & 1 & 0 & 1 & 1 & 0 & 1 \\ 0 & 0 & 1 & 1 & 1 & 1 & 0 \end{bmatrix} = [1_3 \mid A].$$

No column interchanges were necessary here, so

$$G = G' = [-A^T \mid 1_4] = \begin{bmatrix} 0 & 1 & 1 & 1 & 0 & 0 & 0 \\ 1 & 1 & 1 & 0 & 1 & 0 & 0 \\ 1 & 0 & 1 & 0 & 0 & 1 & 0 \\ 1 & 1 & 0 & 0 & 0 & 0 & 1 \end{bmatrix}$$

is a generator matrix for C. The rows of G form a basis for the linear code C.

A useful feature of the check matrix is that from it one can read off the minimum distance of the code.

(12.4.10) *Let H be a check matrix for a linear code C. Then the minimum distance of C equals the largest integer m such that every set of $m - 1$ columns of H is linearly independent.*

Proof. Let d be the minimum distance of C and note that d is the minimum weight of a non-zero codeword, say $d = wt(c)$. Then $cH^T = 0$, which implies that there exist d linearly dependent columns of H. Hence $m - 1 < d$ and $m \leq d$. Also by maximality of m there exist m linearly dependent columns of H, so $wH^T = 0$ where w is a non-zero word with $wt(w) \leq m$. But $w \in C$; thus $d \leq m$ and hence $d = m$. □

Example (12.4.5) Consider the code C of Example (12.4.4). Every pair of columns of the check matrix H is linearly independent, i.e., the columns are all different. On the other hand, columns 1, 4 and 5 are linearly dependent since their sum is zero. Therefore $m = 3$ for this code and the minimum distance is 3 by (12.4.10). Consequently C is a 1-error correcting code.

Using the check matrix to correct errors. Let C be a linear q-ary code with length n and minimum distance d and let H be a check matrix for C. Note that by (12.4.3) C is e-error correcting where $e = \left[\frac{d-1}{2}\right]$. Suppose that a codeword c is transmitted and received as a word w and that *at most e errors* in the entries have been made. Here is a procedure to correct the errors and recover the original codeword c.

Write $w = u + c$ where u is the error; thus $wt(u) \leq e$. Now $|H_n(q) : C| = q^{n-k}$ where $k = \dim(C)$. Choose a transversal to C in $H_n(q)$, say $v_1, v_2, \ldots, v_{q^{n-k}}$, by requiring that v_i be a word of *smallest length* in its coset $v_i + C$. For any $c_0 \in C$ we have $(v_i + c_0) H^T = v_i H^T$, which depends only on i. Now suppose that w belongs to the coset $v_i + C$. Then $wH^T = v_i H^T$, which is called the *syndrome* of w. Writing $w = v_i + c_1$ with $c_1 \in C$, we have $u = w - c \in v_i + C$, so that $wt(v_i) \leq wt(u) \leq e$ by choice of v_i. Hence $w = u + c = v_i + c_1$ belongs to $H_e(c) \cap H_e(c_1)$. But this implies that $c = c_1$ since C is e-error correcting. Therefore $c = w - v_i$ and the transmitted codeword has been identified.

In summary here is the procedure to identify the transmitted codeword c. It is assumed that the transversal $\{v_1, v_2, \ldots, v_{q^{n-k}}\}$ has been chosen as described above.

(i) Suppose that w is the word received with at most e errors; first compute the syndrome wH^T.

(ii) By comparing wH^T with the syndromes $v_i H^T$, find the unique i such that $wH^T = v_i H^T$.

(iii) Then the transmitted codeword was $c = w - v_i$.

12.4 An introduction to error correcting codes

Example (12.4.6) The matrix

$$H = \begin{bmatrix} 1 & 0 & 1 & 1 & 0 \\ 0 & 0 & 1 & 1 & 1 \\ 1 & 1 & 0 & 1 & 1 \end{bmatrix}$$

determines a linear binary code C with length 5 and dimension $5 - 3 = 2$; here H is a check matrix for C. Clearly C has minimum distance 3 and so it is 1-error correcting. Also $|C| = 2^2 = 4$ and $|H_5(2) : C| = 2^5/4 = 8$. By reducing H to reduced row echelon form as in Example (12.4.4) we find a generator matrix for C to be

$$G = \begin{bmatrix} 0 & 1 & 1 & 1 & 0 \\ 1 & 0 & 1 & 0 & 1 \end{bmatrix}.$$

Thus C is generated by $(0\,1\,1\,1\,0)$ and $(1\,0\,1\,0\,1)$, so in fact C consists of $(0\,0\,0\,0\,0)$, $(0\,1\,1\,1\,0)$, $(1\,0\,1\,0\,1)$ and $(1\,1\,0\,1\,1)$.

Next we enumerate the eight cosets of C in $H_5(2)$:

$$C = C_1 = \{(\mathbf{0\,0\,0\,0\,0}), (0\,1\,1\,1\,0), (1\,0\,1\,0\,1), (1\,1\,0\,1\,1)\}$$
$$C_2 = \{(\mathbf{1\,0\,0\,0\,0}), (1\,1\,1\,1\,0), (0\,0\,1\,0\,1), (0\,1\,0\,1\,1)\}$$
$$C_3 = \{(\mathbf{0\,1\,0\,0\,0}), (0\,0\,1\,1\,0), (1\,1\,1\,0\,1), (1\,0\,0\,1\,1)\}$$
$$C_4 = \{(\mathbf{0\,0\,1\,0\,0}), (0\,1\,0\,1\,0), (1\,0\,0\,0\,1), (1\,1\,1\,1\,1)\}$$
$$C_5 = \{(\mathbf{1\,1\,0\,0\,0}), (1\,0\,1\,1\,0), (0\,1\,1\,0\,1), (0\,0\,0\,1\,1)\}$$
$$C_6 = \{(0\,1\,1\,0\,0), (\mathbf{0\,0\,0\,1\,0}), (1\,1\,0\,0\,1), (1\,0\,1\,1\,1)\}$$
$$C_7 = \{(1\,0\,1\,0\,0), (1\,1\,0\,1\,0), (\mathbf{0\,0\,0\,0\,1}), (0\,1\,1\,1\,1)\}$$
$$C_8 = \{(1\,1\,1\,0\,0), (\mathbf{1\,0\,0\,1\,0}), (0\,1\,0\,0\,1), (0\,0\,1\,1\,1)\}.$$

Choose a word of minimum weight from each coset; these are printed in bold face. The coset syndromes are computed as $(0\,0\,0)$, $(1\,0\,1)$, $(0\,0\,1)$, $(1\,1\,0)$, $(1\,0\,0)$, $(1\,1\,1)$, $(0\,1\,1)$, $(0\,1\,0)$.

Now suppose that the word $w = (1\,1\,1\,1\,1)$ is received with at most one error in its entries. The syndrome of w is $wH^T = (1\,1\,0)$, which is the syndrome of elements in the coset C_4, with coset representative $v_4 = (0\,0\,1\,0\,0)$. Hence the transmitted codeword was $c = w - v_4 = (1\,1\,0\,1\,1)$.

Hamming codes. Let C be a linear q-ary code of length n and dimension k. Assume that the minimum distance of C is at least 3, so that C is 1-error correcting. A check matrix H for C has size $\ell \times n$ where $\ell = n - k$, and by (12.4.10) no column of H can be a multiple of another column.

Now consider the problem of constructing such a linear code which is as large as possible for given q and $\ell > 1$. So H should have as many columns as possible, subject to no column being a multiple of another one. Now there are $q^\ell - 1$ non-zero ℓ-column vectors over $GF(q)$, but each of these is a multiple of $q - 1$ other columns. So the maximum possible number of columns for H is $n = \frac{q^\ell - 1}{q - 1}$. Note that the columns of the identity $\ell \times \ell$ matrix can be included among those of H, which shows that H has rank ℓ. It follows that the matrix H determines a linear q-ary code C of length

$$n = \frac{q^\ell - 1}{q - 1}.$$

The minimum distance of H is at least 3 by construction, and in fact it is exactly 3 since we can include among the columns of H three linearly dependent ones, $(1\,0\ldots 0)^T$, $(1\,1\,0\ldots 0)^T$, $(0\,1\,0\ldots 0)^T$. Thus C is 1-error correcting: its dimension is $k = n - \ell$ and its order is q^n. Such a code is known as a *Hamming code*. It is not surprising that Hamming codes have optimal properties.

(12.4.11) *Hamming codes are perfect.*

Proof. Let C be a q-ary Hamming code of length n constructed from a check matrix with ℓ rows. Then

$$|C| = q^{n-\ell} = q^n/q^\ell = q^n/(1 + n(q-1))$$

since $n = \frac{q^\ell - 1}{q - 1}$. Hence C attains the Hamming upper bound of (12.4.5) and so it is a perfect code. □

Example (12.4.7) Let $q = 2$ and $\ell = 4$. Then a Hamming code C of length $n = \frac{2^4 - 1}{2 - 1} = 15$ can be constructed from the 4×15 check matrix

$$H = \begin{bmatrix} 1 & 0 & 0 & 0 & 1 & 1 & 1 & 0 & 0 & 0 & 0 & 1 & 1 & 1 & 1 \\ 0 & 1 & 0 & 0 & 1 & 0 & 0 & 1 & 1 & 0 & 1 & 0 & 1 & 1 & 1 \\ 0 & 0 & 1 & 0 & 0 & 1 & 0 & 1 & 0 & 1 & 1 & 1 & 0 & 1 & 1 \\ 0 & 0 & 0 & 1 & 0 & 0 & 1 & 0 & 1 & 1 & 1 & 1 & 1 & 0 & 1 \end{bmatrix}.$$

Here $|C| = 2^{n-\ell} = 2^{11} = 2048$.

Similarly by taking $q = 2$ and $\ell = 5$ we obtain a perfect linear binary code of length 31 and dimension 26.

Perfect codes. We now carry out an analysis of perfect codes which will establish the unique position of the Hamming codes.

(12.4.12) *Let C be a perfect q-ary code where $q = p^a$ and p is a prime. Assume that C is 1-error correcting. Then:*

(i) *C has length $\frac{q^s - 1}{q - 1}$ for some $s > 1$;*

(ii) *if C is linear, it is a Hamming code.*

Proof. (i) Let C have length n. Then $|C| = \frac{q^n}{1+n(q-1)}$ since C is perfect and 1-error correcting. Hence $1 + n(q-1)$ divides q^n, so it must be a power of p, say $1 + n(q-1) = p^r$. Now by the Division Algorithm we can write $r = sa + t$ where $s, t \in \mathbb{Z}$ and $0 \le t < a$. Then $1 + n(q-1) = (p^a)^s \, p^t = q^s \, p^t = (q^s - 1)p^t + p^t$. Therefore $q - 1$ divides $p^t - 1$. However $p^t - 1 < p^a - 1 = q - 1$, which means that $p^t = 1$ and $1 + n(q-1) = q^s$. It follows that $n = \frac{q^s - 1}{q - 1}$.

(ii) Now assume that C is linear. Since $|C| = \frac{q^n}{1+n(q-1)}$ and we have shown that $1 + n(q-1) = q^s$, we deduce that $|C| = q^n/q^s = q^{n-s}$. Hence $\dim(C) = n - s$ and a check matrix H for C has size $s \times n$. The number of columns of H is $n = \frac{q^s-1}{q-1}$, the maximum number possible, and no column is a multiple of another one since C is 1-error correcting and thus has minimum distance ≥ 3. Therefore C is a Hamming code. □

Almost nothing is known about perfect q-ary codes when q is not a prime power. Also there are very few perfect linear q-ary codes which are e-error correcting with $e > 1$. Apart from binary repetition codes of odd length – see Exercise (12.4.3) below – there are just two examples, a binary code of length 23 and a ternary code of length 11. These remarkable examples are called the *Golay codes;* they are of great importance in algebra.

Exercises (12.4)

1. Give an example of a code for which the minimum distance is different from the minimum weight of a non-zero codeword.

2. Find the number of q-ary words with weight in the range i to $i + k$.

3. Let C be the set of all words $(a\, a \ldots a)$ of length n where $a \in \mathrm{GF}(q)$.
    (i) Show that C is a linear q-ary code of dimension 1.
    (ii) Find the minimum distance and error correcting capacity of C.
    (iii) Write down a generator matrix and a check matrix for C.
    (iv) Show that when $q = 2$, the code C is perfect if and only if n is odd.

4. Let C be a q-ary code of length n and minimum distance d. Prove that

$$|C| \le q^{n-d+1},$$

(the *Singleton upper bound*). [Hint: two codewords with the same first $n - d + 1$ entries are equal].

5. If C is a linear q-ary code of length n and dimension k, the minimum distance of C is at most $n - k + 1$.

6. Let C be a linear q-ary code of length n and dimension k. Suppose that G is a generator matrix for C and that $G' = [I_k \mid A]$ is the reduced row echelon form of G. Prove that there is a check matrix for C of the form $[-A^T \mid I_{n-k}]$ up to a permutation of columns.

7. A linear binary code C has basis $\{(1\,0\,1\,1\,1\,0), (0\,1\,1\,0\,1\,0), (0\,0\,1\,1\,0\,1)\}$. Find a check matrix for C and use it to determine the error-correcting capacity of C.

8. A check matrix for a linear binary code C is

$$\begin{bmatrix} 1 & 1 & 0 & 1 & 1 \\ 0 & 1 & 0 & 0 & 1 \\ 1 & 1 & 1 & 0 & 1 \end{bmatrix}.$$

(i) Find a basis for C;
(ii) find the minimum distance and error correcting capacity of C;
(iii) if a word $(0\,1\,1\,1\,1)$ is received and at most one entry is erroneous, use the syndrome method to find the transmitted codeword.

9. (*An alternative decoding procedure*). Let C be a linear q-ary code of length n with error correcting capacity e. Let H be a check matrix for C. Suppose that a word w is received with at most e errors. Show that the following procedure will find the transmitted codeword.

(i) Find all words u in $H_n(q)$ of weight $\leq e$, (these are the possible errors).
(ii) Compute the syndrome uH^T of each word u from (i);
(iii) Compute the syndrome wH^T and compare it with each uH^T: prove that there is a unique word u in $H_n(q)$ of with weight at most e such that $uH^T = wH^T$.
(iv) The transmitted codeword was $w - u$.

(Notice that the number of possible words u is $\sum_{i=0}^{e} \binom{n}{i}(q-1)^i$, which is polynomial in n.)

10. Use the method of Exercise (12.4.9) to find the transmitted codeword in Exercise (12.4.8).

11. (*Dual codes*). Let C be a linear q-ary code of length n and dimension k. Define the *dot product* of two words v, w in $H_n(q)$ by $v \cdot w = \sum_{i=1}^{n} v_i w_i$. Then define $C^\perp = \{w \in H_n(q) \mid w \cdot c = 0, \forall c \in C\}$.

(i) Show that $C^\perp$ is a linear q-ary code of length n – it is called the *dual code* of C.
(ii) Let G and H be corresponding generator and check matrices for C. Prove that G is a check matrix and H a generator matrix for $C^\perp$.
(iii) Prove that $\dim\left(C^\perp\right) = n - k$ and $|C^\perp| = q^{n-k}$.

12. Let C be a binary Hamming code of length 7. Find a check matrix for the dual code $C^\perp$ and show that its minimum distance is 4.

Bibliography

[1] Adamson, I. T., Introduction to Field Theory, 2nd ed., Cambridge University Press, Cambridge, 1982.

[2] Cameron, P. J., Combinatorics, Cambridge University Press, Cambridge, 1994.

[3] Dummit, D. S. and Foote, R. M., Abstract Algebra, 2nd ed., Prentice Hall, Upper Saddle River, NJ, 1999.

[4] Goldrei, D., Classic Set Theory, Chapman and Hall, London, 1996.

[5] Halmos, P. R., Naive Set Theory, Undergrad. Texts in Math., Springer-Verlag, New York, NY, 1974.

[6] Janusz, G.J., Algebraic Number Fields, 2nd ed., Amer. Math. Soc., Providence, RI, 1996.

[7] van Lint, J. H., Introduction to Coding Theory, Grad. Texts in Math. 86, 3rd rev. and exp. ed., Springer-Verlag, Berlin 1999.

[8] Niven, I. M., Irrational Numbers, Carus Math. Monogr. 11, Wiley, New York, 1956.

[9] Robinson, D. J. S., A Course in Linear Algebra with Applications, World Scientific, Singapore, 1991.

[10] Robinson, D. J. S., A Course in the Theory of Groups, 2nd ed., Grad. Texts in Math. 80, Springer-Verlag, New York, 1996.

[11] Rose, J. S., A Course on Group Theory, Dover, New York, 1994.

[12] Rotman, J. J., Galois Theory, Universitext, Springer-Verlag, New York, 1990.

[13] Rotman, J. J., An Introduction to the Theory of Groups, 4th ed., Grad. Texts in Math. 148, Springer-Verlag, New York, 1995.

[14] Rotman, J. J., A First Course in Abstract Algebra, Prentice Hall, Upper Saddle River, NJ, 1996.

[15] van der Waerden, B., A History of Algebra, Springer-Verlag, Berlin, 1985.

[16] Wilson, R. J., Introduction to Graph Theory, 4th ed., Longman, Harlow, 1996.

Index of notation

$A, B, \ldots$	Sets
$a, b, \ldots$	Elements of sets
$a \in A$	a is an element of the set A
$|A|$	The cardinal of a set A
$A \subseteq B, A \subset B$	A is a subset, proper subset of B
$\emptyset$	The empty set
$\mathbb{N}, \mathbb{Z}, \mathbb{Q}, \mathbb{R}, \mathbb{C}$	The sets of natural numbers, integers, rational numbers, real numbers, complex numbers
$\cup, \cap$	Union and intersection
$A_1 \times \cdots \times A_n$	Set product
$A - B, \bar{A}$	Complementary sets
$P(A)$	The power set
$S \circ R$	The composite of relations or functions
$[x]_E$	The E-equivalence class of x
$\alpha : A \to B$	A function from A to B
$\mathrm{Im}(\alpha)$	The image of the function α
id_A	The identity function on the set A
α^{-1}	The inverse of the bijective function α
$\mathrm{Fun}(A, B)$	The set of all functions from A to B
$\mathrm{Fun}(A)$	The set of all functions on A
gcd, lcm	Greatest common divisor, least common multiple
$a \equiv b \pmod{m}$	a is congruent to b modulo m
$[x]_m$ or $[x]$	The congruence class of x modulo m
$\mathbb{Z}_n$	The integers modulo n
$a \mid b$	a divides b
ϕ	Euler's function
μ	The Möbius function
$\langle X \rangle$	The subgroup generated by X
$|x|$	The order of the group element x
XY	A product of subsets of a group
$H \leq G, H < G$	H is a subgroup, proper subgroup of the group G
$\simeq$	An isomorphism
$[x, y]$	The commutator $xyx^{-1}y^{-1}$

Index of notation

sign(π)	The sign of a permutation π
$(i_1 i_2 \ldots i_r)$	A cyclic permutation
Fix	A fixed point set
Sym(X)	The symmetric group on a set X
S_n, A_n.	The symmetric and alternating groups of degree n
Dih($2n$)	The dihedral group of order $2n$
$\text{GL}_n(R)$, $\text{GL}_n(q)$	General linear groups
$\text{SL}_n(R)$, $\text{SL}_n(q)$	Special linear groups
$\lvert G : H \rvert$	The index of H in G
$N \triangleleft G$	N is a normal subgroup of the group G
G/N	The quotient group of N in G
Ker(α)	The kernel of the homomorphism α
$Z(G)$	The center of the group G
G'	The derived subgroup of the group G
$\varphi(G)$	The Frattini subgroup of the group G
$N_G(H)$, $C_G(H)$	The normalizer and centralizer of H in G
$\text{St}_G(x)$	The stabilizer of x in G
$G \cdot a$	The G-orbit of a
Aut(G), Inn(G)	The automorphism and inner automorphism groups of G
Out(G)	The outer automorphism group of G
$U(R)$ or R^*	The group of units of R, a ring with identity
$R[t_1, \ldots, t_n]$	The ring of polynomials in $t_1, \ldots, t_n$ over the ring R
$F\{t_1, \ldots, t_n\}$	The field of rational functions in $t_1, \ldots, t_n$ over a field F
deg(f)	The degree of the polynomial f
f'	The derivative of the polynomial f
$M_{m,n}(R)$	The ring of $m \times n$ matrices over R
$F(X)$	The subfield generated by X and the subfield F
GF(q)	The field with q elements
$(E : F)$	The degree of E over F
$\text{Irr}_F(f)$	The irreducible polynomial of $f \in F[t]$
dim(V)	The dimension of the vector space V
$[v]_\mathcal{B}$	The coordinate vector of v with respect to $\mathcal{B}$
$C[a, b]$	The vector space of continuous functions on the interval $[a, b]$
$L(V, W)$ $L(V)$	Vector spaces of linear mappings

$\langle\,,\,\rangle$, $\|\ \|$	Inner product and norm
$S^\perp$	The orthogonal complement of S
$G^{(i)}$	The i-th term of the derived series of the group G
$Z_i(G)$	The i-th term of the upper central series of the group G
$\mathrm{Gal}(E/F)$, $\mathrm{Gal}(f)$	Galois groups
Φ_n	The cyclotomic polynomial of order n
$\mathrm{Fix}(H)$	The fixed field of a subgroup H of a Galois group
$\langle X \mid R \rangle$	A group presentation
$H_n(q)$	Hamming n-space over a set with q elements
$B_n(v)$	The n-ball with center v
$d(a,b)$	The distance between points a and b
$wt(v)$	The weight of the word v

Index

Abel, Niels Henrik, 40
abelian group, 40
abelian groups
    number of, 180
algebra of linear operators, 154
algebra over a field, 154
algebraic closure of a field, 237
algebraic element, 187
algebraic number, 189
algebraic number field, 190
algebraically closed field, 237
alternating group, 37
    simplicity of, 166
angle between vectors, 158
antisymmetric law, 5
ascending chain condition on ideals, 120
associate elements in a ring, 115
associative law, 40
    generalized, 43
associative law for composition, 8
automaton, 11
automorphism, 70
    inner, 71
    outer, 72
automorphism group, 70
    outer, 72
automorphism group of a cyclic group, 73
automorphism of a field, 213
Axiom of Choice, 239

ball
    r-, 257
basis
    change of, 142, 153
    existence of, 140
    ordered, 140
    orthonormal, 160
    standard, 140
basis of a vector space, 140
    existence of, 140, 236
Bernstein, Felix, 14
bijective function, 10
binary operation, 39
binary repetition code, 258
Boole, George, 4
Boolean algebra, 4
Burnside p-q Theorem, 174
Burnside's Lemma, 83
Burnside, William, 83

cancellation law, 108
canonical homomorphism, 67
Cantor, Georg, 14
Cantor–Bernstein Theorem, 14
Cardano's formulas, 243
Cardano, Gerolamo, 243
cardinal, 13
cardinal number, 13
cartesian product, 3
Cauchy's formula, 37
Cauchy's Theorem, 89
Cauchy, Augustin Louis, 37
Cauchy–Schwartz Inequality, 157
Cayley's theorem, 80
Cayley, Arthur, 80
center of a group, 61
central series, 174
centralizer of a subset, 85
centralizer of an element, 82
chain, 7, 235
    upper central, 174
characteristic function of a subset, 10
characteristic of a domain, 111
check matrix of a code, 260
check matrix used to correct errors, 262
Chinese Remainder Theorem, 27

Index

circle group, 63
class, 1
    equivalence, 5
class equation, 85
class number, 82
classification of finite simple groups, 169
code, 255
    e-error correcting, 255
    e-error detecting, 255
    q-ary, 255
    binary repetition, 258
    dual, 266
    Hamming, 264
    linear, 258
    minimum distance of, 255
    perfect, 258
codeword, 255
coefficients of a polynomial, 100
collineation, 169
column echelon form, 142
common divisor, 20
commutative ring, 99
commutative semigroup, 40
commutator, 61, 172
commutator subgroup, 62
complement, 2
    relative, 2
complement of a subgroup, 181
complete group, 74, 87
complete set of irreducibles, 121
composite of functions, 10
composite of relations, 8
composition factor, 165
composition series, 165
congruence, 24
    linear, 26
congruence arithmetic, 25
congruence class, 24
congruence classes
    product of, 25
    sum of, 25
conjugacy class, 82

conjugacy classes of the symmetric group, 85
conjugate elements in a field, 216
conjugate of a group element, 60
conjugate subfield, 222
conjugation homomorphism, 71
constructible point, 191
constructible real number, 191
construction of a regular n-gon, 191, 225
content of a polynomial, 124
coordinate column vector, 140
Correspondence Theorem for groups, 64
Correspondence Theorem for rings, 106
coset
    left, 52
    right, 52
countable set, 15
crossover diagram, 36
cubic equation, 242
cycle, 33
cyclic group, 47, 49
cyclic permutation, 33
cyclic subgroup, 47
cyclotomic number field, 221
cyclotomic polynomial, 132
    Galois group of, 220
    irreducibility of, 220

Dedekind, Richard, 58
defining relator of a group, 251
degree of a polynomial, 100
degree of an extension, 186
del Ferro, Scipione, 243
De Morgan's laws, 3
De Morgan, Augustus, 3
derangement, 38, 200
derivation, 182
derivative, 128
derived chain, 172
derived length, 171

derived subgroup, 62, 172
dihedral group, 42
dimension of a vector space, 141
direct product
    external, 65
    internal, 64
direct product of latin squares, 203
direct sum of subspaces, 145
Dirichlet, Johann Peter Gustav Lejeune, 29
discriminant of a polynomial, 240
disjoint union, 4
distance between words, 254
Division Algorithm, 20
division in rings, 115
division in the integers, 19
division ring, 108
domain, 108
double dual, 151
dual code, 266
dual space, 151
duplication of the cube, 191

edge function of a graph, 96
edge of a graph, 95
Eisenstein's Criterion, 132
Eisenstein, Ferdinand Gotthold Max, 132
element of a set, 1
elementary abelian p-group, 146
elementary symmetric function, 232
elementary vector, 137
empty word, 245
equal sets, 2
equation of the fifth degree, solvability of, 228
equipollent sets, 13
equivalence class, 5
equivalence relation, 5
Euclid of Alexandria, 21
Euclid's Lemma, 22
Euclid's Lemma for rings, 119
Euclidean n-space, 135

Euclidean Algorithm, 21
Euclidean domain, 116
Euler's function, 27
Euler, Leonhard, 27
evaluation homomorphism, 127
even permutation, 35
exact sequence, 73
exact sequence of vector spaces, 155
extension field, 186
    algebraic, 188
    finite, 186
    Galois, 214
    radical, 228
    separable, 210
    simple, 186
external direct product, 65

factor set, 181
faithful group action, 79
faithful permutation representation, 79
Feit–Thompson Theorem, 174
Fermat prime, 24, 195, 226
Fermat's Theorem, 26, 54, 60
Fermat, Pierre de, 26
Ferrari, Lodovico, 243
field, 108
    algebraic number, 190
    algebraically closed, 237
    finite, 146, 195
    Galois, 197
    intermediate, 221
    perfect, 210
    prime, 185
    splitting, 130
field as a vector space, 135
field extension, 186
field of fractions, 112
field of rational functions, 113
finite p-group, 87
finite abelian group, 178
finite dimensional vector space, 141
finitely generated vector space, 136
fixed field, 221

fixed point set, 83
formal power series, 103
fraction over a domain, 112
Frattini argument, 176
Frattini subgroup, 175, 240
Frattini subgroup of a finite p-group, 176
Frattini, Giovanni, 175
free group, 244
 projective property of, 253
 structure of, 249
free groups
 examples of, 249
function, 9
 bijective, 10
 characteristic, 10
 elementary symmetric, 232
 identity, 10
 injective, 10
 inverse, 11
 one-one, 10
 onto, 10
 surjective, 10
Fundamental Theorem of Algebra, 128, 225
Fundamental Theorem of Arithmetic, 22, 166
Fundamental Theorem of Galois Theory, 221

Galois correspondence, 222
Galois extension, 214
Galois field, 197
Galois group, 213
Galois group of a polynomial, 214
Galois groups of polynomials of degree ≤ 4, 242
Galois Theory
 Fundamental Theorem of, 221
Galois' theorem on the solvablility by radicals, 229
Galois, Évariste, 213
Gauss's Lemma, 125, 132

Gauss, Carl Friedrich, 24
Gaussian elimination, 139
Gaussian integer, 117
general linear group, 41
generator matrix of a code, 259
generators and defining relations of a group, 251
generic polynomial, 233
Gödel–Bernays Theory, 235
Gram, Jorgen Pedersen, 161
Gram–Schmidt process, 161
graph, 95
graphs
 number of, 97
graphs, counting, 95
greatest common divisor, 20
greatest common divisor in rings, 119
greatest lower bound, 8
group
 abelian, 40
 alternating, 37
 circle, 63
 complete, 74, 87
 cyclic, 47, 49
 dihedral, 42
 elementary abelian p-, 146
 finite p-, 87
 finite abelian, 178
 free, 244
 general linear, 41
 Klein 4-, 44, 45
 metabelian, 171
 nilpotent, 174
 permutation, 78
 simple, 61
 solvable, 171
 special linear, 61
 symmetric, 31, 32, 41
group actions, 78
group extension, 170
group of prime order, 54
group of units in a ring, 102

Hall π-subgroup, 183
Hall's theorem on finite solvable
 groups, 182
Hall, Philip, 183
Hamilton, Rowan Hamilton, 109
Hamming code, 264
Hamming space, 254
Hamming upper bound, 257
Hamming, Richard Wesley, 254
Hasse, Helmut, 6
Hölder, Otto, 165
homomorphism, 67, 104
 canonical, 67
 conjugation, 71
 trivial, 67
homomorphism of rings, 104
homomorphisms of groups, 67

ideal, 104
 irreducible, 240
 left, 114
 maximal, 110, 236
 prime, 110
 principal, 104, 118
 right, 114
identity element, 40
identity function, 10
identity subgroup, 47
image of a function, 9
image of an element, 9
Inclusion–Exclusion Principle, 38
index of a subgroup, 53
infinite set, 15
injective function, 10
inner automorphism, 71
inner product, 156
 complex, 161
 standard, 156
inner product space, 156
input function, 12
inseparability, 209
insolvability of the word problem, 251
integers, 17

integral domain, 108
Intermediate Value Theorem, 225
internal direct product, 64
intersection, 2
inverse, 40
inverse function, 11
irreducibility
 test for, 132
irreducible element of a ring, 116
irreducible ideal, 240
irreducible polynomial, 116, 187
isometry, 42
isomorphic series, 163
isomorphism of algebras, 154
isomorphism of graphs, 96
isomorphism of groups, 45
isomorphism of rings, 104
isomorphism of vector spaces, 148
Isomorphism Theorems for groups, 69
Isomorphism Theorems for rings, 106
Isomorphism Theorems for vector
 spaces, 149

Jordan, Camille, 165

kernel of a homomorphism, 68, 105
kernel of a linear mapping, 149
Kirkman, Thomas Penyngton, 207
Klein 4-group, 44, 45
Klein, Felix, 44

label, 92
labelling problem, 92
Lagrange's Theorem, 53
Lagrange, Joseph Louis, 53
latin square, 45, 199
latin squares
 number of, 200
 orthogonal, 201
lattice, 8
lattice of subgroups, 48
Law of Trichotomy, 15
Laws of Exponents, 46
least common multiple, 24

least upper bound, 8
left coset, 52
left ideal, 114
left regular action, 79
left regular representation, 79
left transversal, 52
linear code, 258
linear combination, 136
linear equations
    system of, 138
linear fractional transformation, 249
linear functional, 151
linear mapping, 147
linear mappings and matrices, 152
linear operator, 148
linear order, 7
linear transformation, 147
linearly dependent subset, 138
linearly independent subset, 138
linearly ordered set, 7

mapping, 9
    linear, 147
mapping property of free groups, 244
mathematical induction, 18
Mathieu group, 170
Mathieu, Émile Léonard, 170
maximal p-subgroup, 240
maximal element in a partially ordered
    set, 235
maximal ideal, 110, 236
maximal ideal of a principal ideal
    domain, 120
maximal normal subgroup, 64
maximal subgroup, 176
metabelian group, 171
minimum distance of a code, 255
Modular Law, Dedekind's, 58
Möbius, August Ferdinand, 199
Möbius function, 199, 218
monic polynomial, 120
monoid, 40
monoid, free, 41

monster simple group, 170
Moore, Eliakim Hastings, 197
multiple root
    criterion for, 129
multiple root of a polynomial, 128
multiplication table, 45

next state function, 12
nilpotent class, 174
nilpotent group, 174
    characterization of, 175
non-generator, 176
norm, 156, 158
normal closure, 61, 250
normal core, 80
normal extension, 208
normal form of a matrix, 141
normal form of a word, 247
normal subgroup, 60
normalizer, 82
normed linear space, 158
n-tuple, 3
null space, 136

odd permutation, 35
1-cocycle, 182
one-one, 10
one-one correspondence, 10
onto, 10
orbit of a group action, 82
order of a group, 45
order of a group element, 49
ordered basis, 140
orthogonal complement, 159
orthogonal vectors, 156
orthonormal basis, 160
outer automorphism, 72
outer automorphism group, 72
output function, 12

partial order, 5
partially ordered set, 6
partition, 6
Pauli spin matrices, 109

Pauli, Wolfgang Ernst, 109
perfect code, 258
perfect field, 210
permutation, 31
    cyclic, 33
    even, 35
    odd, 35
    power of, 34
permutation group, 78
permutation matrix, 77
permutation representation, 78
permutations
    disjoint, 34
π-group, 183
Poincaré, Henri, 59
Polya's Theorem, 93
Polya, George, 93
polynomial, 100
    cyclotomic, 132, 218
    generic, 233
    irreducible, 116
    monic, 120
    primitive, 124
polynomial in an indeterminate, 101
polynomial not solvable by radicals, 230
poset, 6
power of a group element, 46
power of a permutation, 34
power set, 3
presentation of a group, 251
prime
    Fermat, 24, 195, 226
prime field, 185
prime ideal, 110
prime ideal of a principal ideal domain, 120
prime number, 22
primes
    infinity of, 23, 29
primitive n-th root of unity, 217

Primitive Element
    Theorem of the, 212
primitive polynomial, 124
principal ideal, 104, 118
principal ideal ring, 118
Principle of Mathematical Induction, 18
    Alternate Form, 19
product of subgroups, 57
product of subsets, 57
projective general linear group, 168
projective groups and geometry, 169
projective space, 169
projective special linear group, 168
proper subset, 2

quartic equation, 242
quasigroup, 199, 207
quaternion, 109
quotient, 20
quotient group, 62
quotient ring, 106
quotient space, 145
    dimension of, 145

radical extension, 228
rank of a matrix, 141
reduced word, 246
refinement of a series, 163
Refinement Theorem, 164
reflexive law, 5
regular, 82
relation, 4
relation on a set, 5
relatively prime, 21
relatively prime elements of a ring, 119
remainder, 20
Remainder Theorem, 127
right coset, 52
right ideal, 114
right regular representation, 81
right transversal, 53
ring, 99
    commutative, 99

division, 108
ring of polynomials, 100
ring of quaternions, 109
ring with identity, 99
root of a polynomial, 127
root of multiplicity n, 128
root of unity, 217
    primitive n-th, 217
row echelon form, 139
RSA cryptosystem, 29
Ruffini, Paulo, 228
ruler and compass construction, 190

scalar, 134
Schmidt, Erhart, 161
schoolgirl problem
    Kirkman's, 207
Schreier, Otto, 164
Schur's theorem, 181
Schur, Issai, 181
Schwartz, Hermann Amandus, 157
semidirect product
    external, 75
    internal, 75
semigroup, 40
semiregular, 82
separable element, 210
separable extension, 210
separable polynomial, 209
series
    central, 174
    composition, 165
    factors of, 163
    length of, 163
    terms of, 163
series in a group, 163
set, 1
    countable, 15
    infinite, 15
set operations, 3
set product, 3
sign of a permutation, 36
simple extension

structure of, 187
simple group, 61
    sporadic, 169
simple group of Lie type, 169
simple groups
    classification of finite, 169
Singleton upper bound, 265
solvable group, 171
solvablility by radicals, 228
special linear group, 61
splitting field, 130
    uniqueness of, 197
splitting field of $t^q - t$, 196
splitting theorem, 181
squaring the circle, 191
standard basis, 140
Steiner triple system, 204
Steiner, Jakob, 204
subfield, 130, 185
subfield generated by a subset, 185
subgroup, 46
    cyclic, 47
    Frattini, 175, 240
    Hall π-, 183
    identity, 47
    maximal, 176
    normal, 60
    proper, 47
    subnormal, 163
    Sylow p-, 88
    trivial, 47
subgroup generated by a subset, 48
subgroup of a cyclic group, 55
subgroup of a quotient group, 63
subnormal subgroup, 163
subring, 103
    zero, 103
subset, 2
    proper, 2
subspace, 136
    zero, 136
subspace generated by a subset, 136
sum of subspaces, 144

surjective function, 10
Sylow p-subgroup, 88
Sylow's Theorem, 88
Sylow, Peter Ludwig Mejdell, 88
symmetric function, 231
Symmetric Function Theorem, 232
symmetric group, 31, 32, 41
symmetric law, 5
symmetry, 42
symmetry group, 42
symmetry group of the regular n-gon, 42
syndrome, 262

Tarry, Gaston, 204
Tartaglia, Niccolo, 243
Theorem
    Cantor–Bernstein, 14
    Cauchy, 89
    Cayley, 80
    Fermat, 26, 54, 60
    Lagrange, 53
    Polya, 93
    Schur, 181
    Sylow, 88
    von Dyck, 251
    Wedderburn, 110
    Wilson, 50
Thirty Six Officers
    problem of the, 204
transcendent element, 187
transcendental number, 189
transfinite induction, 239
transition matrix, 143
transitive action, 82
transitive law, 5
transitive permutation representation, 82
transpositions, 33
transversal
    left, 52
    right, 53
Triangle Inequality, 158

triangle rule, 135, 156
Trichotomy
    Law of, 15, 239
trisection of an angle, 191
trivial homomorphism, 67
trivial subgroup, 47
2-cocycle, 181

union, 2
    disjoint, 4
unique factorization domain, 122, 125
unique factorization in domains, 121
unit in a ring, 102
unitriangular matrices
    group of, 175, 178
upper bound, 235
upper central chain, 174

value of a polynomial, 127
Varshamov–Gilbert lower bound, 257
Varshamov–Gilbert lower bound
    for linear codes, 258
vector, 134
    column, 135
    elementary, 137
    row, 135
vector space, 134
    dimension of, 141
    finite dimensional, 141
    finitely generated , 136
vector space of linear mappings, 150
vertex of a graph, 95
Von Dyck's theorem, 251
von Dyck, Walter, 251
von Lindemann, Carl Ferdinand, 194

Wedderburn's theorem, 110
Wedderburn, Joseph Henry Maclagan, 110
weight of a word, 255
well order, 7
Well-Ordering
    Axiom of, 239
Well-Ordering law, 18

Wilson's Theorem, 50
Wilson, John, 50
word, 41, 245, 254
    empty, 245
    reduced, 246
word in a generating set, 245

Zassenhaus's Lemma, 164
Zassenhaus, Hans, 164
zero divisor, 107
zero subring, 103
zero subspace, 136
Zorn's Lemma, 235